SOLID STATE PHYSICS

LICENSE, DISCLAIMER OF LIABILITY, AND LIMITED WARRANTY

By purchasing or using this book (the "Work"), you agree that this license grants permission to use the contents contained herein, but does not give you the right of ownership to any of the textual content in the book or ownership to any of the information or products contained in it. *This license does not permit uploading of the Work onto the Internet or on a network (of any kind) without the written consent of the Publisher.* Duplication or dissemination of any text, code, simulations, images, etc. contained herein is limited to and subject to licensing terms for the respective products, and permission must be obtained from the Publisher or the owner of the content, etc., in order to reproduce or network any portion of the textual material (in any media) that is contained in the Work.

MERCURY LEARNING AND INFORMATION ("MLI" or "the Publisher") and anyone involved in the creation, writing, or production of the companion disc, accompanying algorithms, code, or computer programs ("the software"), and any accompanying Web site or software of the Work, cannot and do not warrant the performance or results that might be obtained by using the contents of the Work. The author, developers, and the Publisher have used their best efforts to insure the accuracy and functionality of the textual material and/or programs contained in this package; we, however, make no warranty of any kind, express or implied, regarding the performance of these contents or programs. The Work is sold "as is" without warranty (except for defective materials used in manufacturing the book or due to faulty workmanship).

The author, developers, and the publisher of any accompanying content, and anyone involved in the composition, production, and manufacturing of this work will not be liable for damages of any kind arising out of the use of (or the inability to use) the algorithms, source code, computer programs, or textual material contained in this publication. This includes, but is not limited to, loss of revenue or profit, or other incidental, physical, or consequential damages arising out of the use of this Work.

The sole remedy in the event of a claim of any kind is expressly limited to replacement of the book, and only at the discretion of the Publisher. The use of "implied warranty" and certain "exclusions" vary from state to state, and might not apply to the purchaser of this product.

SOLID STATE PHYSICS

*From the Material Properties
of Solids to Nanotechnologies*

DAVID S. SCHMOOL

MERCURY LEARNING AND INFORMATION
Dulles, Virginia
Boston, Massachusetts
New Delhi

Reprint and Revision Copyright ©2017 by MERCURY LEARNING AND INFORMATION LLC.
All rights reserved.

Original title and copyright: *Introductory Solid State Physics: From the Material Properties of Solids to Nanotechnologies.* Copyright © 2014 by David S. Schmool. ISBN 978-0-9926368-6-9
All rights reserved. Published by The Pantaneto Press.

This publication, portions of it, or any accompanying software may not be reproduced in any way, stored in a retrieval system of any type, or transmitted by any means, media, electronic display or mechanical display, including, but not limited to, photocopy, recording, Internet postings, or scanning, without prior permission in writing from the publisher.

Publisher: David Pallai
MERCURY LEARNING AND INFORMATION
22841 Quicksilver Drive
Dulles, VA 20166
info@merclearning.com
www.merclearning.com
1-800-232-0223

This book is printed on acid-free paper.

David S. Schmool. *Solid State Physics. From the Material Properties of Solids to Nanotechnologies.*
ISBN: 978-1-942270-77-5

The publisher recognizes and respects all marks used by companies, manufacturers, and developers as a means to distinguish their products. All brand names and product names mentioned in this book are trademarks or service marks of their respective companies. Any omission or misuse (of any kind) of service marks or trademarks, etc. is not an attempt to infringe on the property of others.

Library of Congress Control Number: 2016935950

161718321 Printed in the United States of America

Our titles are available for adoption, license, or bulk purchase by institutions, corporations, etc. For additional information, please contact the Customer Service Dept. at 800-232-0223(toll free).

All of our titles are available in digital format at *authorcloudware.com* and other digital vendors. The sole obligation of MERCURY LEARNING AND INFORMATION to the purchaser is to replace the book, based on defective materials or faulty workmanship, but not based on the operation or functionality of the product.

CONTENTS

Preface	*xiii*
Chapter 1 Introduction to Solid State Physics	**1**
1.1 Introduction	1
1.2 Electronic Structure of the Atom	4
1.2.1 Electron Orbits	5
1.2.2 The Bohr Model of the Atom	6
1.2.3 Electron Filling, Quantum Theory and Quantum Numbers	11
1.3 The Periodic Table	16
1.4 Interatomic Bonding	18
1.4.1 Ionic Bonding	21
1.4.2 Covalent Bonding	23
1.4.3 Mixed Covalent and Ionic Bonding	26
1.4.4 Metallic Bonding	26
1.4.5 Hydrogen Bonding	28
1.4.6 Van der Waals Bonding	29
1.5 Summary	29
References and Further Reading	31
Basic Texts	31
Advanced Texts	31
Exercises	32
Notes	33
Chapter 2 Crystallinity in Solids	**35**
2.1 Introduction	35
2.2 Aspects of Symmetry in Crystalline Materials	36
2.2.1 Translational Symmetry	37
2.2.2 The Basis and the Unit Cell	38
2.2.3 Elements of Symmetry	41
2.3 Bravais Lattices	43
2.4 Crystal Planes and Axes: The Miller Indices	44
2.5 Common Crystalline Structures	51
2.6 Atomic Packing	55
2.7 Summary	57
References and Further Reading	59
Basic Texts	59
Advanced Texts	59
Exercises	59
Note	60

Chapter 3 Crystal Structure Determination   **61**
3.1 Introduction   61
3.2 The Reciprocal Lattice   63
3.3 Diffraction of Waves by Crystals   66
    3.3.1 Bragg's Law   68
    3.3.2 The Von Laue Approach   70
    3.3.3 Reconciling the Bragg and von Laue Approaches   72
    3.3.4 The Ewald Sphere Construction   74
3.4 The Atomic Form Factor   76
3.5 The Structure Factor   77
3.6 Diffraction Methods for Structure Determination   80
    3.6.1 X-Ray Diffraction   80
    3.6.2 Electron Diffraction   85
    3.6.3 Neutron Diffraction   88
3.7 Summary   90
References and Further Reading   91
    Basic Texts   91
    Advanced Texts   91
Exercises   92
Notes   93

Chapter 4 Imperfections in Crystalline Order   **95**
4.1 Introduction   95
4.2 Point Defects   96
    4.2.1 Types of Point Defect   96
    4.2.2 Thermodynamics of Defect Density   99
    4.2.3 Diffusion in Crystals   100
    4.2.4 Color Centers   103
4.3 Dislocations   104
    4.3.1 Edge Dislocations   104
    4.3.2 Screw Dislocations   105
    4.3.3 The Burgers Vector   105
    4.3.4 Dislocations and Mechanical Properties of Solids   108
    4.3.5 Dislocation Energy   115
    4.3.6 Interactions between Dislocations   117
4.4 Planar Defects   121
    4.4.1 Grain Boundaries   121
    4.4.2 Tilt Boundaries   122
    4.4.3 Twin Boundaries   123
4.5 Non-Crystalline Materials   124
4.6 Summary   125
References and Further Reading   127

Basic Texts	127
Advanced Texts	128
Exercises	128

Chapter 5 Lattice Vibrations 131

5.1 Introduction 131
5.2 Vibrational Modes of a Monatomic Lattice 132
    5.2.1 One-Dimensional Chain 132
    5.2.2 Extension to Three-Dimensions 138
    5.2.3 Number of Modes: Density of States 139
5.3 Vibrational Modes of a Diatomic 1D Lattice 143
5.4 Thermal Properties of Solids 149
    5.4.1 Classical Specific Heat: Dulong and Petit's Law 150
    5.4.2 Einstein's Model 150
    5.4.3 The Debye Model 152
5.5 Anharmonic Effects 155
    5.5.1 Thermal Expansion 157
    5.5.2 Thermal Conduction 158
    5.5.3 Umklapp Processes 160
5.6 Summary 162
References and Further Reading 163
    Basic Texts 163
    Advanced Texts 163
Exercises 164
Notes 165

Chapter 6 Free Electrons in Metals 167

6.1 Introduction 167
6.2 Metallic Behavior 169
6.3 The Maxwell - Boltzmann Velocity Distribution 171
6.4 The Drude Theory 172
6.5 Fermi - Dirac Statistics of an Electron Gas 174
6.6 The Sommerfeld Model 177
6.7 The Density of States 178
6.8 Specific Heat of an Electron Gas 182
6.9 Pauli Paramagnetism 184
6.10 High Frequency Response and Optical Properties 188
6.11 Summary 192
References and Further Reading 193
    Basic Texts 193
    Advanced Texts 193
Exercises 193
Notes 194

Chapter 7 Band Theories of Solids 197
7.1 Introduction 197
7.2 The Periodic Potential 199
7.3 The Bloch Theorem and Functions 201
7.4 The Schrödinger Equation in a Periodic Potential 203
7.5 Brillouin Zones and the Fermi Surface 205
7.6 The Kronig - Penney Model 207
7.7 Free Electrons in a Periodic Potential 211
7.8 The Nearly Free Electron Model 216
7.9 The Tight - Binding Model 218
7.10 Other Models: Potentials and Wave-Functions 222
7.11 Metals, Semiconductors, and Insulators 224
7.12 Summary 227
References and Further Reading 228
  Basic Texts 228
  Advanced Texts 229
Exercises 229
Notes 230

Chapter 8 Electron Dynamics and Transport Phenomena 231
8.1 Introduction 231
8.2 Electron Dynamics in Crystals 233
8.3 The Effective Mass 237
8.4 The Fermi Surface 240
8.5 Positive Charge Carriers: Holes 243
8.6 Drift and Diffusion of Charge Carriers 247
8.7 Electron Scattering in Bands 251
8.8 Magnetic Field Effects 253
  8.8.1 The Hall Effect 254
  8.8.2 Cyclotron Resonance 256
  8.8.3 Magnetoresistance 260
  8.8.4 Magnetic Sub-Bands and Oscillatory Phenomena in Solids 263
  8.8.5 The Quantum and Fractional Quantum Hall Effects 266
8.9 Summary **269**
References and Further Reading 270
  Basic Texts 270
  Advanced Texts 270
Exercises 271
Notes 272

Chapter 9 Semiconductors 273
9.1 Introduction 273
9.2 Semiconducting Materials 276

9.3	Equilibrium Statistics: Electrons and Holes	279
	9.3.1 Intrinsic Semiconductors	279
	9.3.2 The Law of Mass Action	283
	9.3.3 Extrinsic Semiconductors: Doping	284
	9.3.4 Compensated Semiconductors	290
9.4	Non-Equilibrium Distributions	292
	9.4.1 Carrier Injection: Injection Levels	292
	9.4.2 Generation and Recombination Processes	294
	9.4.3 The Continuity Equations	298
9.5	The p - n Junction	301
	9.5.1 Thermal Equilibrium	301
	9.5.2 The Depletion Zone	304
	9.5.3 Junction Capacitance	306
	9.5.4 Current - Voltage Characteristics	306
9.6	Heterostructures and Quantum Wells	310
9.7	Summary	315
References and Further Reading		317
	Basic Texts	317
	Advanced Texts	317
Exercises		317
Notes		320

Chapter 10 Magnetic Materials and Phenomena   321

10.1	Introduction	321
10.2	The Atomic Magnetic Moment	324
	10.2.1 Orbital and Spin Angular Momenta	324
	10.2.2 Hund's Rules and the Ground State	326
	10.2.3 Moments and Energies	330
10.3	Diamagnetism	331
10.4	Paramagnetism	334
	10.4.1 Classical Treatment	334
	10.4.2 Quantum Mechanical Treatment	336
	10.4.3 Van Vleck Paramagnetism	338
10.5	Interactions, Exchange, and Magnetic Order	339
	10.5.1 Dipolar Interaction	339
	10.5.2 Exchange Interactions	341
10.6	Ferromagnetic Order	344
	10.6.1 Mean Field Theory	344
	10.6.2 Itinerant Ferromagnetism	349
10.7	Antiferromagnetic Order	352
10.8	Ferrimagnetic Order	356
10.9	Magnetic Anisotropies	361
	10.9.1 Shape Anisotropy	361
	10.9.2 Magnetocrystalline Anisotropy	363

x • Contents

10.10	Magnetic Domains, Domain Walls, and Hysteresis	365
10.11	Spin Waves	371
10.12	Giant Magnetoresistance and Spintronics	376
10.13	Spin Dynamics	379
10.14	Summary	386

References and Further Reading — 389
   Basic Texts — 389
   Advanced Texts — 389
Exercises — 390
Notes — 392

Chapter 11 Superconductivity — 393

11.1	Introduction	393
11.2	Phenomena Related to Superconductivity	396
	11.2.1 Zero-Resistivity/Infinite Conductivity and Persistent Currents	396
	11.2.2 Meissner-Ochsenfeld Effect	398
	11.2.3 Perfect Diamagnetism	399
	11.2.4 Critical Fields and Critical Current	401
11.3	Thermodynamics of the Superconducting Transition	405
	11.3.1 Phase Stability of the Superconducting State	405
	11.3.2 Heat Capacity of a Superconductor	407
11.4	The London Equations	408
11.5	Ginzburg - Landau Model	412
11.6	Elements of the BCS Theory of Superconductivity	416
	11.6.1 Electron - Phonon Coupling and Cooper Pairs	419
	11.6.2 The BCS Ground State	422
	11.6.3 Outcomes of the BCS Theory	425
11.7	Josephson Effects	430
11.8	High-Temperature Superconductors	437
11.9	Summary	441

References and Further Reading — 443
   Basic Texts — 443
   Advanced Texts — 443
Exercises — 444
Notes — 445

Chapter 12 Dielectric Materials — 447

12.1	Introduction	447
12.2	Some Basic Properties of Dielectric Materials	449
	12.2.1 Electrical Conductivity	449
	12.2.2 Ionic Conduction	449
	12.2.3 Dielectric Breakdown	450

12.3	Electrostatics and the Maxwell Equations	451
12.4	The Local Field Approximation	455
12.5	The Dielectric Function	458
	12.5.1 Electronic Polarization	460
	12.5.2 Ionic Polarization	462
	12.5.3 The Total Dielectric Function	467
12.6	Ferroelectrics	468
12.7	Piezoelectrics	472
12.8	Multiferroic Materials	472
12.9	Optical Properties of Solids	473
	12.9.1 The Wave Equation	474
	12.9.2 Transmission and Reflection Coefficients	475
	12.9.3 Absorption of Electromagnetic Waves	477
	12.9.4 Optical Properties of Dielectrics	478
12.10	Summary	481
References and Further Reading		483
Basic texts		483
Advanced texts		483
Exercises		483
Note		484

Chapter 13 Nanotechnologies and Nanophysics 485

13.1	Introduction	485
13.2	The Physics of Surfaces	489
	13.2.1 Surface Structure	493
	13.2.2 Surface Composition and Excitation States	497
13.3	Low Dimensional Systems	503
13.4	Electronic and Optical Properties of Nanostructures	507
	13.4.1 Size Reduction and Energy Quantization	509
	13.4.2 Quantum Point Contacts	513
	13.4.3 The Insulating Barrier and Tunnel Junctions	519
	13.4.4 Single Electron Transport: Quantum Dots and Coulomb Blockade	523
	13.4.5 Resonant Tunneling	527
	13.4.6 Single Electron Transistor (SET)	529
	13.4.7 Optical Properties of Nanostructures	534
13.5	Aspects of Nanomagnetism	539
	13.5.1 Magnetic Length Scales	540
	13.5.2 The Stoner - Wohlfarth Model	541
	13.5.3 Superparamagnetism and Ferromagnetic Nanoparticles	545
	13.5.4 Magnetic Thin Films and Multilayers	550
	13.5.5 Magnetic Nanostructures	553
13.6	Summary	559

References and Further Reading 564
  Basic Texts 564
  Advanced Texts 564
Exercises 565
Notes 569

Appendix A **571**

Appendix B **575**

Appendix C **577**

Index **579**

PREFACE

This book is principally intended as a basic introduction to the topic of Solid State Physics for undergraduate students. While aimed primarily as an introductory text for Physics students, it will also be of use to undergraduates who require an understanding of solid state physics for chemistry, materials science and engineering based courses. The project began as a work for the Pantaneto Introductory Physics Series and has leant heavily on my teaching of many topics related to the subject. Over the past 13 years, I have lectured in the Department of Physics and Astronomy at the University of Porto (Portugal), where I have taught undergraduate, Masters, and postgraduate courses in such diverse topics as: Magnetic Materials and Applications, Semiconductors and Devices, Materials Science, Introduction to Modern Physics, Functional Materials and Applications, Nanotechnologies, Characterization Techniques for Materials, and several others, including a range of laboratory classes. Many of the topics in the current textbook have been broadly based on some of my lectures notes for these courses.

My own personal interest in the solid state physics began over 20 years ago when I began my doctoral studies at the University of York (UK), where I researched structural and spin wave resonance properties in magnetic multilayers. I have since worked in many aspects of the preparation, characterization, and study of magnetic and spin dynamic properties of low dimensional magnetic systems. This experience has allowed me to develop my interest in many areas of solid state physics. Solid state physics is a popular and enormously rewarding area of research and one which has provided the basis for a huge number of devices and applications and interestingly, many Nobel Prize winners in Physics have worked in the field.

The book is organized in a chapter sequence that I find logical to the introduction of the various topics of solid state physics. As such, I start by exposing some basics concepts of quantum mechanics and the bonding mechanisms between atoms. This is fundamental to the formation of solids and rightly belongs at the beginning. The next chapter deals with the all important subject of crystalline structures. This follows on nicely from the introduction; since once the bonds are formed between atoms, they can condense into the solid structures. These will naturally occur in ordered arrays of atoms, i.e., crystals, depending on the ambient conditions for equilibrium structures. Rapid quenching from the melt, for example, can produce non-equilibrium solid structures, such as found in amorphous materials. Chapter 3 deals with some fundamental aspects of crystal structure determination and is an essential topic for the budding experimentalist. This mainly treats some of the basic concepts of crystal diffraction theory as well as outlining some of the main experimental techniques available. Sadly, nothing is perfect and Chapter 4 approaches the subject of crystalline impurities; defects and imperfections which I have dealt with in terms of the solids dimensionality. The motion of atoms in the form of vibrations or phonons forms the subject of Chapters 5. In Chapter 6, the subject of metallic behavior and how the free electrons in such materials are so important to their unique properties is introduced. The formal crystalline properties of materials has a profound effect on the way electrons and charge carriers move. In Chapter 7, some theoretical aspects which allow us to treat the complex behavior of charge carriers in solids in a simple manner are discussed. Also included, are some extended introductions to some of the simpler models for the calculation of energy bands in solids. I have here tried to demystify the origin of the complicated looking band structures. The specifics of how electrons move in energy bands are dealt with in the following chapter (Chapter 8). Here, we introduce some further concepts which aid our description of the movement of charge carriers and some of the simpler transport phenomena encountered in materials. Of particular importance are the reaction of charge carriers to the application of electric and magnetic fields. The remaining five chapters deal with specific properties of solids. In Chapter 9, semiconductors are introduced. Semiconducting materials form such an important class of materials in the electronics industry. Introduced are the basic concepts that are necessary to understand the principal behavior and the different types of semiconductor. Also introduced, in some detail, is the p-n junction, which forms the fundamental unit of a majority of semiconducting devices as well as approaching the more recent works based heterostructures and quantum wells. In Chapter 10, I tried to bring together the principal elements required for a reasonable understanding of magnetic materials and how solids behave under the influence of a magnetic field. Magnetic materials have a number of important technological applications. My aim was to give a broad overview of its fundamentals, as well as introducing the different classes of magnetic material. An introduction to some more modern aspects of the subject, such as spintronics and spin dynamics, is provided. Superconductivity is a complex subject and forms the topic of Chapter 11. The topic is approached so as to highlight the basic phenomena associated with such behavior, while giving an in-depth account of the fundamental theories of superconductivity. Also included in this chapter, is a fundamental outline of some of the applications, which are mainly based on the Josephson effects and junction. The chapter is completed with an overview of some of the more important concepts of high-temperature superconductors. In Chapter 12, the subject of dielectric properties of solids is dealt with, including some general introductions to the oxide materials with ferroelectric, piezoelectric, and multiferroic properties, which have important applications. Additionally, an outline of the principal optical properties of insulating materials is given. (The optical properties of metallic materials are discussed in Chapter 6.) The final chapter consists of a detailed overview of some of the more recent work in Nanotechnologies. I am of the opinion that this subject belongs as an integral part of solid state science. This is a rather long chapter and includes some extended discussions on the more recent works based on the electronic, optical, and magnetic properties of nanostructured materials. I have discussed this subject under a personal perspective, relating it to the modification of the fundamental properties of solids in low dimensional systems, broadly classing these as the effects of surfaces and confinement effects. Integral with this description, is the introduction to the fundamental length scales of different physical properties. A brief introduction to surface physics and low dimensionality is given. It is hoped that this chapter will serve as an introduction to this important development in modern solid state science. Of course, it could be no more than an introduction, since Nanotechnologies as a subject encompasses many other areas of scientific study and research, such as chemistry, biology, and medicine. My discussion is limited only to the pertinent areas of condensed matter physics. In each of the chapters, a general introduction to the specific topic discussed is provided as well as an extended summary at the end to highlight the main points. At the end of each chapter, I list some of the references used while also indicating others in the text, which are generally primary sources. Finally, a number of exercises are included for the student to test their grasp of each subject area.

There are a large number of excellent textbooks available that deal with solid state physics, from basic introductions to advanced texts. Probably the best known are those by Kittel, probably known to all students who have some across the subject, and Ashcroft and Mermin. The former is an excellent book, which provides a broad overview of the subject, while the latter gives a more in depth study of the subject, particularly on subjects related to the electronic properties of solids. I believe the current book provides a broad general introduction to the subject, covering all the major topics that most undergraduate courses require. The book is intended to provide a text which is both approachable to the newcomer and provides some deeper insights to the student who has some basic knowledge of the subject. In particular, the final chapter on Nanotechnology and Nanophysics should allow the interested reader to see how the extension to the main topics of solid state physics have developed in to this enormously important area of study with its plethora of applications. Indeed, this is a subject rarely dealt with in textbooks on solid state physics and I hope it will provide a stimulating introduction.

I am very grateful to my colleague Dr. David Navas for proofreading the whole text, highlighting some areas that were unclear and needed some clarification. I, of course, take full responsibility for any errors that remain. I would also like to acknowledge many of my students, who over the years have helped me to become a better teacher. I have been very lucky to have made the acquaintance of some very talented and distinguished scientists, who have helped shape my thoughts on many areas of my research and those related to some of the subjects in this book. I am privileged to be able to call them colleagues and friends. Among them, I would like to mention the following: Austin Chambers, Jeremy Whiting, Bret Heinrich, Bob Stamps, Marcel Guyot, Niels Keller, Hamid Kachkachi, José Javier Sáiz Garitaonandia. I have many colleagues at the University of Porto and I would like to thank them all for the professional and pleasant working environment and many fond memories. In particular, I would like to mention: João Bessa Sousa, Helder Crespo, Orfeu Bertolami, and João Lopes dos Santos. I should add to this list some people who were also fundamentally important in my development. I am forever indebted to Hank and Alex Kahney, Alex West, and Peter Kenny, as well as many others. An enormous "Merci" to François Vernay. I would like to extend my deep appreciation to my partner Ana, who has had to put up with me while this book was being written: I hope I can make it up to you. I am sad that both my parents are not alive to have seen this book in print; I know they would have been proud.

D. S. Schmool
Porto and Perpignan
July 2016

CHAPTER 1

INTRODUCTION TO SOLID STATE PHYSICS

"The future is unwritten"

—Joe Strummer

1.1 INTRODUCTION

All substances that we observe in our daily lives are made up of atoms. These are assemblies of smaller, fundamental, particles: protons, neutrons and electrons. The protons, which have a positive charge (by convention) and neutrons form what we call the nucleus, which concentrates at the center of the atom and contains a vast majority of its mass. The electrons, which have a negative charge equal in magnitude to that of the proton, orbit the nucleus, occupy specific shells and are in equal numbers to the protons. This means that an isolated atom with be electrically neutral since the neutrons do not have electrical charge. Each element (atom) is distinguished by the number of protons it has in its nucleus, i.e., the *atomic number.* The nucleus for a certain atomic species can have differing numbers of neutrons, these are called *isotopes,* and occur in specific abundances, where there is usually a dominant isotope. For example, there are two principal or primordial isotopes for uranium; ^{235}U and ^{238}U, both have 92 protons, but the latter has three more neutrons than the former and has a larger natural abundance (around 99.2%). Atoms can be isolated in gaseous form or joined to other atoms to form a molecule, which can also exist in the gaseous phase. The phase or state of matter depends largely on the ambient conditions of temperature and pressure. In general conditions of high pressure and low temperature favor the condensation of a gas to form a liquid, and with further increase of pressure or reduction in temperature, a solid will by formed. The specifics of the changes of phase are crucially dependent on the atomic species and the number of electrons it contains.

Solid state physics is a broad area of modern physics which deals with the fundamental physical properties of materials in their frozen phase. Materials come in a variety of forms and the way we classify substances depends on what properties we are interested in. For example, when we discuss electrical properties of materials, we separate them into groups depending on whether they conduct well (e.g. metals), poorly (insulators) or somewhere in between (semiconductors). In terms of magnetism, we divide materials into those with differing magnetic order; ferromagnetic, ferrimagnetic, paramagnetic etc. It is one of the main concerns of this book to outline this differentiation of material solids, as well as to give an overview of the relation between atomic species, atomic ordering and the physical phenomena and properties of solid matter.

At the root of all solid state physics are two basic aspects: chemical composition and their spatial arrangement. In simple terms, what we have and how they arrange themselves with respect to one another. Atomic species will determine how they bond together (if at all). The bonding mechanism will frequently determine the spatial arrangement of atoms, thus giving a specific crystalline structure (taken under equilibrium conditions). The resulting physical properties are principally determined by these two factors. For example, in metals, atoms are held together by the interaction between the ionic cores and free electrons, typically in some form of close packed structure. The mechanical properties are directly related to the bond strength between the atoms, which in turn depends on the type of atom (number of electrons). The structure will determine how the mechanical properties vary with direction, i.e., the anisotropy in the elastic constants of the material. It is worth noting, for example, that the melting temperature of a solid is a good measure of the interatomic bond strength, but will also depend on the number of nearest neighbor (*nn*) bonds or coordination number. The electronic properties of a solid will generally depend on the number of free electrons per unit volume (electron density) and the physical structure of the material. Magnetic phenomena in solids similarly depend on the atomic species and the distribution of electrons around the atoms as well as the way in which the electron orbits interact, and will in the case of ordered magnetic materials, govern the magnetic anisotropies that are evident in these solids. The properties we are discussing here are referred to as intrinsic properties, and depend strictly on the atomic species and their arrangement. On the other hand, extrinsic properties can arise from other factors, such as the shape of a solid, which can also affect the physical response of a materials to an external stimulus, such as an applied field or mechanical force.

As a general note, the measurement of physical properties typically follows the basic principle of observing the reaction of a material to the application of an external force or stimulus. This can be a mechanical force (compressive, tensile, or shear) and measuring the change in shape of the solid. Alternatively, if we are interested in electronic properties, we may apply an electric field (potential) to the material and measure the flow of current (i.e., number of electrons passing per unit time), which will allow us to assess important parameters such as electrical resistance etc. In a similar way, if we want to assess the magnetic properties of a substance, we generally see how it behaves in a magnetic field. Another important class of measurement consists in subjecting the material or sample to some form of radiation, such as x-ray, electron and neutron beams. The intensity and spatial distribution of the transmitted or reflected radiation can provide valuable information on a broad range of physical characteristics, such as crystalline order, electronic structure and magnetic properties. Some of these measurement techniques will be outlined in later chapters. In particular we shall discuss some aspects of diffraction techniques in the assessment of the crystalline properties of materials (Chapter 3).

In this chapter, we will outline some of the basic concepts which are necessary to understand what atoms are and how they form solids. We will touch on the basic ideas of the structure of the atom, mainly concerning the Bohr atomic model. We will then discuss some basic principles of quantum theory, which enables us to build up a picture of how electrons are ordered in orbitals, which ultimately determines their chemical and bonding properties. Finally in this chapter, we will introduce the fundamental principles of the bonding mechanism between atoms, which form the basis of the crystallization or condensation of atoms in a solid.

1.2 ELECTRONIC STRUCTURE OF THE ATOM

Since all solid matter is comprised of atoms frozen in some form of three-dimensional order, we will commence by considering the basic internal structure of the atom and see how this structure affects the chemical properties of the element. We will consider the atom to be a stable entity or building block for our solid. This is of course not entirely true, but once our atom in the solid has found its equilibrium, it is will not generally change its internal structure. Of course it may be mobile inside the solid, but the basic structure (atom) will, in general, be considered as a stable body. Since we are not considering radioactive phenomena, this is a valid assumption.

Atoms are agglomerates of more elemental particles; protons and neutrons, which make up the massive nucleus and surrounded by an electron cloud. It is this latter which is essentially responsible for the chemical properties of the atom and will determine how it interacts with other atoms in its vicinity. Despite this, the nucleus of the atom makes up virtually all the atomic mass. The number of electrons in an atom is equal to that of the protons, thus maintaining charge neutrality. The positive charge of the nucleus balances that of the negative electrons, where the unit of this charge is equal to 1.60219×10^{-19} C. As we have stated above, the chemical behavior of an atom depends on the number of electrons it contains, which is a direct consequence of the number of protons in its nucleus. However, this is only part of the story and to really understand the chemical behavior of the atom we need to consider how the electrons are held in place. We will now review some basic concepts of quantum theory which describe electronic states in atoms, from which we can then infer the basic chemical properties of atoms.

1.2.1 Electron Orbits

One of the early models of the atom was envisaged by Ernest Rutherford (1911), and is based on the experimental study of the scattering of alpha and beta particles, though it has its origins in the Nagaoka *planetary model* (1904). This model describes the atom as having a tiny but massive positively charged nucleus around which electrons move in dynamically stable orbits. The simplest case is that of the hydrogen atom (atomic number $Z = 1$), having a single proton in the nucleus and orbited by one electron. The centripetal force for the electron in a stable circular orbit (though an elliptical orbit would also be acceptable) with mass, m_e, and velocity, v, can be written as:

$$F_c = \frac{m_e v^2}{r} \tag{1.1}$$

where r represents the radius of the orbit, i.e., the distance between the electron and the nucleus. The electron is held in its orbit by the electrostatic force between itself and the proton and is given by:

$$F_e = \frac{e^2}{4\pi\epsilon_0 r^2} \tag{1.2}$$

Here ϵ_0 is the permittivity of free space (or vacuum), having a numerical value of 8.854×10^{-12} Fm^{-1} and e is the electronic charge. We can obtain the condition for this dynamically stable orbit by setting these two forces equal, from which we write:

$$\frac{m_e v^2}{r} = \frac{e^2}{4\pi\epsilon_0 r^2} \tag{1.3}$$

Rearranging Equation (1.3) we can obtain an expression for the electron velocity:

$$v = \frac{e}{\sqrt{4\pi\epsilon_0 m_e r}} \tag{1.4}$$

Thus the kinetic energy of the electron will take the form:

$$K = \frac{m_e v^2}{2} = \frac{e^2}{8\pi\epsilon_0 r} \tag{1.5}$$

The total energy of the electron will be the sum of both kinetic (K) and potential (V) energies and takes the form:

$$T = K + V = \frac{e^2}{8\pi\epsilon_0 r} - \frac{e^2}{4\pi\epsilon_0 r} = -\frac{e^2}{8\pi\epsilon_0 r} \tag{1.6}$$

The negative value is interpreted as being the binding energy of the electron to the proton (in the formation of an atom). In the case of the hydrogen atom, experiments show that the energy required to separate the electron and the proton is 13.6 eV[1] (2.179×10^{-18} J = 2.179 aJ). It is then a simple matter to estimate the orbital radius of the hydrogen atom from Equation (1.6), which gives a value of 5.29×10^{-11} m or 0.53 Å. This quantity is known as the *Bohr radius*. Such an electron in this orbit would, according to Equation (1.4), have a corresponding orbital velocity of 2.19×10^6 ms^{-1}.

1.2.2 The Bohr Model of the Atom

The principal weakness of the Rutherford model of the atom is the fact that from classical electromagnetic theory, moving charges will emit electromagnetic radiation. In doing so the kinetic energy of the orbiting electron will diminish in an amount equal to the energy of the electromagnetic radiation produced. Such a reduction of kinetic energy will mean that the orbital radius will gradually reduce and the electron will eventually spiral down and collapse into the nucleus. Since this is not experimentally observed, we must conclude that the model is deficient and provides no explanation of the orbital stability of the electrons. To be fair, neither does the Bohr model, which introduces this as a supposition to the model.

The model of the atom, postulated in 1913 by Niels Bohr, provides a stunning early success for the old quantum theory and is based on some basic premises, which are given without justification. While this may seem a little unsatisfactory, the agreement between the Bohr theory and spectroscopic data was astoundingly good and gave much support for the model and is worth adding here since it gives a simple manner in which to think of the atom. In fact, the Bohr model is one that is still frequently used to envisage the structure of atoms. (Later some support for the Bohr model was provided, as we shall shortly see, by the de Broglie hypothesis.) The postulates of the Bohr theory can be expressed as follows:

1. The electrons in an atom can only have certain specific orbits, termed *stationary orbits*, with discrete values of their radius and hence energy.

2. The energy of the orbit is related to its size. The lowest energy is found in the smallest orbit, i.e., that closest to the nucleus.

3. These discrete orbital energies are the only ones the electrons can exist in. Therefore to change orbit and thus energy, the electron must gain or emit discrete energies which correspond to the following rule:

$$\delta E = E_f - E_i = h\nu \qquad (1.7)$$

where E_i and E_f represent the initial and final energies, h is Planck's constant ($h = 6.626 \times 10^{-34}$ Js) and ν is the frequency of the emitted/absorbed electromagnetic radiation.

There are a number of consequences to these postulates. Since we now have discrete electron orbits, the angular momentum will also be discrete and is given by:

$$L = \frac{m_e v_n}{r_n} = n\left(\frac{h}{2\pi}\right) = n\hbar \qquad (1.8)$$

where we see that we have labelled the orbital velocity and radius with n, to denote the discreteness of the electron orbits. We note that n is an important quantity, termed the *principal quantum number*, whose significance will be seen shortly. The quantization of orbital angular momentum here is seen as n, an integer, multiplied by the Planck constant divided by 2π. From Equation (1.3) we can relate the velocity and radius of the orbits and further divide by Equation (1.8) and obtain the velocity of the electron in the nth orbit as:

$$v_n = \frac{Ze^2}{2nh\epsilon_0} \qquad (1.9)$$

Z being the atomic number. It is then a simple matter to obtain the corresponding radius:

$$r_n = \frac{n^2 h^2 \epsilon_0}{\pi m_e Z e^2} \qquad (1.10)$$

This then leads to the energy associated with this nth orbit as:

$$E_n = -\left(\frac{m_e Z^2 e^4}{8 \epsilon_0^2 n^2 h^2}\right) \qquad (1.11)$$

The negative sign here is interpreted as the binding energy of the electron by the positive potential of the nucleus and that work must be done to remove it. We note that as n increases, E_n becomes less negative. We therefore understand that the orbit closest to the nucleus ($n = 1$) has a larger binding energy. This is logical, since it feels a stronger Coulomb (attractive) force from the nucleus. We can now calculate the energy associated with the discrete jumps between the different orbits (which are labelled as i and j) with the corresponding frequency of the radiation, ν_{ij}:

$$\begin{aligned} h\nu_{ij} = E_i - E_j &= -\left(\frac{m_e Z^2 e^4}{8 \epsilon_0^2 n_i^2 h^2}\right) + \left(\frac{m_e Z^2 e^4}{8 \epsilon_0^2 n_j^2 h^2}\right) \\ &= \frac{m_e Z^2 e^4}{8 \epsilon_0^2 h^2}\left(\frac{1}{n_j^2} - \frac{1}{n_i^2}\right) \end{aligned} \qquad (1.12)$$

This process is schematically illustrated in Figure 1.1. A negative value corresponds to absorption, while a positive value to emission, N.B. $E_{ij} = -E_{ji}$.

While being enormously successful as a scientific model, the Bohr theory suffers from some shortcomings. The main success of the Bohr atom is its ability to accurately predict the frequencies of the emission spectra of the hydrogen atom, giving a theoretical basis for the previously empirical Balmer formula and giving the correct value for its Rydberg constant. Further support came in the form of the de Broglie's hypothesis (1924), which proposed a wave formulation to quantum theory, where electrons and particles in general are seen as wave-like entities. Central to this hypothesis is the relation

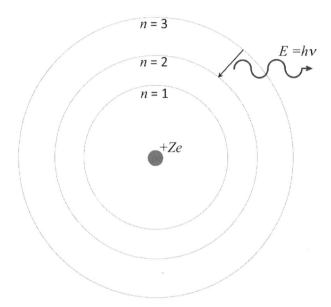

FIGURE 1.1: Bohr orbits.

between the electron (particle) wavelength and its linear momentum (p), which is expressed as:

$$\lambda = \frac{h}{p} = \frac{h}{m_e v} \qquad (1.13)$$

The Bohr radii are then simply generated by the use of the principle of wave interference and the fact that allowed standing waves occur for integral half wavelengths in the permitted orbits, as given by:

$$2\pi r_n = n\lambda \qquad (1.14)$$

We note that any other condition will produce a destructive interference of the wave and would therefore not be a stable state for the electron-wave. Substituting (1.14) in Equation (1.13) we find:

$$2\pi r_n = n\frac{h}{m_e v} \qquad (1.15)$$

From which we obtain:

$$r_n = \frac{nh}{2\pi m_e v} = \frac{nh}{2\pi m_e} \cdot \frac{2nh\epsilon_0}{Ze^2}$$
$$= \frac{n^2 h^2 \epsilon_0}{\pi m_e Z e^2} \tag{1.16}$$

which is the same as given by Equation (1.10). This thus provides a rationale for the existence of the stable Bohr orbits. A schematic illustration shows how this works, see Figure 1.2. From the figure, we see that the number of nodes is equal the the orbital number. It will be noted that for the case of the hydrogen atom ($Z = 1$), in the ground state ($n = 1$), the above equation gives a value for the radius equal to the Bohr radius, as given above.

Despite this, the model is ultimately unsatisfactory; its failings to fulfill some important physical principles mean that it is somewhat limited in its potential use. We can outline the main shortcomings as follows:

- The model fails to predict the ground state orbital angular momentum.

- It violates the Heisenberg uncertainty principle since the electrons have a known radius and orbit.

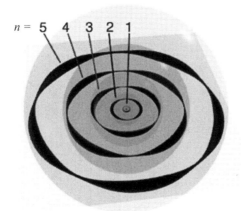

FIGURE 1.2: de Broglie matter waves in Bohr orbits ($n = 1 - 5$).

- The relative intensities of the spectral are not accounted for.
- The Bohr atom does not explain the fine structure and hyperfine structure in spectral lines.
- The model can only be used for the hydrogen atom.
- No explanation is given for the Zeeman effect.

1.2.3 Electron Filling, Quantum Theory and Quantum Numbers

A more precise description of the electronic behavior in atoms is provided by quantum mechanics, which takes into account the *Heisenberg uncertainty principle* as well as the probabilistic and wave natures of particles. While being a rather complex many-body problem, which is beyond the scope of our description, we can learn much with a brief overview.

The Heisenberg uncertainty principle is a quantum phenomenon which reflects the probabilistic nature of fine particulate behavior. Essentially it states that the position and momentum of a particle (such as an electron) cannot be explicitly known. This is often expressed mathematically with the relation:

$$\delta x \delta p \geq \frac{\hbar}{2} \qquad (1.17)$$

where $\hbar = h/2\pi$. The uncertainty relation can also be expressed in terms of time and energy as:

$$\delta E \delta t \geq \frac{\hbar}{2} \qquad (1.18)$$

Both forms express the limit of determinism that can be attributed to the physical quantities which define the electron states in an atom, for example. A full discussion of the consequences of the Heisenberg uncertainty principle are beyond the scope of this section, for further information the reader is referred to texts on quantum theory, see for example Bransden and Joachain (2003 2e).

The wave nature of particles can be conveniently expressed in terms of the Schrödinger equation, which expressed in the time independent form can be written as:

$$\left[\frac{-\hbar^2}{2m_e}\nabla^2 + V(\mathbf{r})\right]\psi(\mathbf{r}) = E\psi(\mathbf{r}) \quad (1.19)$$

This is a very important mathematical representation of the total energy of the electron. The first term in brackets corresponds to a mathematical operator of the kinetic energy, while the second term represents the potential in which the electron finds itself. The overall form the the equation is called an *eigenvalue equation,* where the operator represented in the square brackets operates on the (wave) function, $\psi(\mathbf{r})$, thus returning the eigen or characteristic energy values, which the electron can take given the energy constraints. We will further consider this equation when we discuss the electron energies in a solid, which will be the subject of Chapter 7. The form of the wave function is important since it will also limit the way in which the electron can behave. It is instructive to consider a very simple example. Consider a free electron which is moving in the x-direction of our coordinate system. We note that the mathematical representation of *free,* means that we can write $V(\mathbf{r}) = 0$, where there is no potential to restrict the electron motion, as described by the kinetic energy. The form of the wave function is that of a plane wave, which can be represented in the form: $\psi = Ae^{ikx}$. Since the second partial derivative of this can be written as $-k^2\psi$, we obtain the free electron energy as a parabolic function of the wave vector, k, as:

$$E = \frac{\hbar^2 k^2}{2m_e} \quad (1.20)$$

We note that the wave vector is related to the momentum of the electron through the expression: $p = \hbar k$, and can also be expressed in vectorial form as: $\mathbf{p} = \hbar \mathbf{k}$. This comes from the realization that the de Broglie relation for the wavelength of a particle can be written as: $\lambda = h/p$, from which we can express the wave vector as: $k = 2\pi/\lambda$. An electron in an atom is subject to the Coulomb potential of the nucleus and thus will be very different from this result. The nuclear potential will severely restrict the movement of the electron. In atoms with more that one electron, the potential will be further complicated by their charges, shifting the movement and energies available to them due to the combination of electronic and nuclear charges.

At this point we can state that the wave function for an electron in an atom is called an *atomic orbital.* This atomic orbital will describe the region of space, in the vicinity of the atomic nucleus, where it is most probable to find the electron. The mathematical probability of finding the electron, with wave function, $\psi(\mathbf{r})$, at a position $\mathbf{r}$, is given by the probability density function, $P_n(\mathbf{r}) = |\psi_n(\mathbf{r})|^2$. The subscript n refers to the specific electron orbit in question. We can picture the change of energy of an electron in the atom as a shift in the wave pattern, and is typically associated with the absorption or emission of radiation, as discussed in the Bohr model.

Since electrons can only occupy restricted energies in the atom, we are presented with the problem of defining how this occurs. We must of course always bear in mind that the electron is subject to forces or potentials which limit how they can move in the atom itself. This is a complex situation and we will limit ourselves to the basic rules which govern the way electrons occupy the available energy levels in the atom. We saw earlier that we can define the basic energy levels based on the quantization given by n, see Equation (1.11). This quantity is called the *principal quantum number,* taking discrete values; 1, 2, 3, ... and defines the energy (shell) of the electron. In fact, the electron states in an atom are described by four quantum numbers, three associated with the orbit (n, l, m_l), with the fourth, (m_s), specifying how many electrons can occupy that particular orbital. We can list the four quantum numbers as follows:

1. **Principle quantum number**, $n : n = 1, 2, 3, \ldots$. This determines the energy of the electron as well as defining its orbital size or radial probability distribution. All electrons with the same value of n are said to occupy the same *shell* or *level* and have the same energy. Such energies (of different electrons in the same shell) are said to be degenerate. In the case of the hydrogen atom, $n = 1$ defines the so-called *ground state* of the atom, and for $n > 1$, the atom is said to be *excited.* The total number of orbitals for any given n is n^2. Shells are usually labelled as: K $(n = 1)$, L $(n = 2)$, M $(n = 3)$, ...

2. **Angular momentum quantum number**, $l : l = 0, 1, \ldots, n - 1$. This quantity specifies the orbital shape and essentially divides the shells into sub-shells or sublevels. The designation of the sub-levels derives from spectroscopic nomenclature, given as follows: s: $l = 0$; p: $l = 1$; d: $l = 2$; f: $l = 3$; g: $l = 4$;

h: $l = 5$; ... The meanings for the first designations are, s - sharp, p - principal, d - diffuse, f - fundamental.

3. **Magnetic quantum number**, m_l : $m_l = -l, ..., 0, ..., +l$. The magnetic quantum number specifies the spatial orientation of an orbital for a given energy (n) and shape (l). This quantum number divides the sub-shell into the electron orbits, of which there will be $2l + 1$ in each sub-shell. With the application of a magnetic field, these ordinarily degenerate levels can be separated (i.e., degeneracy is lifted), giving rise to further spectral lines in emission spectra. This is known as the *Zeeman effect*.

4. **Spin quantum number**, m_s : $m_s = +1/2$ or $-1/2$. The final quantum number indicates the orientation of the spin axis for that particular electron. So for each electron configuration of n, l and m_l, there are two possible states; $+1/2$ (*spin-up*) or $-1/2$ (*spin-down*).

Electrons are particles with spin half (as indicated above) and are labelled as *fermions*, particles which obey Fermi - Dirac statistics. We will briefly touch on this in Chapter 6. Fermions are also subject to the *Pauli exclusion principle*, which states that no two fermions in the same system (atom) can have the exact same set of quantum numbers. This means that within an atom, there is only space for a maximum of two electron in any given combination of (n, l, m_l). These must have opposite spin to be accommodated. The Pauli exclusion principle can be thought to arise due to the Coulomb repulsion between the electrons, hence the opposite spin keeps the electrons as far away from each other as much as possible in any particular orbital, since they move in opposite senses around the nucleus. The Pauli exclusion principle has a number of important consequences and in effect brings about very specific properties of the elements. For example, for two electrons in the same orbital, the spins must be opposite to each other; the spins are said to be *paired*. Substances in which all the electrons occur in such pairs are not attracted by magnets and are said to be *diamagnetic*. Atoms with more electrons that spin in one direction than another contain unpaired electrons. These substances are weakly attracted to magnets and are said to be *paramagnetic*. We will have more to say about these and other magnetic properties in Chapter 10.

TABLE 1.1: Allowed quantum numbers and orbital designations

n	l	m_l	No. of orbitals	Orbital name	No. of electrons
1	0	0	1	1s	2
2	0	0	1	2s	2
	1	−1, 0, +1	3	2p	6
3	0	0	1	3s	2
	1	−1, 0, +1	3	3p	6
	2	−2, −1, 0, +1, +2	5	3d	10
4	0	0	1	4s	2
	1	−1, 0, +1	3	4p	6
	2	−2, −1, 0, +1, +2	5	4d	10
	3	−3, −2, −1, 0, +1, +2, +3	7	4f	14

It is useful to see some of the patterns emerging from this picture of the electronic states in an atom as governed by the quantum numbers. Table 1.1 above shows this to good effect.

For a particular atom in its ground state, the distribution of electrons in their orbitals is referred to as the *electronic configuration*. The order of filling of electrons generally occurs according to the *aufbau principle* ("building-up"). This is illustrated in Figure 1.3. The arrows show the order of filling of the various orbitals.

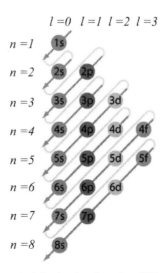

FIGURE 1.3: The Aufbau principle, showing the order of filling of the electronic shells.

1.3 THE PERIODIC TABLE

Once we have understood the basics of quantum numbers and the electronic configurations of the various atoms, we are in a position to understand the shape and structure of the periodic table of elements. As previously noted, the number of electrons in an atom must be equal to the number of protons in its nucleus, thus in its atomic elemental form, the atom maintains charge neutrality. For example, the He atom has two protons and two electrons, while beryllium has four of each. Once we have established the number of electrons in an element/atom, we can use the aufbau principle to determine its electronic configuration. So for the case of beryllium, its four electrons are distributed with 2 electrons in the 1s state and two in the 2s. In each of these sub-shells the electrons are accommodated with opposite spin, as dictated by the Pauli exclusion principle. We can write this electronic configuration as follows: $1s^2 2s^2$. A more complex example would be sodium, which has 11 electrons; its ground state electronic configuration is: $1s^2 2s^2 2p^6 3s^1$. We should add a word of warning here that the aufbau principle is a rough rule of thumb and there are important exceptions to the rule, which occur in the heavier elements. Each atom in elemental order will have one more electron than the previous, thus gradually filling up the available electronic orbitals and being designated with the sequential order of quantum numbers. This is reflected in the the periodic table of elements, as shown in Figure 1.4.

This particular example of the periodic table indicates the electronic configurations for the outer electron shells, the inner shells being full. There are a number of things that should be pointed out from the general structure of this table and its relation to chemical properties and position. Firstly, we note that the noble gases have completely full electronic shells and are located in the right most column of the periodic table. These are chemically inert and do not form chemical bonds under normal conditions of temperature and pressure, and are hence gases. Elements to the left hand side are more reactive chemically. This can be conveniently seen if we group the elements in terms of the shells and orbitals, see Figure 1.5.

Here we clearly see that the s-block of elements correspond to the alkali metals, the d-block corresponds to the transition metals.

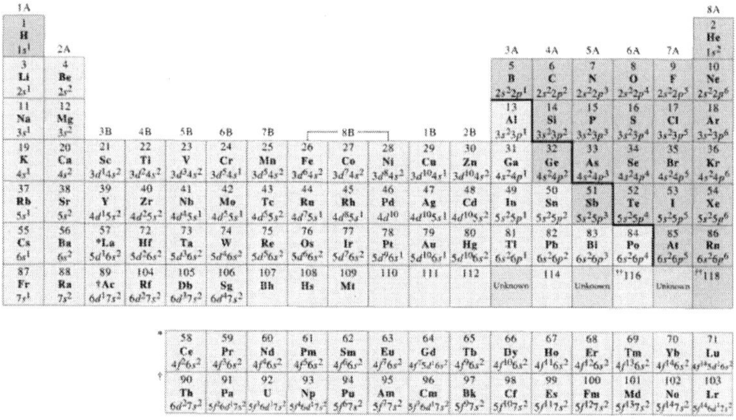

FIGURE 1.4: The periodic table of elements indicating the outer electron configurations.

The p-block terminate in the noble gases and the f-blocks correspond to the lanthanides (rare-earths) and actinide series. While the final column in the p-block corresponds to the noble (inert) gases, the previous column are known as halogens, and are also chemically active. There are many other properties to this table and we will be referring to some of these later in the book. What is essential to note is the similarity of the electronic configurations in any one group

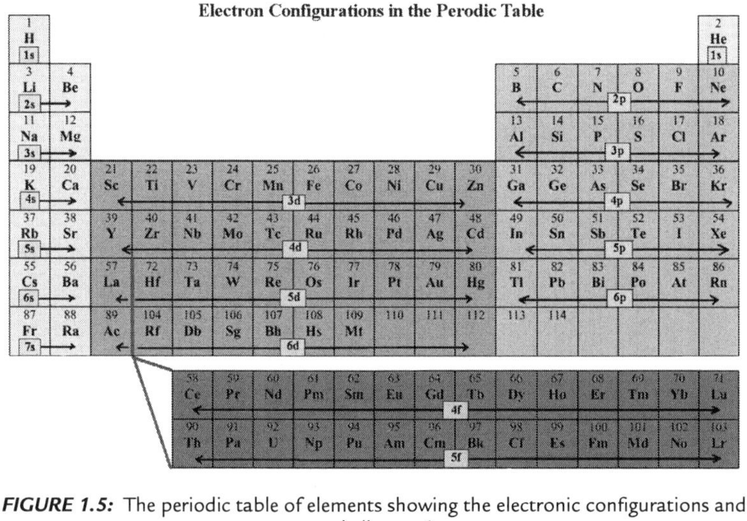

FIGURE 1.5: The periodic table of elements showing the electronic configurations and outer shell groupings.

(column) of the periodic table. It is from this relation that elements in a particular column have similar chemical properties, as indicated in their name groupings; e.g. alkalis, halogens etc. For example, they have the same valence electron states and can form very similar types of compounds as a result. We will consider the example of group 6A (16), to illustrate the point. The electronic configurations of this group of elements is shown in Table 1.2 below.

TABLE 1.2: Electronic configurations for the group 6A elements

Element	Configuration
O	[He] $2s^2 2p^4$
S	[Ne] $3s^2 3p^4$
Se	[Ar] $3d^{10} 4s^2 4p^4$
Te	[Kr] $4d^{10} 5s^2 5p^4$
Po	[Xe] $4f^{14} 5d^{10} 6s^2 6p^4$

Here we note that the electronic configurations are indicated in terms of those after the preceding noble gas element. In all cases the outer shells have two s-electrons and four p-electrons. They display very similar chemical and physical properties. Unfortunately a full description of the periodic table would be the subject of a book in itself, though we get a flavour for the relation between the position of an element in the periodic table and the electronic structure or configuration.

1.4 INTERATOMIC BONDING

The manner in which the different elements interact and bond together to form molecules and solids is intimately related to their electronic structure. Of course to form any type of molecule or solid, there must exist an attractive force between atoms. However, there are also repulsive forces which prevent the atoms from getting too close together. It is the balance of these attractive and repulsive forces that give rise to an equilibrium configuration between a conglomerate of atoms in the formation of a three dimensional spatial arrangement of the atoms in the solid. For simplicity we can consider two identical atoms that are gradually brought closer together. For large separations the forces between the atoms will be negligible (i.e., the interaction potential energy can be said to be zero). As the atoms draw closer together, the interaction potential will increase and attractive forces come into play drawing them closer together. We note that in mathematical terms, this energy is negative, as with the binding energies of the electrons in the atoms, and work must be done on the atoms to separate them. When the atomic separation is of the order of a few atomic diameters, repulsive short range forces come into play, which prevent crystals collapsing in on themselves. The potential energy here is positive meaning that work must be done to bring the atoms even closer. It is the balance between these opposing forces which leads to a final equilibrium. As a matter of fact, the attractive forces between the atoms draw them together until the electron clouds begin to overlap and then strong repulsive forces arise to comply with the Pauli exclusion principle. The equilibrium separation, r_0, between the atoms occurs where the two forces are of equal magnitude, giving a stable configuration of minimum potential energy. We can represent this situation in mathematical form as a simple sum of attractive and repulsive forces as expressed by the relation:

$$U(r) = -\frac{A}{r^n} + \frac{B}{r^m} \qquad (1.21)$$

where r is the distance between the atom centers and A, B, n and m are constants characteristic for these particular atoms. The form of this equation is known as the *Mie potential*. The interaction force between the atoms can be obtained from:

$$F = -\frac{dU(r)}{dr} = -\frac{nA}{r^{n+1}} + \frac{mB}{r^{m+1}} \qquad (1.22)$$

This situation is illustrated in Figure 1.6.

We note that the equilibrium occurs at the point where the potential energy is a minimum and the interatomic force passes through zero. Furthermore, noting that at equilibrium ($r = r_0$) $F = 0$ and $U(r) = $ min., we can obtain the equilibrium separation of the atoms as:

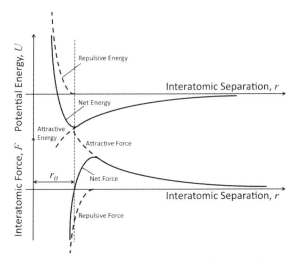

FIGURE 1.6: Variation of (a) potential energy and (b) force as a function of the interatomic separation. Attractive and repulsive regions are shown. The depth of the minimum of the potential energy corresponds to the total binding energy between the atoms. The equilibrium separation is indicated with r_0.

$$r_0 = \left(\frac{mB}{nA}\right)^{1/(m-n)} \tag{1.23}$$

Substituting this into the potential energy, we can find the equilibrium state potential as:

$$U(r_0) = -\frac{A}{r_0^n}\left(1 - \frac{n}{m}\right) \tag{1.24}$$

This value is negative as appropriate to a bound state. This would be the energy required to separate the two atoms. We can see that while at equilibrium the attractive and repulsive forces must be equal, however, their corresponding energies are not; $m \neq n$. In fact, stable aggregations of atoms can only occur for the condition $m > n$. This is another way of saying that repulsive forces must be of shorter range than attractive ones. We can show this by employing the condition:

$$\left[\frac{d^2 U(r)}{dr^2}\right]_{r=r_0} > 0 \tag{1.25}$$

This condition leads directly to $m > n$. Of course this should be evident from the form of Equation (1.24), since only then would $U(r_0)$ remain negative for the bound state condition. A very specific and well known example of the above uses $m = 12$ and $n = 6$ and is known as the *Lennard - Jones (6 - 12) potential*.

Chemists are concerned with the way in which small numbers of atoms bind together to form a molecule, in solid state physics we are concerned with essentially the same problem, though with more atoms involved. In the remainder of this section we will outline the main physical mechanism by which this occurs. Here we will draw on our knowledge of physical forces and the periodic table. Before discussing these mechanisms, we note an important distinction in types of bonding: strong bonds occur when electrons are exchanged in some form (as in ionic, covalent and metal bonds), while weak bonding occurs when electrons are unevenly distributed in atoms to form dipole moment which interact via electrostatic forces (hydrogen and van der Waals bonds). The former are termed *chemical bonds* while the latter are called *physical bonds*. A similar distinction is made for atoms on a surface, which can *chemisorb* or *physisorb* depending on whether there is an exchange of electrons or not.

1.4.1 Ionic Bonding

This is probably the simplest form of bonding. This occurs in compounds whose elements readily form ions, which is an excess or deficiency of electrons from their normal neutral state. A common example is for rock salt, NaCl. It will be seen from their positions in the periodic table, that Na and Cl either have one electron in the outer shell (Na) or one missing in the outer shell (Cl). Once these atoms are close enough, it becomes energetically favorable for the excess electron in the Na to fill the outer shell of the Cl, thus leaving both in an ionic state; Na^+, Cl^-, and in much more stable configurations since they both now have filled outer electron shells. The pair of atoms have the same number of excess/deficient electrons in their outer shell, so pure ionic bonding will only occur in these cases. Other examples of such pairs are NaF, KCl, CsCl. This type of bonding cannot occur between atoms of the same type for this reason. A consequence of the electron exchange is that the atoms are no longer neutral entities and attract one another via the Coulomb interaction. This process of electron transfer can be otherwise seen as the effect of the small ionization potential of the alkali metal (Na) and the large electron affinity of the halogen (Cl). The Coulomb potential has the form:

$$U_{Coul}(r) = -\frac{\nu e^2}{4\pi\epsilon_0 r} \tag{1.26}$$

where $\nu = 1$ for singly ionized species as is the case for NaCl. The energy associated with repulsion is frequently given in the form of an exponential, which is an alternative form of the repulsion energy given above. This can be written as: $B \exp - (r/\rho)$. Here ρ plays the role of a repulsion exponent (having the units of length) and determines the distance at which the repulsive forces become important. The total energy per molecule of an ionic crystal can be written as:

$$U_{ion}(r) = -\frac{\alpha \nu e^2}{4\pi\epsilon_0 r} + B \exp - (r/\rho) \tag{1.27}$$

The additional constant here, α, is known as the *Madelung constant* and depends on the geometrical arrangement of the ions. (We note here that for the NaCl structure this has a value of 1.74756, while for zincblende it is 1.63805 and for the CsCl type structure we have a value of 1.76267.) We can eliminate one of the unknowns in the Equation (1.27) by recognizing that the first derivative with respect to r will be zero at the equilibrium separation, r_0. From this we find:

$$B = \left(\frac{\alpha \nu e^2}{4\pi\epsilon_0 r_0^2}\right) \exp\left(\frac{r_0}{\rho}\right) \tag{1.28}$$

We can now rewrite the energy as:

$$U_{ion}(r) = -\left(\frac{\alpha \nu e^2}{4\pi\epsilon_0 r_0}\right)\left[1 - \frac{r\rho}{r_0^2}\exp\left(\frac{r-r_0}{\rho}\right)\right] \tag{1.29}$$

From this it is a simple task to evaluate the *cohesive energy* (potential energy at equilibrium):

$$U_{ion}(r_0) = -\left(\frac{\alpha \nu e^2}{4\pi\epsilon_0 r_0}\right)\left[1 - \frac{\rho}{r_0}\right] \tag{1.30}$$

Typically ρ is only a small percentage of r_0 and the cohesive energy will be dominated by the Madelung term. The ionic bonding mechanism can be visualized as shown schematically in Figure 1.7.

1.4.2 Covalent Bonding

Elements in the more central region of the periodic table are not so easily reduced to closed electron shell structures since it would require excessive energies to transfer electrons between species and therefore the ionic type bonding is unlikely. Instead bonding can occur via the sharing of valence electrons, where each atom contributes one of more electron in the sharing. The mechanism of electron sharing is termed *covalent bonding*. In its simplest form, for the hydrogen molecule, H_2, the two hydrogen atoms each contribute their $1s^1$ electron to form an electron-pair covalent bond.

The proximity of the two H atoms will cause an interaction between their respective electrons, and split into two states of differing energy. One is referred to as the *bonding state* and has an even (symmetric) orbital wave function, whose overall energy is lowered from the isolated electron energy, thus bringing about the bond. The symmetric solution requires the electron charge density, $-e|\psi|^2$, to

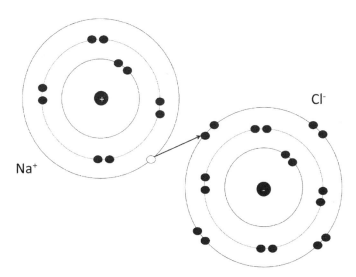

FIGURE 1.7: The ionic bond: An Na^+ cation (left) and Cl^- anion (right) are formed by the electron transfer from the Na to Cl atom and are held together by Coulomb forces.

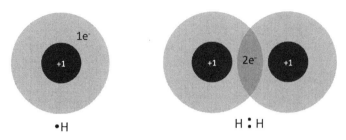

FIGURE 1.8: The covalent bond in its simplest form between a pair of H atoms to form the H_2 molecule.

be concentrated in the region between the nuclei. A requirement of the Pauli exclusion principle is that the total wave function be anti-symmetric, and will be satisfied for the 1s state with the electrons aligning with anti-parallel spins. The other state, called the *anti-bonding* state, has the two electrons with parallel spins and is a purely repulsive. Since this latter state has a higher energy and the two electrons can be accommodated in the 1s state, it is the formed, bonding state which will be realized and a strong bong is formed. It will be noted that this situation could not occur for example with two He atoms, since all the s - states are occupied and would only result in an increase of the overall total energy.

It is clear that for the covalent bonding to exist, the valence electron shells cannot be fully occupied, as in He. For heavier atoms, higher orbitals are occupied. These orbitals, unlike the s-orbitals have a very directional character, as illustrated in Figure 1.9. This directional character of the orbitals arises from energy considerations and the Pauli principle which means that the electrons try to avoid each other forming orbitals which are symmetrically distributed in space. The consequence of this is that covalent bonds formed by the overlap of partially vacant orbitals are also highly directional. For example, the covalent bonds of carbon atoms gives rise to the tetrahedral unit of the diamond structure, which will be discussed in Chapter 2. The electronic configuration for C is $1s^2 2s^1 2p_x^1 2p_y^1 2p_z^1$. In this ground state configuration we have four unpaired electrons, which are hybridized to form the four equally spaced lobes (tetrahedron) of the sp^3 hybrid orbitals with angles of 109.5° between them, see Figure 1.10. We note here that the hybridization of states (in this particular case

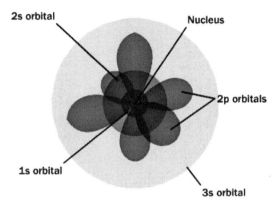

FIGURE 1.9: Lower level orbitals showing the directional character of the p-orbitals and the symmetric distribution of the s-orbitals

we are considering s and p states) involves a mixing of the electronic states between more than one atom. This gives rise to different energy levels which are more favorable since there are more states available for the electrons and bind the atoms together. The combination will have a lower overall energy compared to the isolated atoms. A full discussion of hybridization is beyond the scope of the current chapter, see for example Sutton (1996) for further details.

The covalent mechanism is responsible for the bonds in most of the organic compounds. It is also the dominant pairing for halogen atoms. The covalent bond is also very important in a majority of semiconducting materials. The sp hybridization (whether type 2 or 3) are important in the formation not only of diamond (sp^3),

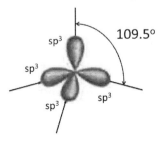

Tetrahedral

FIGURE 1.10: Spatial distribution of the sp^3 - orbitals forming the basis of a tetrahedron.

but also in the much more common allotrope graphite form, which results from sp² type bonds, forming an atomically thin carbon sheet (graphene). This structure is essentially a 2D hexagonal formation, where in graphite the van der Waals interaction between the graphene sheets forms the 3D structure. The difference between the diamond and graphite forms of carbon illustrate the importance of structure on physical properties: while diamond is an electrical insulator, graphite is a good conductor. The weak interlayer bonds in graphite allow it to be flaked or even to produce graphene sheets using sticky tape. The covalent bond can be extremely strong and is responsible for the legendary hardness of diamond, for example.

1.4.3 Mixed Covalent and Ionic Bonding

Exclusively covalent or ionic bonding is not always possible, and the two mechanism compete for dominance. In such cases, mixed bonds can result. In fact, there is a continuous progression from a purely ionic character to a purely covalent one as we consider compounds in which the electronegativity gradually changes. The partial tendency towards electron sharing arises from a resonance between the ionic and covalent configurations. The time-averaged wave function for a bonding electron can be expressed as:

$$\psi = \psi_{cov} + \lambda \psi_{ion} \qquad (1.31)$$

where ψ_{cov} and ψ_{ion} are the normalized electron bonding wave functions for covalent and ionic forms, λ is a parameter which expresses the degree of ionicity, defined as:

$$\text{percent ionicity} = \left(\frac{100\lambda^2}{1+\lambda^2} \right) \qquad (1.32)$$

The appropriate value of λ will be determined by quantum-mechanical calculations based on the most stable (i.e., equilibrium) configuration.

1.4.4 Metallic Bonding

The metallic bond is a more complex mechanism and we need to consider the solid as a whole instead of the individual interactions between atoms. The atoms in a metal readily release their valence electrons which are then shared in the solid. The easiest way of thinking about the metallic bond is by considering the positive ions (i.e., the atoms which have released their valence electrons) as being held together by the "free" electrons, which move around in the spaces between the ions. This is a rather crude image and in reality the situation is somewhat more complex. The electrons are partially or weakly attached to the atoms and move from one atom to another, and as such are *delocalized,* as opposed to the localized electrons in the ionic or covalent bonding mechanisms. The materials most commonly formed by the metallic bonds fall into the d-block elements (see Figure 1.5), though this is not exclusively the case. It is these outer d-electrons which form the "cloud" or "sea" of electrons which act as the "glue" for the positive ions. This picture is illustrated in Figure 1.11. The core electrons are still localized on the atoms.

The description of the free electrons will be via wave functions, which have more in common with free electrons in a weak periodic potential (formed by the positive ions). This will be discussed in more detail when we consider the electronic properties of solids and we will introduce the concepts related to the periodic potential, see Chapter 7. It would be incorrect not to mention that while there is a net attractive force at play, the positive ions also exert repulsive forces between themselves. We end this section by noting that the

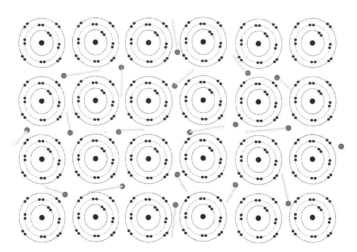

FIGURE 1.11: The metallic bond results from the attraction between the "free" electrons and the ion cores of the atoms.

metallic bond gives rise to some of the most important close-packed structures (face-centered cubic and hexagonal close-packed), which is a consequence of the lack of directionality of the metallic bond.

1.4.5 Hydrogen Bonding

Since the hydrogen atom has only one electron, it can only be covalently bonded to one other atom. In forming the covalent bond, the electron distribution can be very asymmetric, giving rise to an electric dipole moment. An example of this is the water molecule, where two hydrogen atoms covalently bond to an oxygen atom (H_2O). The directionality of these covalent bonds then gives rise to the polar water molecule, which can now, via electrostatic forces, attract and bond to other such molecules. This is illustrated in Figure 1.12, which shows the formation of ice crystals, which can take on a structure similar to wurtzite (see Chapter 2), having hexagon symmetry.

The relatively weak H-bonding provided by this type of electrostatic force allows the ice structure to flip into a variety of forms, depending on the prevailing conditions of pressure and temperature. In addition to water, the H-bonds are also responsible for the formation of many molecules, such as HF, HCN and NH_4F. The hydrogen bond plays an enormously important role in biological materials, such being the forces holding together the two strands of the double helix of the DNA molecule. It is the relatively weak H-bonds between the polynucleotide chains which allow it to "unzip" for gene replication. While the individual bonds between the base pairs is

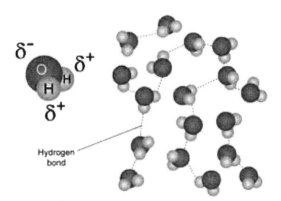

FIGURE 1.12: Hydrogen bonds between water molecules.

relatively weak, the structure as a whole is very stable since there are many such base pairs and hence bonds in the DNA molecule. It is important that this bond strength is significantly weaker than the interatomic bonds between the atoms within the bases, otherwise the molecule could not function biologically.

1.4.6 Van der Waals Bonding

The dipole bonds as found in hydrogen bonding are directional. However, weak dipole bonds that are nondirectional can also be formed between neutral atoms and molecules. These fluctuating bonds are termed van der Waals bonds and are due to the weak attractive forces that can exist when there are (temporarily) more electrons on one side of an atom than another. The centers of positive and negative change momentarily have different positions giving a weak net dipole moment. Interactions can thus take place between such atoms or molecules in a gas, leading to deviations from the ideal gas equation. It is also this force which allows inert gas atoms to condense a very low temperatures and give rise to structures of a close-packed type, typically face-centered cubic.

1.5 SUMMARY

The internal structure of an atom plays a crucial role in how atoms interact with other atoms. This ultimately depends on how many protons it has. Since the charges on the proton and electron are of the same magnitude, the nucleus essentially attracts an equal number of electrons to make the atom a neutral and stable entity. However, the complex behavior of electrons in their interaction with both the nucleus and other electrons, means they form shell structures. Inner (core) electrons being more tightly bound than outer (valence) electrons, which are shielded from the full nuclear charge by the core electrons. While the Bohr model of the atom improved on the previous Rutherford picture, quantum mechanics was ultimately more successful in bringing out a more detailed and accurate picture of how the electrons behave in the presence of the nuclear potential, binding them to it. We designate four (quantum) numbers to each electron, which fundamentally determine how the are located in the

TABLE 1.3: Summary of the classification of bonding types in solids. (Table adapted from Blakemore 1985.)

Bond Type	Material	Crystal Structure	Binding Energy (eV/molecule)	Nearest neighbor (Å)	Characteristic Properties
van der Waals	Ar	fcc	0.1	3.76	Low melting and boiling temperatures.
	Cl	Tetragonal	0.3	4.34	
	H	hcp	0.01	3.75	Electrically insulating.
Covalent	Si	Diamond	3.7	2.35	Hard. High melting temperature.
	InSb	Zinc blende	3.4	2.80	Insulator or semiconductor.
	Mg_2Sn	Cubic	1.0	2.92	
Ionic	KCl	Rock salt (cubic)	7.3	3.14	Brittle and fairly high melting temperature.
	AgBr	Rock salt (cubic)	5.4	2.88	Electrically insulating.
Metallic	Na	bcc	1.1	3.70	High conductivity.
	Ag	fcc	3.0	2.88	Close-packing.
	Ni	fcc	4.4	2.48	
Hydrogen	Ice	Hexagonal	0.5	1.75	Low melting temperature and insulating.

atom. The realization that the physical and chemical properties of the elements depends directly on the electronic structure of atoms was a major advance in modern science. It is from here that we can understand the shape and structure of the periodic table of elements. From the electron filling of the available states, which importantly adheres to the Pauli exclusion principle, we can determine how atoms, via electron sharing and electrostatic forces, interact with one another allowing molecules and solids to form. Directionality, in many cases, will determine the physical distribution of these atoms into three-dimensional structures. We will discuss the arrangement of these atoms in the next chapter. In Table (1.3), we can summarize the main properties of the bonding mechanisms that we have outlined in this chapter. Although this is not an exhaustive comparison, it does provide a good indication of the strengths and properties of bonding types giving some examples.

REFERENCES AND FURTHER READING

Basic Texts

- B. H. Bransden and C. J. Joachain, *Quantum Mechanics*, 2e Prentice-Hall, Essex (2000)

- H. M. Rosenberg, *The Solid State: An Introduction to the Physics of Crystals for Students of Physics, Materials Science and Engineering*, Oxford University Press, Oxford (1978)

- J. S. Blakemore, *Solid State Physics*, Cambridge University Press, Cambridge (1985)

- H. P. Myers, *Introductory Solid State Physics*, Taylor and Francis, London (1998)

Advanced Texts

- A. P. Sutton, *Electronic Structure of Materials*, Oxford University Press, Oxford (1996)

- B. H. Bransden and C. J. Joachain, *Physics of Atoms and Molecules*, Longman, London (1986)

- M. A. Wahab, Solid State Physics: Structure and Properties of Materials, Alpha Science International Ltd., Harrow (2007)

- N. W. Ashcroft and N. D. Mermin, *Solid State Physics*, Saunders College, Philadelphia (1976)

EXERCISES

Q1. Derive Equations (1.9) through (1.11).

Q2. Calculate the first four lines (wavelengths and frequencies) of the emission spectrum for atomic hydrogen.

Q3. Determine the radii of the first four electronic orbits for the hydrogen atom. Quote your answers in terms of the Bohr radius.

Q4. What are the orbital periods and frequencies of electrons in the $n=1$, 2 and 3 states of a hydrogen atom?

Q5. What is the binding energy of an electron in the 1s state of singly ionized helium?

Q6. What are the ground state electronic configurations of P, Ga^{3+}, and Cl^{2-}?

Q7. Show that for the potential given by Equation (1.21), it is necessary for the condition $m > n$ to have a stable binding state.

Q8. Use the Lennard-Jones (6-12) potential to evaluate the binding energy between a K and a Cl ion. ($A = 1.78$).

Q9. The potential energy for an ionic crystal of the rock salt structure containing N ions of each type can be expressed as:

$$U(r) = -N\left(\frac{\alpha e^2}{4\pi\epsilon_0 r} - \frac{B}{r^n}\right) \qquad (1.33)$$

where α is the Madelung constant and B a constant. Derive an expression for the bulk modulus of compressibility, $V(\mathrm{d}V/\mathrm{d}P)$.

Q10. Prove Equation (1.30)

NOTES

[1] Note, the energy unit eV is equivalent to the numerical value of the fundamental electronic charge in Joules; i.e., 1 eV = 1.60219×10^{-19} J

CHAPTER 2

CRYSTALLINITY IN SOLIDS

"The person who never made mistakes, never tried anything new"
—Albert Einstein

2.1 INTRODUCTION

In the previous chapter, we discussed the various mechanisms for atomic bonding. In doing so we noted that in certain cases this bonding has a directional character, while in others it does not. The consequence of directionality, or not, has a profound effect on the way solids are formed. This gives rise to the variety of crystalline order found in solids. For example, in the case of the covalent bonding in carbon, the directionality of the sp hybridization produces a tetragonal sub-unit (sp^3) in the diamond structure, or a three-fold symmetry (sp^2) in the bond directions, as is observed in graphite or graphene. The prevailing structure will depend on the crystallization conditions of temperature and pressure. In the case of metals, the homogeneous nature of the bond and the delocalized nature of the electrons frequently gives rise to some form of close packed structure, for example Ni, which has a face-centered cubic (fcc) structure.

In this chapter, we will be concerned with the description of the solid formation into regular periodic structures which result from the interatomic bonding between various atomic species. In solid state physics, we aim to describe the material properties in terms of the crystalline structures of the constituent elements. Therefore we need a rigorous way in which to characterize the three-dimensional spatial distribution of atoms. We do this by describing the symmetry relations of the particular crystalline arrangement. By this we mean the long range periodicity of the atomic array. Clearly, real structures are never perfect and defects in this ordering can result from a number of origins. Defects are an extremely important part of the description of real crystals and materials and can have a profound effect on the overall properties of the solid, a discussion of imperfections to the crystalline order will be given in Chapter 4. In this chapter, we will consider only perfect arrays of atoms or molecules, which produces the principal physical properties in solid materials. The description of crystalline symmetries allows us to describe very large collections of atoms (of the order of 10^{24} in a typical solid crystal), which would be impossible otherwise. It is this regular structure which allows us to reduce the material to its basic constituent components (which could be an atom, a molecule or a group of atoms or molecules). This is repeated many times through the solid and related by some form of symmetry operation. By reducing the problem to the minimum repeat structure, with appropriate boundary condition, we can simplify the description of the solid and model the physical properties of the crystal. It is therefore essential for us to describe the symmetry and crystalline structure in a formal way, which will then allow us to consider the specific physical properties of the material.

2.2 ASPECTS OF SYMMETRY IN CRYSTALLINE MATERIALS

An ideal single crystal will have an *infinite* three-dimensional repetition of the fundamental constituents or building blocks, each with the same spatial orientation. This building block is called the *basis*, and is made of an atom or a group of atoms. The basis is therefore the smallest repeat unit, which makes up the *primitive unit cell*, (we will return to discuss the choice of *unit cells* shortly). This is a 3D geometrical form that is then spatially translated in discrete steps in three-dimensional space to fill all our crystal. We will not consider the surfaces of the crystal, which will have different coordinations of atoms and neighbors, resulting in different physical properties. Usually the surface region is a very small fraction of the crystal which has a negligible effect on the bulk properties. Cases where this is not the case concern ultrathin films and nanocrystals, and will be discussed in the final chapter of this book, when we consider aspects of *nanotechnologies*.

2.2.1 Translational Symmetry

The *translational symmetry* is the simplest form of symmetry operation which is required to describe the crystalline structure of a solid. We define this quantity in terms of three basic translation vectors, $\mathbf{a}_1$, $\mathbf{a}_2$ and $\mathbf{a}_3$. The importance of this translation vector is that when the operation is made inside the solid, the atomic environment remains unchanged (or invariant), and any number of repeats of this operation leave us in identical positions inside the crystal. We define the translation operation as:

$$\mathbf{T} = n_1 \mathbf{a}_1 + n_2 \mathbf{a}_2 + n_3 \mathbf{a}_3 = \sum_i n_i \mathbf{a}_i \qquad (2.1)$$

where the n_i are integers and $\mathbf{a}_i$ define the edges of a parallelopiped of the unit cell. We note that the $\mathbf{a}_i$ vectors are not necessarily orthogonal and are chosen, depending on the type of crystalline order, for convenience. The vector $\mathbf{T}$, as defined above, is also known as a *lattice vector*. It is important to note that the translational symmetry extends beyond just describing the atomic environment and refers to any location in the crystal, which may or may not coincide with the atomic positions. Thus any point described by a general vector $\mathbf{r}$, in the crystal, can be translated by the vector $\mathbf{T}$ and the local environment will be identical. This means that any physical environment, such as charge density or local internal magnetic field, will remain the same under the action of this operation, i.e., invariant. We thus write:

$$\mathbf{r}' = \mathbf{r} + \mathbf{T} \qquad (2.2)$$

where $\mathbf{r}'$ is a general vector and $\mathbf{r}$ is a vector within the unit cell. The set of operations described by **T** define the *space lattice* or *Bravais lattice* of the crystal. These are geometrical concepts which aid our classification of crystal symmetries. The real crystal results when we define a basis around any geometric point in the Bravais lattice. There are five possible Bravais lattices in two dimensions and fourteen in three-dimensional space. These latter will be defined in the next section.

Only in special cases can lattices be defined in which the three primitive vectors, $\mathbf{a}_i$, be chosen to be of equal length and more rarely as being mutually perpendicular. We note that the term *primitive* in our context means that the vector must be the smallest possible vector to satisfy the repeat unit. These vectors can be used to define the *primitive unit cell*. Other unit cells (i.e., non-primitive cells) can be defined using other non-primitive vectors to define a convenient form of repeat unit.

2.2.2 The Basis and the Unit Cell

In order to correctly define a lattice, we must define the basis, or repeat unit, as well as the symmetry of that repeat unit. In the case of a primitive unit cell, we usually need less atoms in the basis than a non-primitive choice of unit cell. For some simple monoatomic crystals, this will be a single atom, e.g. Na and Fe, which form body-centered cubic lattices. Other more complex structure require more than one atom in the basis, such as Si, which crystallizes in the diamond structure and has two atoms in the basis. More complex bases are required for diatomic crystals where the basis must contain at least one atom of each type, such as GaAs, which has an atom of each type in the basis, forming a zinc-blende like crystal. Crystals of organic compounds can have many atoms in the basis.

The choice of unit cell can be quite arbitrary and is frequently made as a matter of convenience. In Figure 2.1 below, we show a 2D lattice with different, and importantly, equally valid examples of unit cells for the same crystalline structure. It would be more common to choose the simplest unit cell, such as choice (i). In cases where the primitive vectors can be drawn at right angles, these would be the most convenient and more common choice.

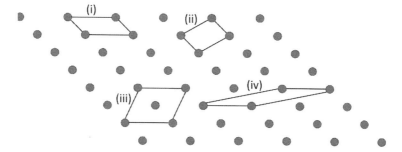

FIGURE 2.1: Some examples of units cells in a 2D lattice.

In some cases where the primitive vectors are not orthogonal, the choice of unit cell vectors can be made to define a unit cell which is different from the primitive case. Note that choice (iii) in Figure 2.1 is still a valid unit cell, though it is not a primitive unit cell.

It is useful to distinguish between the unit cell and the primitive unit cell. This is usually done by using using a different notation. In the above we have use the symbol $\mathbf{a}_i (i = 1, 2, 3)$ for the primitive vectors, which define the primitive unit cell. It is convention to use $\mathbf{a}$, $\mathbf{b}$ and $\mathbf{c}$ to define the vectors of the conventional unit cell.

We can define the volume of a unit cell, in three dimensions, using the lattice vectors which define the unit cell. We note that this volume is called the *primitive unit cell volume* if the vectors used are those of the primitive vectors, as defined by the three vectors, $\mathbf{a}_1$, $\mathbf{a}_2$ and $\mathbf{a}_3$. Otherwise it is simply referred to as the volume of the unit cell, as defined by the three vectors $\mathbf{a}$, $\mathbf{b}$ and $\mathbf{c}$. The volume for the primitive unit cell is expressed as:

$$V_{puc} = (\mathbf{a}_1 \times \mathbf{a}_2) \cdot \mathbf{a}_3 \tag{2.3}$$

The volume of the conventional unit cell can be similarly expressed in terms of the unit cell vectors as:

$$V_{uc} = (\mathbf{a} \times \mathbf{b}) \cdot \mathbf{c} \tag{2.4}$$

One special way of choosing a primitive unit cell was suggested by E. P. Wigner and F. Seitz in 1933. The construction of the so-called Wigner-Seitz cell has its origin in a lattice point, see Figure 2.2. The method of construction is as follows: (i) Select a lattice point in the crystalline structure and draw lines to the nearest lattice points. (ii) Bisect each of the lines from the origin to the nearest lattice points with a plane such that they overlap. The Wigner-Seitz cell is the defined as the smallest area (in two dimensions) or volume (in three dimensions) that is enclosed by these planes. Figure 2.2 shows this for a 2D lattice array.

Some three-dimensional Wigner-Seitz structures are shown in Figure 2.3. These constructions turn out to be very useful in the description of electronic states in solids and we will have recourse to their use in Chapter 7, when we describe electron dynamics in solids. Despite the usefulness of the Wigner-Seitz method in electronic structure calculations, it is frequently more convenient to choose a unit cell which has orthogonal axes, as illustrated in Figure 2.4 for the fcc structure. The main difference between the formal choice of the unit cell is that the primitive unit cell will contain a single basis unit per unit cell, while other unit cell choices may have more. Note, for the fcc structure illustrated in Figure 2.4,

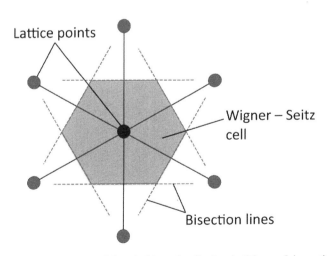

FIGURE 2.2: Construction of the primitive unit cell using the Wigner - Seitz method.

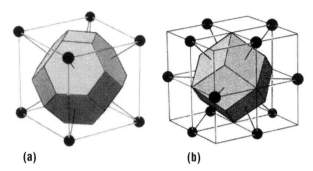

FIGURE 2.3: Wigner - Seitz cell for the bcc (left) and fcc (right) structures.

the primitive unit cell has only one atom, while the conventional unit cell has four atoms.

2.2.3 Elements of Symmetry

Symmetry operations are of fundamental importance when we consider crystalline structures. The symmetry of a crystal dictates the number of allotropic forms that can exist for a specific substance,

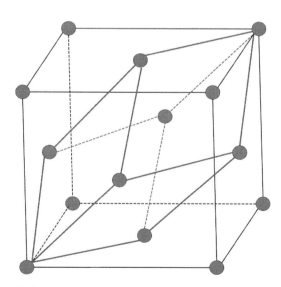

FIGURE 2.4: The face-centered cubic (fcc) structure illustrating the conventional unit cell and the primitive unit cell (inner structure).

though often only one exists. This results from the symmetry of the basis itself (point group) having to be consistent with the specific lattice (Bravais). A rigorous treatment of lattice symmetry operations uses group theory[1]. However, we will only give a brief overview of some of the main symmetry operations that occur in crystalline solids. Further to the translational symmetry that we have already discussed, there are other symmetry operations that leave the original point of consideration in the crystal *invariant*. These symmetry operations can be summarized as: i) Reflection or mirror symmetry, ii) Rotation about an axis (this can be 1, 2, 3, 4 or 6-fold), iii) Inversion through a point, iv) Glide (reflection plus translation), v) Screw (rotation plus translation). The latter two are compound operations, which require the application of more than one of the previous operations. Inversion symmetry is also a complex operation, which can also be represented as a compound operation, being equivalent to a reflection followed by a rotation of 180°. The degree of rotational symmetry can be expressed as follows: a crystal has an n-fold axis of rotation if we can rotate it through an angle of $2\pi/n$ and leave the crystal lattice invariant. In Figure 2.5, we illustrate some of the main symmetry operations.

By considering the combinations of symmetries we can generate the various *point groups* and *space groups* that can be used to define the specific crystal type. A full description of the three-dimensional forms leads to *seven* crystal systems, 32 point groups and 230 space

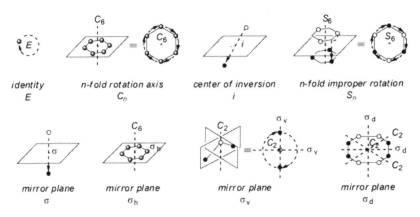

FIGURE 2.5: Main symmetry operations.

groups. The full set of 230 space groups and their relation to the 32 point groups are given in Appendix A.

2.3 BRAVAIS LATTICES

The *Bravais lattice* is a designation of crystal types, which take into account the specific symmetry of the crystalline structure. In all Bravais lattices, the lattice points are generated by the discrete set of translation vectors, as defined in Equation (2.1), where the $\mathbf{a}_i$ are primitive unit vectors of the crystalline array. It is important to note that the Bravais lattice is closed under the addition and subtraction of the lattice vectors. The designation of the Bravais lattice will be unique and two Bravais lattices are often considered equivalent if they have isomorphic symmetry groups. In two dimensions all periodic structures of lattice points can be reduced to one of the five possible Bravais lattices, while in three dimensions there turn out to be 14 possible Bravais lattices (see Section 13.2.1 for a discussion of surface structures). Since we are concerned here with three dimensional solids, we will only consider these.

It is important that we reconcile the fourteen lattice structures with the 32 point groups and the 230 space groups. This can be understood in that while there are various permutations of the symmetry operations within the crystal structure, the way in which we repeat the periodic structure in three-dimensional space can be reduced to just fourteen lattices. This is illustrated in Appendix A, where we see how the seven crystal systems are divided into the various symmetry groups. Clearly there is much more that can be said of the different notations and divisions in this scheme, but this would involve a disproportionate discussion on crystallography. It is sufficient for the purposes of this book to only take into account the Bravais lattices in three dimensions since it allows us to provide a concise and accurate description of three-dimensional crystal lattices. The quantities a_1, a_2 and a_3 are called the lattice parameters. In the case of cubic structures there is just one lattice parameter; $a_1 = a_2 = a_3 = a$. In the Figure (2.6) and Table (2.1) below we illustrate and define these fourteen lattices (with the seven crystal systems).

44 • Solid State Physics

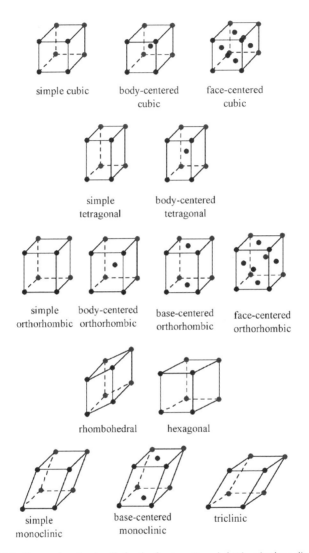

FIGURE 2.6: Conventional unit cells for the fourteen Bravais lattices in three dimensions.

2.4 CRYSTAL PLANES AND AXES: THE MILLER INDICES

It is often necessary to define a particular lattice direction or plane in a three-dimensional crystal. It is therefore important that we follow a specific definition to make sure we know exactly what

TABLE 2.1: The seven crystal systems and fourteen Bravais lattices in three-dimensional structures.

Crystal system	Axial distances	Axial angles	Bravais lattices in system
Triclinic	$a_1 \neq a_2 \neq a_3$	$\alpha \neq \beta \neq \gamma \neq 90°$	Primitive (P)
Monoclinic	$a_1 \neq a_2 \neq a_3$	$\alpha = \gamma = 90° \neq \beta$	Primitive (P) Base centered (C)
Orthorhombic	$a_1 \neq a_2 \neq a_3$	$\alpha = \beta = \gamma = 90°$	Primitive (P) Base-centered (C) Body-centered (I) Face-centered (F)
Tetragonal	$a_1 = a_2 \neq a_3$	$\alpha = \beta = \gamma = 90°$	Primitive (P) Body-centered (I)
Trigonal (Rhombohedral)	$a_1 = a_2 = a_3$	$120° > \alpha = \beta = \gamma \neq 90°$	Primitive (P)
Hexagonal	$a_1 = a_2 \neq a_3$	$\alpha = \beta = 90°, = 120°$	Primitive (P)
Cubic	$a_1 = a_2 = a_3$	$\alpha = \beta = \gamma = 90°$	Primitive (P) Body-centered (I) Face-centered (F)

we mean. This is extremely important in solid state physics, since many physical properties are dependent on the direction in the crystal lattice that we are considering. Very rarely are physical properties homogeneous or isotropic, in scientific terms, we say that the properties of the crystal are *anisotropic*. The internationally agreed system for defining crystal directions and planes are using the *Miller indices*, which are obtained in the following manner:

- Consider a three-dimensional crystal lattice with primitive translation vectors $\mathbf{a}_1$, $\mathbf{a}_2$ and $\mathbf{a}_3$, which are not necessarily of the same length and not necessarily orthogonal. This is illustrated in Figure 2.7.
- The positions of O, A, B and C coincide with lattice points.
- Then vectors OA, OB and OC correspond to the directions of the primitive vectors.
- Distances $|OA|$, $|OB|$ and $|OC|$ will be multiples of n_1, n_2 and n_3 for the respective primitive unit vector lengths.

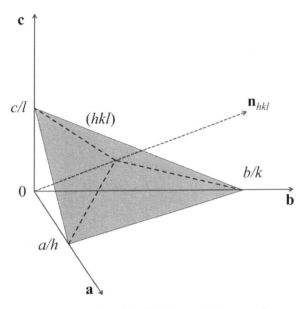

FIGURE 2.7: Intersection of the (*hkl*) plane with the *x*, *y*, and *z* axes.

In this way we have defined the plane corresponding to ABC as $(n_1|\mathbf{a}_1|, n_2|\mathbf{a}_2|, n_3|\mathbf{a}_3|)$. However, for purposes of conciseness and convenience and without loss of generality, this can be written as $(n_1 n_2 n_3)$. The generally accepted convention for expressing crystalline planes is in the format (*hkl*), where the *hkl* are integer related to n_1, n_2, n_3 as:

$$h : k : l = n_1^{-1} : n_2^{-1} : n_3^{-1} \tag{2.5}$$

This reciprocal relationship has important implications which will become apparent when we discuss the *reciprocal lattice* and *reciprocal space* in the next chapter. A plane satisfying Equation (2.5) is said to have Miller indices (*hkl*), where the choice of *h*, *k* and *l* are such that they are the smallest numerical integer values which satisfy this equation. Figure 2.8 illustrates the Miller indices for some low index planes for cubic lattices. In addition to describing crystal planes, the Miller indices are also used to express directions in a crystal, as shown in Figure 2.9.

Crystallinity in Solids • 47

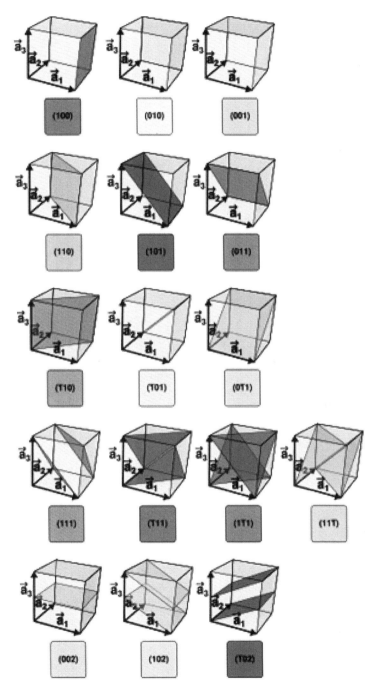

FIGURE 2.8: Low index planes for cubic lattices showing the Miller indices.

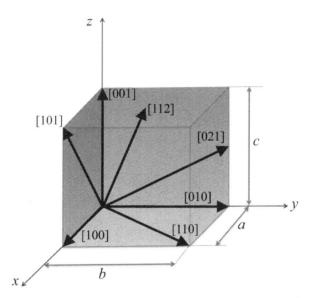

FIGURE 2.9: Miller indices for some principal axes in a cubic crystal.

The convention for the designation of planes and directions in crystallography are given in Table 2.2. When we refer to families of planes and directions, we mean the equivalent types of planes of directions. For example, the (100) plane is equivalent to the (010) and (001) planes in a cubic system, thus we can write this as {100} family. In terms of directions, we can say that the directions [100], [010], [001], [$\bar{1}$00], [0$\bar{1}$0] and [00$\bar{1}$] are equivalent and can be jointly represented as ⟨100⟩. We note that there is no need to designate bar signs for the planes since the (100) and ($\bar{1}$00) planes are essentially the same, however the directions [100] and [$\bar{1}$00] point in opposite directions.

We should add a word of caution here, as it would be easy to assume that a particular direction [hkl] is normal to the corresponding

TABLE 2.2: Convention for brackets in the use of Miller indices

Bracket notation	Meaning
(hkl)	Planes
[hkl]	Directions
{hkl}	Family of planes
⟨hkl⟩	Family of directions

plane (hkl). While this may be the case for materials which crystallize in the cubic system, this is not the general case and can cause considerable confusion if care is not taken. A crystal direction is expressed as the smallest set of integers which are proportional to the magnitudes of the vectors between lattice points of a specific orientation. Any two planes which are parallel will have the same set of Miller indices.

In addition to the three parameter notation (hkl) used for the Miller indices in three-dimensional crystals, the trigonal and hexagonal crystal classes are frequently quoted using four Miller indices. In this convention, the first three indices refer to the three coplanar directions in the basal plane and the fourth indicates the uniaxial axis or *c-axis*, as illustrated in Figure 2.10.

Once we have a knowledge of the crystalline planes and directions we can derive the interplanar spacings for specific consecutive parallel planes in the crystal. These are useful for evaluating the lattice parameters from diffraction patterns. We will consider here the

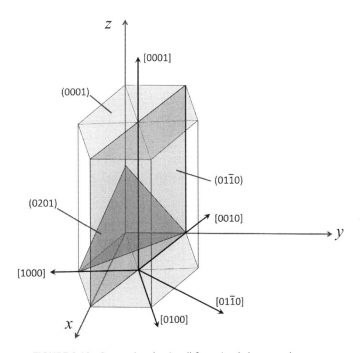

FIGURE 2.10: Conventional unit cell for a simple hexagonal structure

simplest case for crystal systems with orthogonal principal axes and assume that the plane (hkl) passing through the origin is parallel to the plane with intercepts at a_1/h, a_2/k and a_3/l, see Figure 2.7. The nearest distance between two planes (hkl), which we write as d_{hkl}, will be along the line which is perpendicular to the planes.

From the triangle made up or the line from the origin to the central point in the plane then to the intersect on the x-axis back to the origin, we can establish the relation:

$$\cos \alpha = \frac{d_{hkl}}{(a_1/h)} \tag{2.6}$$

Similar equations can be written analogous triangles from the corresponding to those taken from the origin to the central point of plane (hkl) and the intersects on the y and z axes. From this we can write:

$$\cos \beta = \frac{d_{hkl}}{(a_2/k)} ; \cos \gamma = \frac{d_{hkl}}{(a_3/l)} \tag{2.7}$$

Making use of the cosine rule:

$$\cos^2 \alpha + \cos^2 \beta + \cos^2 \gamma = 1 \tag{2.8}$$

and substituting in Equations (2.6) and (2.7) we obtain:

$$\frac{d_{hkl}^2}{(a_1/h)^2} + \frac{d_{hkl}^2}{(a_2/k)^2} + \frac{d_{hkl}^2}{(a_3/l)^2} = 1 \tag{2.9}$$

It is now possible to for us to write the general interplanar spacing for crystalline systems with orthogonal axes (orthorhombic), as:

$$d_{hkl} = \left(\frac{h^2}{a_1^2} + \frac{k^2}{a_2^2} + \frac{l^2}{a_3^2} \right)^{-1/2} \tag{2.10}$$

It is a simple matter to show that from this, the corresponding relation for the tetragonal system; $a_1 = a_2 \neq a_3$, Equation (2.9) reduces to:

$$d_{hkl} = \left(\frac{h^2 + k^2}{a_1^2} + \frac{l^2}{a_3^2} \right)^{-1/2} \tag{2.11}$$

and for the cubic system; $a_1 = a_2 = a_3 = a$, this simplifies to:

$$d_{hkl} = \frac{a}{\sqrt{h^2 + k^2 + l^2}} \qquad (2.12)$$

Crystalline structures with non-orthonormal principal axes will have more complex mathematical relations between the interplanar spacing and the Miller indices.

2.5 COMMON CRYSTALLINE STRUCTURES

In this section, we shall review some of the more common crystalline structures as well as the important close-packed structures. These are of particular significance for metallic solids. If we consider a two dimensional atomic plane in a monatomic solid, then we may expect that the "hard-sphere" depiction would be of a close-packed hexagonal array of touching spheres, as shown in Figure 2.11. Such hexagonal close-packed arrays of atoms can stack to form the closest possible packing structures of all solids. There are two important structures that are formed in this way: (i) hexagonal close-packed (hcp), which arises from stacking sequence ABAB... and (ii) face-centered cubic (fcc), which is formed from the stacking sequence ABCABC... These are illustrated in Figure 2.12, note that the relative positions of A, B and C are shown in Figure 2.11 (right). In the

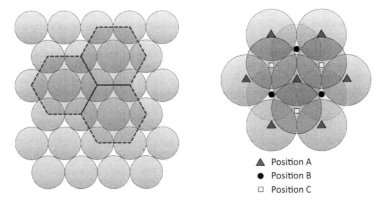

▲ Position A
● Position B
□ Position C

FIGURE 2.11: Close-packed hard spheres in a single atomic plane.

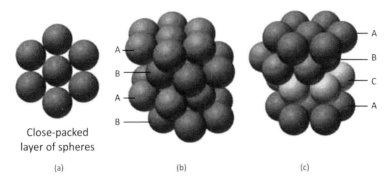

FIGURE 2.12: Stacking sequences for the close-packed structures hcp (b) and fcc (c).

case of the hcp crystal, the close packed plane corresponds to the (0001) plane, while for the fcc structure, the close packed plane is a (111) type plane. Examples of the hcp crystal structure can be found in various elements of the periodic table, such as Be, Mg, Zn and Cd. The fcc structures are common in the transition metals, such as Cu, Ag, Au, Ni, Al and Pb. The more conventional views of the hcp and fcc unit cells are shown in Figure 2.13. It can be seen that the primitive vectors for the fcc lattice can be expressed as:

$$\mathbf{a}_1 = \frac{a}{2}(\hat{\mathbf{y}} + \hat{\mathbf{z}}); \mathbf{a}_2 = \frac{a}{2}(\hat{\mathbf{z}} + \hat{\mathbf{x}}); \mathbf{a}_3 = \frac{a}{2}(\hat{\mathbf{x}} + \hat{\mathbf{y}}). \quad (2.13)$$

Another common crystal in metallic systems is the body-centered cubic (bcc) structure, which is also illustrated in Figure 2.13, with examples being Fe, Cr, Cs and Na. The bcc structure is not a close

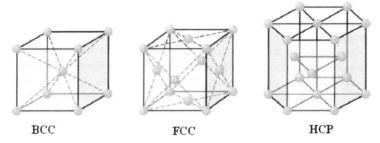

FIGURE 2.13: The most common elemental crystal structures. Body-centered cubic or BCG (left), face-centered cubic or FCC (center) and hexagonal close-packed or HCP (right).

packed crystal, having a more open crystal lattice. Atomic packing will be discussed in the following section for some of the main crystal structures. The primitive vectors for the bcc lattice can be written in the form:

$$\mathbf{a}_1 = \frac{a}{2}(\hat{\mathbf{y}} + \hat{\mathbf{z}} - \hat{\mathbf{x}}); \mathbf{a}_2 = \frac{a}{2}(\hat{\mathbf{z}} + \hat{\mathbf{x}} - \hat{\mathbf{y}}); \mathbf{a}_3 = \frac{a}{2}(\hat{\mathbf{x}} + \hat{\mathbf{y}} - \hat{\mathbf{z}}). \quad (2.14)$$

The cubic systems, apart from being simpler to visualize, form a very important class of structures. Other common crystal systems which form cubic lattices are the rocksalt (NaCl) and CsCl, see Figure 2.14. In the former, the basis has one atom of both Na and Cl at positions (0, 0, 0) and (1/2, 1/2, 1/2), respectively and form an fcc lattice. The formation of this crystal is a result of the anion - cation ratios in the ionic bonding process. There are various alkali halides which form this type of crystal as well as silver halides and lead chalcogenides. The CsCl structure appears to form a bcc like structure, but actually it is based on the simple cubic structure, with the Cs atom at (0, 0, 0) and the Cl at (1/2, 1/2, 1/2), forming the basis of the structure.

In addition to the fcc and bcc structures discussed above, the diamond and zinc-blende crystals are very relevant in the p-block elements, where many of the semiconducting materials are to be found. The cubic unit cells for these two crystals are very similar and are shown in Figure 2.15. The diamond crystal structure is related to the fcc Bravais lattice, in which the basis is formed of two atoms of the same type, with atomic positions (0, 0, 0) and (1/4, 1/4, 1/4).

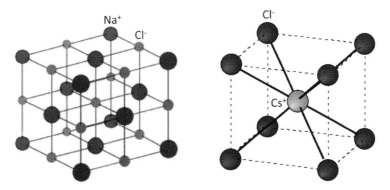

FIGURE 2.14: The NaCl (left) and CsCl (right) crystal structures.

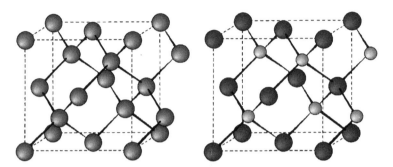

FIGURE 2.15: The diamond (left) and zincblende (right) crystal structures.

The basic unit of the diamond structure can be found in addition to diamond (carbon), in silicon and germanium and derives from the covalent bonding we discussed in Chapter 1. The fundamental unit can be seen to be a tetragonal block from the four equivalent nearest neighbor atoms. The main difference between the diamond and zinc-blende structures is that the basis has two different atomic species in the latter. Materials which crystallize in this structure are GaAs and a broad range of the III - V semiconductors. This forms the main structure of the binaries, ternaries, and even quarternies so common in *band structure engineering* (see Chapter 9). Still in the cubic system, the perovskite structure is quite common, especially with oxides of the ABO_3 type. As the formula unit shows there are three oxygen atoms and one of each type A and B per unit cell, which is clearly seen from the schematic in Figure 2.16.

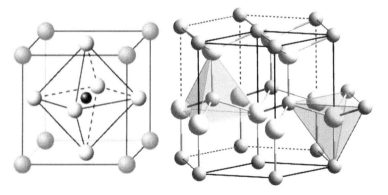

FIGURE 2.16: The perovskite (left) and wurtzite (right) crystal structures.

What diamond is to fcc, wurtzite is to hcp. The wurtzite structure, illustrated in Figure 2.16 (right), is again based on the tetragonal unit with two different atomic species in the basis.

2.6 ATOMIC PACKING

Atomic packing in crystals is based on using a hard-sphere model of atoms and is a useful concept in the consideration of the spatial arrangement of atoms. Central to this concept is the so-called *packing fraction*, which is defined as the ratio of the volume of the atoms (hard-spheres) in the unit cell to the volume of the unit cell itself. This is best demonstrated with an example. Consider the simple cubic structure, which has just one atom per unit cell. For the hard-sphere model, in which the spheres are envisaged to touch, we can calculate the packing fraction as:

$$(PF)_{sc} = \frac{Volume\ of\ atom}{Volume\ of\ unit\ cell} = \frac{(4/3)\pi r^3}{a^3} \qquad (2.15)$$

Now, since $a = 2r$, see Figure 2.17, we obtain

$$(PF)_{sc} = \frac{(4/3)\pi r^3}{8r^3} = \frac{\pi}{6} = 0.5236 \qquad (2.16)$$

Alternatively, we say that there is a packing efficiency of 52.4% for the simple cubic structure. We will now consider the fcc unit cell. The conventional unit cell for this structure has four atoms and in the hard-sphere model, the spheres will touch along the $\langle 110 \rangle$ directions, as illustrated in Figure 2.17. From this we can establish the relation: $\sqrt{2}a = 4r$,. The packing fraction in this case can be expressed as:

$$(PF)_{fcc} = \frac{4(4/3)\pi r^3}{a^3} = \frac{4(4/3)\pi r^3}{16\sqrt{2}r^3} = \frac{\pi}{3\sqrt{2}} = 0.7405 \qquad (2.17)$$

As a final example we will look at the hcp structure. For our purposes, it is sufficient to consider one-third of the conventional unit cell, which is a parallelopiped, with an atom at each apex and

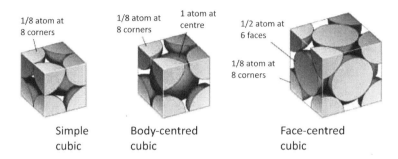

FIGURE 2.17: Hard sphere models for simple cubic, body-centered cubic and face-centered cubic crystalline structures.

one atom at a position $(1/3, 1/3, 1/2)$. The area of the base is $a \times a \sin(60°) = a^2\sqrt{3}/2$. The ideal (c/a) ratio is simple to calculate for the hcp crystal and has a value of 1.633. Since there are two atoms per cell unit, we can calculate the packing fraction as:

$$(PF)_{hcp} = \frac{2(4/3)\pi r^3}{ca^2\sqrt{3}/2} = \frac{8\pi r^3}{3\sqrt{2}a^3} \tag{2.18}$$

Since $a = 2r$, we obtain:

$$(PF)_{hcp} = \frac{\pi r^3}{3\sqrt{2}r^3} = \frac{\pi}{3\sqrt{2}} = 0.7405 \tag{2.19}$$

This is exactly the same value we obtained for the fcc structure. In fact, this represents the maximum packing fraction for hard spheres in any arrangement, and are hence referred to as close packed. All other structures have a lower *PF*.

In addition to the three-dimensional structures, we have considered, we can also consider the packing in particular atomic planes of the crystalline lattices. This turns out to be important when considering defects in solids, where slipping generally occurs along the planes of highest atomic density. We will consider this and other defects in solids in Chapter 4. As an example we will consider the principal planes; (100), (110) and (111), for the simple cubic structure. The density of atoms in a particular plane can be defined as:

$$\delta_{structure}^{(hkl)} = \frac{Number\ of\ atoms\ in\ unit\ cell\ of\ plane\ (hkl)}{Area\ of\ unit\ cell\ in\ (hkl)} \tag{2.20}$$

TABLE 2.3: Atomic packing in some common crystal types.

Crystal structure	Atomic coordination	Packing fraction	Close packed planes	Close packed directions
sc	6	0.524	⟨100⟩	{100}
fcc	12	0.7405	⟨111⟩	{111}
bcc	8	0.680	⟨110⟩	{110}
hcp	12	0.7405	⟨0001⟩	{11$\bar{2}$0}
diamond	4	0.340	⟨111⟩	{111}

From this definition and a projection of each of the planes we can find for the simple cubic structure:

$$\delta_{sc}^{(100)} = \frac{1}{a^2} \qquad (2.21)$$

$$\delta_{sc}^{(110)} = \frac{1}{\sqrt{2a^2}} = \frac{0.7071}{a^2} \qquad (2.22)$$

$$\delta_{sc}^{(111)} = \frac{1}{\sqrt{3a^2}} = \frac{0.5773}{a^2} \qquad (2.23)$$

Thus we conclude that for the sc structure, the {100} planes have the highest atomic density. Furthermore, the close packed direction in this structure will be the ⟨100⟩ directions. Other cubics structures have different close packed planes and directions, as will be seen from Table 2.3.

2.7 SUMMARY

In this chapter, we have outlined some of the principal concepts of crystallography and reviewed some of the more important crystalline structures that occur frequently in solids. This is far from an exhaustive study and the interested reader is referred to the texts outlined in the References given below. Many physical properties depend strongly on the arrangement of atoms in a solid. We commenced this chapter with the consideration of perfectly ordered systems, called crystals. However, not all solids are crystalline. We can distinguish between materials with perfect order (crystals) and those in which there are small regions of perfect order (polycrystalline materials), where there is no correlation to other regions and finally solids in which there is no long range order. In this latter case, such a solid with no long range atomic order, is said to be *amorphous*. Both amorphous and polycrystalline systems generally exhibit homogeneous or isotropic physical properties due to the randomness inherent in their structure. Crystals, however, frequently show anisotropy in many of their physical properties. It is therefore very important that we consider the physical arrangement of atoms in a solid when we discuss their physical properties.

We define crystalline structures in terms of their symmetries and using symmetry operations we can define up to 230 space lattices, which can then be subdivided into 32 point groups, fourteen Bravais lattices and seven crystal systems. The complex inter-relationships being defined by symmetry operations, such as translation, rotation, reflection and inversion. As an aid to our understanding and as a useful tool when considering crystals, we define a unit cell, which is made up of a basis and is attached to each lattice point. Each basis has an identical composition and orientation in the crystalline structure, of which the unit cell defines a sub-unit, which is chosen for convenience. A primitive unit cell should have only one basis or lattice point. It is worth noting that since there is no limit on how large a basis can be, there can be many more crystal structures than space groups; the known crystal structures run into thousands.

When discussing the specifics of crystals, it is useful to be able to refer to certain directions and planes in a given structure. To do this we define planes and directions in terms of the Miller indices, which are three integers specifying these orientations and directions in the crystal and are distinguished by the type of brackets that are used; (hkl) for a plane, [hkl] for a direction, {hkl} for a family of planes and ⟨hkl⟩ for a family of directions.

Specific types of crystal structure are commonly found in nature and we have reviewed some of the more important of these, such as face-centered cubic, body-centered cubic, hexagonal close-packed. These structures are widely found in monatomic metallic systems. Alloys and compounds are also frequently found in cubic phases, though generally with more complex crystal structures. The question of atomic packing can be a useful concept in discussing certain physical properties and we have defined this in terms of directions, plane and three-dimensional structures. The table below (2.3) reviews some of the more important crystal types, giving some of the packing parameters.

REFERENCES AND FURTHER READING

Basic Texts

- J. S. Blakemore, *Solid State Physics*, Cambridge University Press, Cambridge (1985)
- H. P. Myers, *Introductory Solid State Physics*, Taylor and Francis, London (1998)

Advanced Texts

- M. A. Wahab, *Solid State Physics: Structure and Properties of Materials*, Alpha Science International Ltd., Harrow (2007)
- N. W. Ashcroft and N. D. Mermin, *Solid State Physics*, Saunders College, Philadelphia (1976)

EXERCISES

Q1. Calculate the volume of the primitive unit cell as a proportion of the conventional unit cell for the fcc and bcc crystalline structures.

Q2. Consider the perovskite structure, which is illustrated in Figure 2. What is its crystal type and what is the basis for the structure?

Q3. Prove that the Wigner-Seitz cell for a 2D hexagonal array has the same area as the primitive unit cell for this structure.

60 • Solid State Physics

Q4. Show that the ideal c/a ratio in the hcp crystal is $\sqrt{8/3} = 1.633$.

Q5. Demonstrate that the inversion symmetry operation can be produced by the compound operation of reflection and rotation. Describe the *screw* and *glide* as compound operations.

Q6. Illustrate the following planes for a cubic crystal:
a) (110), b) (120), c) (112)

Q7. Explain why, in general, the direction [hkl] is not perpendicular to the plane (hkl). In what cases might this be true?

Q8. Calculate the interplanar separation in copper for the planes (100), (110) and (111). Note, Cu has fcc structure and has a lattice parameter of 3.61 Å.

Q9. Evaluate the primitive vectors for the hcp and diamond lattices and show the form of their primitive unit cells.

Q10. Repeat the above exercise for the NaCl and CsCl structures.

Q11. Calculate the packing fraction of the diamond lattice.

NOTE

[1]See for example, M. M. Woolfson, *An Introduction to X-ray Crystallography*, Cambridge University Press (1997); M. Buerger, *Elementary Crystallography: An Introduction to the Fundamental Geometric Features of Crystals*, MIT Press, (1978)

CHAPTER 3

CRYSTAL STRUCTURE DETERMINATION

"Nothing is ever as good or as bad as it first appears"

—La Rochefoucauld

3.1 INTRODUCTION

As has been stated in previous chapters, the crystalline order in solids has an enormous impact on the physical properties of the material. It is therefore of crucial importance that we are able to relate this structure to the physical properties of interest. It is the principal aim of this chapter to discuss how we can do this. One of the most powerful tools for elucidating crystalline structures is by the diffraction of waves. This is most commonly performed using x-rays and to some extent with electrons, the use of neutrons is also not uncommon for specific cases. These have appropriate wavelengths for diffraction effects to be observed in crystals; i.e., the wavelength of the radiation used is of the order of the atomic separations in solids. We will begin by looking at the basic diffraction phenomenon, which can be viewed as a scattering process due to conditions of constructive and destructive interference of waves. We will not discuss it here, but the student will be aware of the wave - particle duality phenomenon famous from the early progress of quantum theory.

Suffice it to say, all radiation will here be treated as a wave and one of the proofs of the wave-like nature of electrons was found from the diffraction of electrons by NiO crystals (credit for which goes to Davisson and Germer, 1925).

The usefulness of diffraction as a technique for the characterization of materials cannot be overestimated. This was recognized quite early on by Max von Laue in 1912 and independently in 1913 by W. L. Bragg, who independently formulated the basic conditions for diffraction by crystals. Some of the earliest studies from around this time were performed using x-ray radiation, with photon energies in a range from 10 - 100 keV. Such energies are required for diffraction to occur from atoms with separations in the few Å range. Also such radiation can penetrate sufficiently well into the crystal to allow a determination of the bulk structure. More detailed descriptions of diffraction by electrons and x-rays will be discussed in Section 3.7. We will also briefly discuss the use of neutrons in diffraction experiments, since they can provide further information about the ordering in crystals.

The theory of diffraction is outlined in Section 3.3, where we will discuss the approaches to this problem by Bragg and von Laue. It will become clear that these two important models actually come down to the same thing. However, before we do that, it is useful to introduce the concept of the *reciprocal lattice,* which is essentially a geometric construction, but one which significantly aids the interpretation of diffraction patterns and also will be of interest to later discussions of the electronic structure of materials. A further tool in the interpretation of diffraction patterns is provided by the Ewald sphere construction and is used in conjunction with the reciprocal lattice. We will also introduce the so-called *structure factor,* which provides a simple to use tool for constructing *reciprocal space.* Actually we will be working mostly in reciprocal space. This may seem a strange concept to the uninitiated, however, it nothing to be concerned about. It is simply a geometric construction or mathematical tool. It is widely used in diffraction physics, so we will introduce it here. It is simplest just to think of at as normal space, with the proviso that all lengths in reciprocal space have the dimensions of inverse length (1/m), hence the name. To be honest, we have already introduced something in the previous chapter that is related to the reciprocal lattice; the Miller indices, see Equation (2.4).

3.2 THE RECIPROCAL LATTICE

The reciprocal lattice for a real space lattice structure can be fairly easily generated by applying a mathematical procedure. The relation between the real and reciprocal space lattice is analogous to that between frequency and time. This is best described in terms of the sum of the Fourier components. For a time varying quantity, the Fourier components will give the frequency domain spectrum, which will reveal oscillatory-like motion in terms of its frequency. For the case of spatial quantities, the Fourier transform yields the spatial regularities in the form of an inverse separation of reciprocal-space varying function. The general Fourier transform for any function is shown in Appendix B. We will now outline the reciprocal lattice transformation equations in terms of the lattice vectors of the real space lattice. It is worth noting that the reciprocal lattice for a Bravais lattice is also a Bravais lattice. In the definition of the reciprocal lattice, we can choose to apply the unit cell lattice vectors or the primitive unit cell vectors. In the former case the transformation will provide the lattice vectors of the reciprocal lattice and in the latter the primitive vectors of the reciprocal lattice. Thus either one will do. For the primitive lattice we define the primitive reciprocal lattice by:

$$\mathbf{b}_1 = 2\pi \frac{\mathbf{a}_2 \times \mathbf{a}_3}{\mathbf{a}_1 \cdot (\mathbf{a}_2 \times \mathbf{a}_3)}; \mathbf{b}_2 = 2\pi \frac{\mathbf{a}_3 \times \mathbf{a}_1}{\mathbf{a}_1 \cdot (\mathbf{a}_2 \times \mathbf{a}_3)}; \mathbf{b}_3 = 2\pi \frac{\mathbf{a}_1 \times \mathbf{a}_2}{\mathbf{a}_1 \cdot (\mathbf{a}_2 \times \mathbf{a}_3)}. \tag{3.1}$$

The analogous expressions for the reciprocal lattice using the base vectors **a**, **b** and **c** are expressed as:

$$\mathbf{a}^* = 2\pi \frac{\mathbf{b} \times \mathbf{c}}{\mathbf{a} \cdot (\mathbf{b} \times \mathbf{c})}; \mathbf{b}^* = 2\pi \frac{\mathbf{c} \times \mathbf{a}}{\mathbf{a} \cdot (\mathbf{b} \times \mathbf{c})}; \mathbf{c}^* = 2\pi \frac{\mathbf{a} \times \mathbf{b}}{\mathbf{a} \cdot (\mathbf{b} \times \mathbf{c})}. \tag{3.2}$$

The factor 2π is introduced for convenience to match the factors used in *reciprocal space* (also sometimes referred to as *momentum space*).[1] Therefore we maintain the same dimensions as the wavevector. An important property of the real lattice and reciprocal lattice vectors is that they satisfy:

$$\mathbf{b}_i \cdot \mathbf{a}_j = 2\pi \delta_{ij}, \tag{3.3}$$

where δ_{ij} is known as the Kronecker delta, defined as:

$$\delta_{ij} = \begin{cases} 0, & i \neq j; \\ 1, & i = j. \end{cases} \quad (3.4)$$

This shows the orthogonality between the vectors $\mathbf{a}_i$ and $\mathbf{b}_j$, for $i \neq j$. For example, $\mathbf{b}_1 \cdot \mathbf{a}_1 = 2\pi$, while $\mathbf{b} \cdot \mathbf{a}_2 = \mathbf{b}_1 \cdot \mathbf{a}_3 = 0$. This means that vectors $\mathbf{b}_1$ and $\mathbf{a}_1$ are parallel, while $\mathbf{b}_1$ and $\mathbf{a}_2$ etc. are perpendicular.

In the same way we wrote lattice vectors in real space, we can write vectors in reciprocal space as linear combinations of the $\mathbf{b}_i$:

$$\mathbf{k} = k_1 \mathbf{b}_1 + k_2 \mathbf{b}_2 + k_3 \mathbf{b}_3 \quad (3.5)$$

Using the real space vector $\mathbf{T} = n_1 \mathbf{a}_1 + n_2 \mathbf{a}_2 + n_3 \mathbf{a}_3$, where the n_i are integers, it will be evident that we have:

$$\mathbf{k} \cdot \mathbf{T} = 2\pi(k_1 n_1 + k_2 n_2 + k_3 n_3) \quad (3.6)$$

This is a direct consequence of Equation (3.3). We now consider the plane wave $e^{\mathbf{k} \cdot \mathbf{T}}$, where from the previous discussion in Chapter 2, $\mathbf{T}$ is a vector describing the Bravais lattice. For a general $\mathbf{k}$, the plane wave will not have the periodicity of the Bravais lattice, this will only occur for special choices of the wave-vector $\mathbf{k}$. In fact, the set of wave-vectors $\mathbf{K}$ that yield plane waves of the periodicity of a specific Bravais lattice is the reciprocal lattice! Consider the plane wave:

$$e^{\mathbf{K} \cdot (\mathbf{T} + \mathbf{r})} = e^{\mathbf{K} \cdot \mathbf{r}}, \quad (3.7)$$

where there is no restriction on $\mathbf{r}$. For the above to be true, the following condition must be satisfied:

$$e^{i \mathbf{K} \cdot \mathbf{T}} = 1 \quad (3.8)$$

for all $\mathbf{T}$ of the Bravais lattice. In relation to Equation (3.6) above, the reciprocal lattice would require that all coefficients k_i to be integers. Therefore we see that the reciprocal lattice is a Bravais lattice in which the $\mathbf{b}_i$ can be taken as the primitive vectors. Now given that the reciprocal lattice is a Bravais lattice, it is possible again to construct its reciprocal, which will produce the real space or direct lattice. It is convention to term members of the set of translation vectors in reciprocal space in terms of the Miller indices, where we write:

$$\mathbf{G}_{hkl} = h\mathbf{b}_1 + k\mathbf{b}_2 + l\mathbf{b}_3 \qquad (3.9)$$

in which the plane with Miller indices (hkl) is normal to the reciprocal lattice vector, $\mathbf{G}_{hkl}$. Therefore, any vector, $\mathbf{R}$ in the plane (hkl) will satisfy the condition:

$$\mathbf{R} \cdot \mathbf{G}_{hkl} = 0 \qquad (3.10)$$

In Chapter 2, we found a simple expression for the interplanar spacings for cubic and other crystals with orthogonal principal axes. It is possible to construct a compact expression for interplanar distances, d_{hkl}, in terms of the reciprocal lattice vector, $\mathbf{G}_{hkl}$, for any lattice. This can be derived from the scalar products of $\mathbf{G}_{hkl}$ with the vectors $\mathbf{a}_i$. From Equation (3.3) we note that $\mathbf{a}_1 \cdot \mathbf{G}_{hkl} = 2\pi h$, etc. Since the vector $\mathbf{G}_{hkl}$ is perpendicular to the plane (hkl), we can write the unit vector normal for a plane as:

$$\hat{\mathbf{n}}_{hkl} = \frac{\mathbf{G}_{hkl}}{|\mathbf{G}_{hkl}|} \qquad (3.11)$$

If we take an equation of the plane (hkl), we can write:

$$d_{hkl} = \mathbf{r} \cdot \hat{\mathbf{n}}_{hkl} = \mathbf{r} \cdot \frac{\mathbf{G}_{hkl}}{|\mathbf{G}_{hkl}|} \qquad (3.12)$$

where $\mathbf{r}$ can be any vector larger than d_{hkl} taken from the point of origin to the plane (hkl), see Figure 3.1. Considering that $\mathbf{r}$ can be made up from any linear combination of $\mathbf{a}_i$, we can now write the interplanar spacing as:

$$d_{hkl} = \frac{2\pi}{|\mathbf{G}_{hkl}|} \qquad (3.13)$$

It is useful to demonstrate the construction of the reciprocal lattice from the Bravais lattice. We will do this by considering the fcc

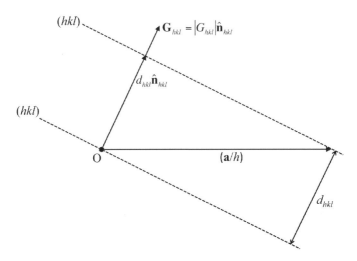

FIGURE 3.1: Schematic diagram of interplanar spacing between two planes (*hkl*).

lattice, with a lattice constant, a. The primitive vectors $\mathbf{a}_i$ were given in Chapter 2:

$$\mathbf{a}_1 = \frac{a}{2}(\hat{\mathbf{y}} + \hat{\mathbf{z}}); \; \mathbf{a}_2 = \frac{a}{2}(\hat{\mathbf{z}} + \hat{\mathbf{x}}); \; \mathbf{a}_3 = \frac{a}{2}(\hat{\mathbf{x}} + \hat{\mathbf{y}}). \quad (3.14)$$

Applying the definition of the reciprocal lattice given in Equation (3.1), we obtain the primitive vectors of the reciprocal lattice as:

$$\mathbf{b}_1 = \frac{4\pi}{a}\frac{1}{2}(\hat{\mathbf{y}} + \hat{\mathbf{z}} - \hat{\mathbf{x}}); \; \mathbf{b}_2 = \frac{4\pi}{a}\frac{1}{2}(\hat{\mathbf{z}} + \hat{\mathbf{x}} - \hat{\mathbf{y}}); \; \mathbf{b}_3 = \frac{4\pi}{a}\frac{1}{2}(\hat{\mathbf{x}} + \hat{\mathbf{y}} - \hat{\mathbf{z}}). \quad (3.15)$$

It is thus found that the reciprocal lattice of an fcc lattice takes on a bcc form (and vice versa). This will be seen by comparing the above result to Equation (3.14), where here the cubic cell has a lattice parameter of $4\pi/a$.

3.3 DIFFRACTION OF WAVES BY CRYSTALS

Diffraction is a wave phenomenon in which the scattering of an incident wave by an object gives rise to destructive and constructive interference. In general, for such effects to be physically observable, the wavelength of the incident radiation must be of the same order of magnitude as the diffracting object. For our purposes, we are concerned with the diffraction of radiation by the atoms in a solid, which typically have separations of the order of a few Å. To be of the correct order of magnitude, electromagnetic radiation should be in the x-ray region of the electromagnetic spectrum. This can be evaluated from the following relationship:

$$E = h\nu = \frac{hc}{\lambda} \qquad (3.16)$$

This gives an energy of about 1.24 MeV (1.9878×10^{-16} J) for a wavelength of 1nm, which is in the x-ray region of the electromagnetic spectrum. For visible light, say green, the wavelength is around 500 nm, which corresponds to an energy of about 2.5 eV.

For electrons, say in an electron microscope, the relevant energies are of the order of 100 keV. This corresponds to the accelerating voltage for the electrons delivered by an electron gun. The relation between the wavelength and the accelerating voltage can be derived from the de Broglie expression, $\lambda = h/p$ and from the kinetic energy equated with the accelerating voltage, $V : eV = mv^2/2 = p^2/2m$. From these we obtain:

$$\lambda = \frac{h}{\sqrt{2meV}} \qquad (3.17)$$

Relativistic corrections can add a small shift in this wavelength, but we shall not concern ourselves with this. From this we can evaluate the accelerating potential required for a particular wavelength. For example, a wavelength of 1 Å corresponds to an energy of 150 eV, i.e., an accelerating potential of 150 V. Alternatively, we can calculate the wavelength corresponding to an accelerating potential; for example, from a 1 kV potential, a wavelength of about 0.4 Å is achieved. This corresponds to an electron velocity of around $1.8 \times 10^7 ms^{-1}$. Diffraction experiments are frequently performed using neutrons, which are about 1000 times heavier than electrons, this means to achieve similar wavelengths, they can be much slower. The 1 Å wavelength will be achieved for neutrons with an energy of only

0.08 eV, corresponding to a velocity of just under 4000 ms^{-1}. These are hence referred to as *slow neutrons*. The lack of electrical charge of the neutron means that they interact more weakly with atoms and can thus penetrate much further into a crystal.

The diffraction of the different types of radiation can differ due to the the specific nature of the radiation itself, and will depend on the *scattering cross-sections* of the radiation with the atoms. We will discuss this in more detail later in this chapter. To understand the diffraction of incident radiation we will consider two important approaches which will be dealt with separately in the following sections. We can note that in both of these approaches, the incident radiation is considered to only scatter *elastically*, which is an essential condition for constructive interference. Elastic scattering means that there is no energy loss (or transfer) in the scattering process, and the wavelength of the incident radiation remains unchanged after interaction. By contrast *inelastic scattering* can occur in a number of processes, which can excite atoms and produce the emission of secondary electrons and characteristic radiation from atoms. Such processes are the basis of many spectroscopic techniques. The inelastically scattered radiation will not contribute to the diffraction image and is thus ignored in the first instance.

We can mathematically express the wave-vectors of the incident and scattered beams as **k** and **k**′, respectively. So for assumption of elastic scattering, we can write $|\mathbf{k}| = |\mathbf{k}'| = (2\pi/\lambda)$. The difference in **k** and **k**′ being only in their direction.

The scattering efficiency depends on a number of factors, including the energy of the incident radiation, but also importantly on the atomic species. This scattering efficiency is later taken into account when we discuss the *atomic form factor*, being proportional to the atomic number.

3.3.1 Bragg's Law

The Bragg law was developed in 1913 by W. L. Bragg as a way to explain the x-rays diffraction patterns observed from crystalline samples, which were measured by W. H. and W. L. Bragg (father and son). The principal idea derives from the consideration of the specular reflection of wavefronts from successive atomic planes in a crystal. The basic construction can be illustrated via the path difference between waves reflected by successive atomic planes, as shown in Figure 3.2. In this approach the first ray reflects from an upper atomic plane (*hkl*), while a second ray reflects from the next plane down in the crystal. The *Bragg condition* to observe *constructive interference* from these two rays will be that the path difference (being longer for the ray reflected by the lower plane and corresponds to the distance from A to B plus B to C in Figure 3.2) between them must be a whole number of wavelengths. This can be expressed mathematically as:

$$n\lambda = 2d_{hkl}\sin\theta_{hkl} \qquad (3.18)$$

The subscript denotes the Miller indices of the atomic plane under consideration. The angle θ_{hkl} corresponds to the angle of incidence of the radiation with respect to the plane (*hkl*), as shown in the figure. The integer n is referred to as the order of the corresponding reflection. By changing the angle of incidence, it is possible to satisfy a Bragg condition from the same planes by increasing the order of reflection. Note that the angle of deflection

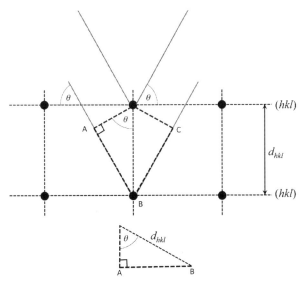

FIGURE 3.2: Schematic diagram of interplanar spacing between two planes (*hkl*).

of the incident beam corresponds to $2\theta_{hkl}$. In this analysis, we have considered that the incident beam is monochromatic, which corresponds to a single wavelength of radiation. For a beam of x-rays which contains a range of wavelengths ("white radiation"), many different reflections will be observed, resulting from the various Bragg conditions being satisfied from higher order reflections and from different crystal planes.

In a real experiment, the x-ray beam, of finite width, will impinge upon many atoms and be scattered by them. It is the scattering from multiple atoms that will give rise to a host of diffraction spots, where the constructive interference occurs, resulting from many Bragg conditions being satisfied by a large number of crystalline planes. The dark regions occur due to destructive interference. Indeed this must be the case since other wise we would expect a diffuse background, which occurs from non-ordered materials and liquids.

3.3.2 The Von Laue Approach

Max von Laue (1912) used a somewhat different way of looking at the problem. There are no assumptions as to the direction of the incident beam with respect to the crystalline planes and no restrictions as to specular reflections. Indeed it is a more general approach, which regards the crystal as an assembly of scattering centers (atoms or ions) being positioned at the sites, **R**, of a Bravais lattice. The radiation then scatters from these sites in all directions. The sharp (diffraction) spots will occur only for specific directions and at wavelengths for which the rays scattered from the lattice sites interfere constructively.

The condition for constructive interference can be found by considering the two points (scatterers) in the Bravais lattice, see Figure 3.3. The incident x-rays of wavelength, λ, have a wave-vector of $\mathbf{k} = 2\pi\hat{\mathbf{n}}/\lambda$, while the (elastically) scattered x-rays have a wave-vector $\mathbf{k}' = 2\pi\hat{\mathbf{n}}'/\lambda$, where $\hat{\mathbf{n}}$ and $\hat{\mathbf{n}}'$ are the vectors that define the directions of the incident and scattered rays. As seen in Figure 3.3, the path difference between the scattered rays from the two scattering centers, separated by d can be expressed as:

$$d\cos\theta + d\cos\theta' = \mathbf{d} \cdot (\hat{\mathbf{n}} - \hat{\mathbf{n}}') \tag{3.19}$$

Therefore the condition for constructive interference will take the form:

$$\mathbf{d} \cdot (\hat{\mathbf{n}} - \hat{\mathbf{n}}') = m\lambda \tag{3.20}$$

where m must be an integer. Multiplying through by $2\pi/\lambda$ allows us to express this condition in terms of the wave-vectors, such that:

$$\mathbf{d} \cdot (\mathbf{k} - \mathbf{k}') = 2\pi m \tag{3.21}$$

or

$$\mathbf{d} \cdot \Delta \mathbf{k} = 2\pi m \tag{3.22}$$

Since our crystal is made up of a large array of such scatterers, in its Bravais lattice, we can replace $\mathbf{d}$ by the Bravais lattice vectors $\mathbf{R}$. The Laue condition is thus expressed as:

$$\mathbf{R} \cdot (\mathbf{k} - \mathbf{k}') = 2\pi m \tag{3.23}$$

For integral m and all Bravais lattice vectors R. Now since m is integer, this can be equally written in the form:

$$e^{i(\mathbf{k} - \mathbf{k}') \cdot \mathbf{R}} = 1 \tag{3.24}$$

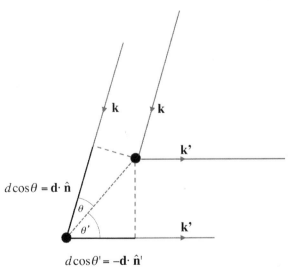

FIGURE 3.3: Illustration of the path difference between two points in the Bravais lattice.

It should now be clear from our definition of the reciprocal lattice, Equation (3.8), that for the Laue condition will be satisfied if the change in wave-vector $\Delta \mathbf{k}$, is a vector of the reciprocal lattice. We can now express this as:

$$\mathbf{k} - \mathbf{k}' = \mathbf{G}_{hkl} \tag{3.25}$$

or

$$\mathbf{k} = \mathbf{G}_{hkl} + \mathbf{k}' \tag{3.26}$$

Squaring both sides yields:

$$G^2 + 2\mathbf{k}' \cdot \mathbf{G} = 0 \tag{3.27}$$

This is illustrated in Figure 3.4, where we see that the difference between the incident and scattered wave-vectors is equal to the reciprocal lattice vector, $\mathbf{G}_{hkl}$. The plane perpendicular to this vector is called the *Bragg plane*. We note that all these vectors are represented in reciprocal or *k-space*.

3.3.3 Reconciling the Bragg and von Laue Approaches

It is a relatively easy task to show the equivalence of the Bragg and von Laue formulations. We can start by using the von Laue

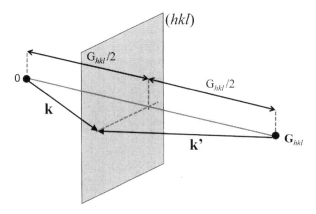

FIGURE 3.4: The von Laue condition: the difference between the wave-vectors of the incident and scattered beams is equal to the reciprocal lattice vector for a Bragg plane (*hkl*).

formula given in Equation (3.25). We substitute in the following expressions: $\mathbf{k} = 2\pi\hat{\mathbf{n}}/\lambda$, $\mathbf{k}' = 2\pi\hat{\mathbf{n}}'/\lambda$ and $\mathbf{G}_{hkl} = 2\pi\hat{\mathbf{n}}_{hkl}/d_{hkl}$. We note that each term includes a unit vector normal to describe its direction from the von Laue formulation. However, referring to the original expression, we note that the direction of $|\mathbf{k} - \mathbf{k}'|$ is parallel to $\mathbf{G}_{hkl}$ and corresponds to nothing more than twice the sine of the angle θ_{hkl}. This should be clear from the illustrations in Figures 3.4 and 3.5. Writing this out in mathematical form we obtain:

$$\mathbf{k} - \mathbf{k}' = \mathbf{G}_{hkl}$$
$$\Rightarrow |\mathbf{k} - \mathbf{k}'| = |\mathbf{G}_{hkl}|$$
$$\Rightarrow 2|\mathbf{k}|\sin\theta_{hkl} = \frac{2\pi}{d_{hkl}} \qquad (3.28)$$

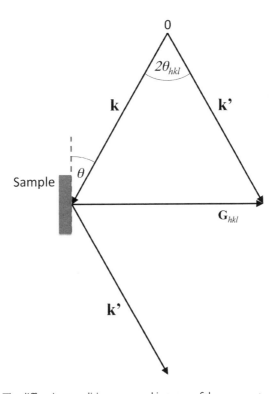

FIGURE 3.5: The diffraction condition expressed in terms of the wave-vector representation.

Therefore we can write:

$$2\left(\frac{2\pi}{\lambda}\right)\sin\theta_{hkl} = \frac{2\pi}{d_{hkl}} \qquad (3.29)$$

or

$$2d_{hkl}\sin\theta_{hkl} = \lambda \qquad (3.30)$$

Which is the expression of Bragg's law. Of course the n (not to be confused with the unit vector normal) is missing from the right hand side, but this is just because we didn't consider the number of wavelengths in the path difference for the constructive interference condition. We can add this in the form, $\mathbf{G}_{hkl} = 2\pi n/d_{hkl}$, to give the more familiar expression:

$$2d_{hkl}\sin\theta_{hkl} = n\lambda \qquad (3.31)$$

Thus we see that both the Bragg and von Laue formulations are indeed expressions of the same phenomenon and arise from the consideration of the condition for the constructive interference of scattered rays.

3.3.4 The Ewald Sphere Construction

The Ewald sphere construction for diffraction from a crystal is a rather nice way of visualizing the process. It is based on the von Laue formulation and effectively considers the compete set of reflections possible from the Bravais lattice. The representation is performed in k-space, where the origin is taken (arbitrarily) on a reciprocal lattice point. The surface of a sphere coincides with this point, the radius of which corresponds to the incident wave-vector $\mathbf{k}$. Therefore the direction of the incident wave with respect to the lattice is also taken into account. Since the Laue condition is for elastic scattering, the scattered ray, $\mathbf{k}'$, with the same magnitude, also coincides with the surface of our sphere and with a point in the reciprocal lattice. This will mean that the vector from the origin of k-space and this reciprocal lattice point will be a reciprocal lattice vector, $\mathbf{G}_{hkl}$. It should now be clear that wherever a reciprocal lattice point coincides with the surface of the Ewald sphere a Laue (that is to say also a Bragg) condition will be satisfied. This is more easily visualized in Figure 3.6.

This construction is a powerful tool in the interpretation of diffraction images, whether that be from x-ray or electron diffraction. We can note that a change in the orientation if the incident beam will be reflected as a rotation of the Ewald sphere in k-space, with the point at the origin being fixed, i.e., the position of the center of the sphere will move according to the direction of the incident beam. A change in energy of the incident beam will cause the radius to change; the larger the energy the larger the radius of the Ewald sphere and thus a larger proportion of the reciprocal lattice will be sampled. Only a monochromatic beam will produce an infinitesimally thin shelled Ewald sphere. Therefore, any energy dispersion of the primary beam can have a significant effect on the resulting diffraction pattern.

Furthermore, the degree of perfection of the crystalline structure will affect the resulting diffraction pattern. For example, only an infinitely large perfectly ordered crystal will have infinitesimally small reciprocal lattice points. Thermal agitation will cause the points to increase in size. The size of crystalline grains will also affect the

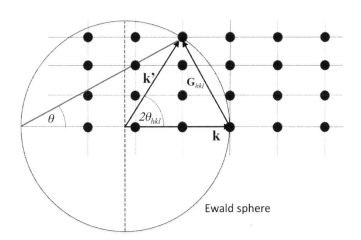

FIGURE 3.6: The Ewald sphere construction showing the simultaneous satisfaction of an array of Laue conditions. The incident k - vector points from the origin to the center of the sphere, having radius k. The scattered ray, $\mathbf{k'}$, goes from a point simultaneously on the surface of the Ewald sphere and the reciprocal lattice, given as a reciprocal lattice vector, $\mathbf{G}_{hkl}$.

spot size. This effect is enhanced in the case of diffraction from the surface of a crystal, where effectively only a single atomic layer contributes to the diffraction pattern. The reciprocal lattice points in this case will be stretched infinitely in k-space in the direction perpendicular to the plane. Now the Ewald construction will be of a sphere intersection will reciprocal lattice rods and can give a streaked image to the diffraction spots. Such a situation is regularly seen in the RHEED (reflection high-energy electron diffraction) technique, as will be discussed in Section 3.6.2.

3.4 THE ATOMIC FORM FACTOR

In our discussion thus far we have made no assumptions as to the strength of the scattering from the atoms in our crystal. Indeed the scattering of radiation is different for different atomic species and also has an angular dependence. We characterize the scattering of x-rays using the atomic form factor, f_j, where subscript j denotes the atomic species. In general the scattering strength increases with the number of electrons in atom, i.e., with the atomic number, Z. Therefore the scattering will vary drastically as we go through the periodic table. The radial dependence of scattering is shown in Figure 3.7 and can be represented mathematically as:

$$f = \sum_{n=1}^{Z} \int_0^\infty \frac{4\pi r^2 \psi_n^2(\mathbf{r}) \sin[4\pi r (\sin\theta)/\lambda] dr}{[4\pi r (\sin\theta)/\lambda]} \quad (3.32)$$

Here $\psi_n(\mathbf{r})$ denotes the radial dependence of the normalized wave-function for the nth electron in the atom of atomic number, Z. Each term in the summation becomes unity as $(\sin\theta)/\lambda$ goes to zero.

The scattering of different types of incident radiation will be different because of their intrinsic nature. The above describes the scattering of x-rays, which is principally due to their interaction with the electrons in the atoms or ions. Electrons also interact strongly with the atomic electrons and will have a similar atomic form factor. Neutrons, on the other hand, are more strongly scattered by the nuclei of atoms rather than electrons and the scattering is virtually isotropic. Furthermore, neutrons have an intrinsic spin and are

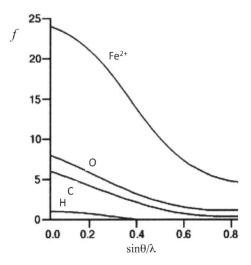

FIGURE 3.7: The atomic form factor, f, as a function of $(\sin\theta)/\lambda$, showing the response for various elements.

sensitive to the magnetic moment of atoms. This will give additional information for magnetically ordered materials.

3.5 THE STRUCTURE FACTOR

In addition to the intensity variation of scattered diffraction spots due to atomic species, variations in intensity can also occur due to the structure itself. The structure factor allows us to evaluate this important piece of information. We note that for a monatomic structure, i.e., one with only one type of atom, we do not need to include the atomic form factor, as it will be the same for all atoms. The structure factor indicates how the Bragg peaks are formed from the change in wave-vector: $\mathbf{k} - \mathbf{k}' = \mathbf{G}_{hkl}$. The path difference due to scattering from different lattice sites, say $\mathbf{d}_i$ and $\mathbf{d}_j$ can be expressed as $\mathbf{G}_{hkl} \cdot (\mathbf{d}_i - \mathbf{d}_j)$, with a phase difference of $e^{i\mathbf{G}_{hkl} \cdot (\mathbf{d}_i - \mathbf{d}_j)}$. We can incorporate this into a neat concise form as:

$$S_{hkl} = \sum_{j=1}^{n} f_j(\mathbf{r}) e^{i2\pi(hx_j + ky_j + lz_j)} \quad (3.33)$$

where the summation is over the unit cell of our structure. The quantity S_{hkl} in Equation (3.33) is called the structure factor. The intensity of diffracted Bragg reflections are given by:

$$I_{hkl} \propto |S_{hkl}|^2 \qquad (3.34)$$

The form of the structure factor given in Equation (3.33) shows the basic Fourier transform relation between the real space lattice and that of the reciprocal lattice. The inclusion of the atomic form factor in Equation (3.33) permits us to distinguish between different atomic scatterers and will introduce an intensity variation into the resulting scattered intensity distribution.

The best way to illustrate the usefulness of the structure factor is by using an example. We will start by considering the bcc lattice, which has a unit cell with just two atoms at positions (000) and (1/2, 1/2, 1/2). Therefore our summation will give rise to only two terms. In the first, centered at (000), the exponential will term will be unity. We can now write the structure factor for the bcc structure as:

$$S_{hkl}^{bcc} = f[1 + e^{i\pi(h+k+l)}] \qquad (3.35)$$

Now since h, k and l are integers, there can only be two possible outcomes; i) when $h + k + l$ is even and ii) when $h + k + l$ is odd. Thus we find:

$$S_{hkl}^{bcc} = \begin{cases} 2f, & \text{when } h+k+l \text{ is even,} \\ 0, & \text{when } h+k+l \text{ is odd.} \end{cases} \qquad (3.36)$$

To construct the reciprocal lattice from this information we should remember that the above analysis is based on using a unit cell for the bcc structure as a simple cubic structure with a two atom basis. The reciprocal lattice will now be a simple cubic structure with points of intensity 0 and $2f$. The result of the structure factor calculation in Equation (3.36) is illustrated in Figure 3.8, where we have effectively an fcc reciprocal lattice of side $4\pi/a$, which is the same result as obtained from the analysis in Section 3.2, where we saw the reciprocal relation between the fcc and bcc structures. It is useful to see how the structure factor works for a polyatomic crystal. Here we

Crystal Structure Determination • 79

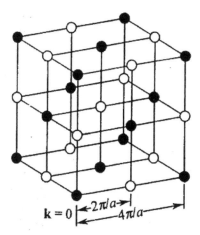

FIGURE 3.8: Reciprocal lattice for the bcc structure as obtained from the structure factor. Open circles have an intensity of zero, while filled circles have an intensity of 2f.

shall consider the example of the CsCl lattice. This has a basis with say the Cs atom at (000) and the Cl atom at (1/2, 1/2, 1/2). We can now obtain the structure factor as:

$$S_{hkl}^{CsCl} = [f_{Cs} + f_{Cl}e^{i\pi(h+k+l)}] \tag{3.37}$$

Again the result depends on whether $h + k + l$ is even or odd and we can express this as follows:

$$S_{hkl}^{CsCl} = \begin{cases} f_{Cs} + f_{Cl}, & \text{when } h+k+l \text{ is even,} \\ f_{Cs} - f_{Cl}, & \text{when } h+k+l \text{ is odd.} \end{cases} \tag{3.38}$$

We note that this result agrees with the bcc lattice above for which $f_{Cs} = f_{Cl}$. The reciprocal lattice has the form of an NaCl type lattice, which consist of an fcc structure with two different type lattice points. In the analogy with the bcc structure, the $h + k + l$ is odd condition gives rise to a zero intensity point.

It will now be clear that the structure factor allows us to assess the form of the reciprocal lattice with the additional information with regards to the relative intensity of the reciprocal lattice points due to the periodic structure of the crystal lattice. The atomic forma factor is also important in polyatomic crystals. It does not, however, provide the full absolute intensity of the Bragg peak. Where it is reliable is when the intensity goes to zero, which occurs from a fully destructive interference condition being produced for a particular **G**. In taking the modulus squared of the structure factor:

$$|S_{hkl}|^2 = \left\{\sum_{j=1}^{n} f_j(\mathbf{r})\cos[2\pi(hx_j + ky_j + lz_j)]\right\}^2$$
$$+ \left\{\sum_{j=1}^{n} f_j(\mathbf{r})\sin[2\pi(hx_j + ky_j + lz_j)]\right\}^2 \qquad (3.39)$$

we lose any phase information regarding the amplitudes of the rays. This means that the intensity does not contain phase, and leads to the well-known *phase problem* in crystallography. There are a number of solutions to this problem, though it is beyond the scope of the present text to deal with this. The interested reader is encouraged to consult a textbook on crystallography, see reference section at end of chapter for some examples.

3.6 DIFFRACTION METHODS FOR STRUCTURE DETERMINATION

There are a relatively large number of experimental techniques based on the diffraction of radiation to elucidate the interior structure of materials. While we cannot provide an exhaustive review, we will discuss the main points of some of the more common techniques, which are based on x-ray, electron and neutron diffraction. We have already outlined some of the main differences in the nature of these types of radiation, and the important thing to remember is that the differences observed in terms of the resulting diffraction patterns arises due to the nature of the interaction of the radiation with the atoms/ions in the solids. This is principally due to their so-called interaction cross-sections, which intimately depend on the velocity, charge and mass of the incident radiation.

3.6.1 X-Ray Diffraction

The production of x-rays for experimental use in diffractometers, is mainly achieved via the collision of electrons with a metallic target. Characteristic radiation is emitted from the target in the form of Bremsstrahlung and from the de-excitation process in the metallic atoms of the metal. The former process arises from the deceleration of the incident electrons, causing the emission of x-radiation, which has a short wavelength cut-off of $\lambda_{min} = hc/eV$, where V is the accelerating potential in volts. This continuous spectrum is of little interest for x-ray diffraction. It is the excitation of core (typically K - level) electrons, which, in the de-excitation process, give rise to characteristic spectra of lines; K_α and K_β lines. The choice of target material is important and is based on good thermal and electrical conductivities, in addition to having a relatively high melting point. Furthermore, for x-ray diffraction capabilities, the emitted radiation should have a wavelength of around 1 Å. The most common choice for target is Cu, which has a K_α line at around 1.54 Å. Since the emission is not monochromatic, filters are applied, usually in the form of a crystal monochromator, though this is not always necessary. The characteristic emission spectra of a range of materials were studied by Moseley in 1913 - 1914, and ultimately led him to provide an explanation for the order of elements in the periodic table in the expression of what is now known as Moseley's law[2].

A more intense source of x-radiation is the synchrotron. This is usually a very large installation where charged particles, usually electrons, are accelerated in a circular path using magnets. The laws of electrodynamics can show that electromagnetic radiation is emitted when charged particles are accelerated. So by sending electrons around a *storage ring*, a stream of high energy and high intensity x-rays can be produced in a direction tangential to the synchrotron ring. This is a rather expensive way to do standard diffraction and is typically only used for samples which require very specific conditions and for dynamic studies for which a high intensity beam is necessary.

X-rays, like most other electromagnetic radiation, mainly interact with the electron clouds of atoms, which occupy the vast majority of the atomic size.

The Laue method. This was the first technique to use x-rays for diffraction from crystals and dates from 1912, when von Laue gave his interpretation of the diffraction phenomenon from crystals. The method is experimentally simple, consisting of a collimated beam of *white* x-rays falling on a single crystal which is mounted on a triple-axis goniometer. Photographic plates can be placed in front or behind the sample to record the diffraction pattern, as illustrated in Figure 3.9(a). A typical image is illustrated in Figure 3.9(b).

Since the technique does not employ a monochromatic x-ray beam, the diffraction pattern consists of the superposition of the reciprocal lattice with a range of Ewald spheres, whose radii are determined by the upper and lower wavelength limits; . As such we sample a large area of reciprocal space, as illustrated in Figure 3.10, as indicated by the shaded region. All the spots in this region will therefore give rise to a spot in the diffraction image, providing it falls onto the photographic plate. By ensuring there is a sufficiently broad range of wavelengths in the incident beam, we can guarantee that diffraction conditions will be satisfied.

The Laue technique is especially well suited to the determination of the crystalline orientation of a sample whose structure is known. This is useful for orienting samples, though is not recommended for the determination of unknown crystalline structures.

The spot distribution can be somewhat complex due to the range of wavelengths used, so each set of Bragg planes can give rise to a range of spots in the diffraction pattern. For this reason, this technique is usually limited to the orientation of known crystal structures using the goniometer.

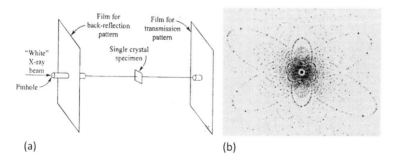

FIGURE 3.9: (a) A flat Laue camera, where diffraction images can be recorded in both the forward and back-reflection geometries. (b) A sample image from a white beam diffraction pattern of Silicon single crystal in the (001) orientation. Pattern taken with large area 2D detector.

Crystal Structure Determination • 83

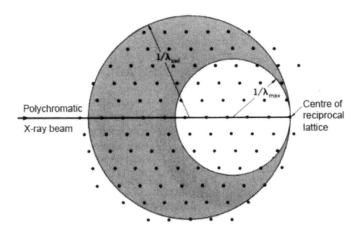

FIGURE 3.10: Ewald construction relative to the Laue diffraction method, where the upper and lower limits of the wavelength of the white x-radiation determine the amount of reciprocal space sampled.

The rotating crystal method. This method employs a monochromatic beam of x-rays in which the sample is rotated as a function of time. The gradual rotation of the sample ensures that various Bragg reflections will be recorded. In terms of the Ewald construction, the incident **k**-vector can be considered to be fixed, while the reciprocal lattice rotates about the point of origin of k-space. The resulting x-ray diffraction pattern is frequently expressed as the intensity of the exiting x-ray beam as a function of the angle, such an image is shown in Figure 3.11. There are several variations of the principle of the rotating crystal method, which can provide a wealth of information for the study of crystalline structures. Further information on these techniques can be found in specialist books on crystallography, as indicated in th bibliography at the end of this chapter.

The Debye - Scherrer method. This is also known as the *powder* method, since this is the nature of the samples studied using this technique. As such we are presented with a polycrystalline sample. Actually, the fact that the sample has multiple crystals provides another way of ensuring that Bragg reflections will occur. In terms of the Ewald construction, we are presented with a single sphere of fixed orientation, while the reciprocal lattice will be made up of all possible orientations of the Bragg planes. This

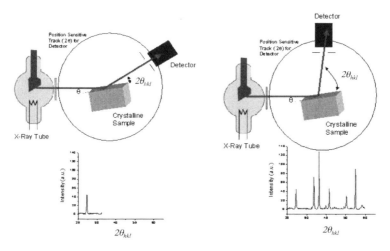

FIGURE 3.11: Experimental set-up for an x-ray diffractometer. Here we see two positions of the apparatus, where the results show the diffracted beam intensity as a function of the angle of diffraction, $2\theta_{hkl}$.

will mean that for each particular reflection (hkl), where for a single crystal we have a single diffraction spot for diffraction angle θ_{hkl}, in the polycrystalline sample, we obtain all possible orientations of this reflection as a cone of corresponding Bragg angle θ_{hkl}. Figure 3.12(a) shows this situation in schematic form. The final diffraction image will then be one of concentric circles with characteristic radii. It is from this that the crystalline structure can be deduced to a good degree of accuracy and allowing the lattice parameters to be evaluated. The photographic film is straightened out after adequate exposure and consists of arcs, as shown in Figure 3.12(b). Measurement is made from the $\theta_{hkl} = 0°$ position to $\theta_{hkl} = 90°$. Interpretation is based on the Bragg equation, which we can express as:

$$\sin^2 \theta_{hkl} = \frac{\lambda^2}{4d_{hkl}^2} \qquad (3.40)$$

We can now substitute in an expression for the interplanar separation, see Chapter 2. In the case of cubic latices Equation (3.40) can be written as:

$$\sin^2 \theta_{hkl} = \frac{\lambda^2}{4a^2}(h^2 + k^2 + l^2) \qquad (3.41)$$

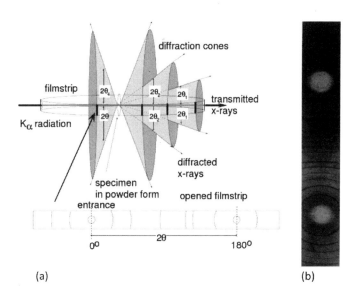

FIGURE 3.12: Experimental set-up for the Debye-Scherrer technique. The monochromatic x-rays impinge on the sample and the diffraction rings are recorded on a strip of film which encircles the sample. On the right hand side we see an actual image of a diffraction pattern.

For specific structure, certain combinations of h, k, l will be absent, as we saw in Section 3.5. Using this fact, we can express the values of $\sin^2 \theta_{hkl}$ as a function of $(h^2 + k^2 + l^2)$, where, according to Equation (3.41), we can evaluate the lattice parameter, a, given that the wavelength is of a known and fixed value.

3.6.2 Electron Diffraction

Electron beams can be produced in a number of ways. The most common method used for electron diffraction experiments is based on thermionic emission of a heated wire. The electrons are then accelerated via a potential difference and collimated before impinging on the sample of interest. The energy of the electrons is usually well defined by this accelerating potential, from which we can evaluate the associated wavelength of the electrons, as expressed by Equation (3.17). We can substitute in the constants to express this in terms of Ångstrom units:

$$\lambda(\text{Å}) \simeq \sqrt{\frac{150}{V}} \qquad (3.42)$$

where V represents the accelerating voltage. For very high energies a relativistic correction can be applied.

While the basic theory of diffraction is equivalent to that expressed above for x-rays, the experimental conditions and the nature of the electrons means that there are some specific limitations to their use. In the first place, electrons will interact strongly with atoms and ions due to their electrical charge, and their transmission through samples is limited to around 50 nm. Another way of expressing this is to say that electrons have a high cross-section for scattering and is about four times greater than that for x-rays. The most common place to find electron diffraction experiments is generally inside a transmission electron microscope (TEM). In fact, the TEM technique relies on the diffraction of the electrons by the sample. The electron beam inside a TEM is controlled and focussed using electric and magnetic fields. There are many textbooks on this important materials characterization technique[3]. The high scattering cross-section means that sample will normally be thinned to allow the electrons to pass through it and hence allow the diffraction pattern and the magnified image to be observed.

Electron diffraction is a commonly used tool in the study of crystalline surfaces and thin film growth. There are two principal techniques used and are defined by the electron energies used and the geometry of the experiment. We briefly mentioned one of these techniques earlier; reflection high-energy electron diffraction (RHEED), the other related technique is called low-energy electron diffraction (LEED). In the former, the electron beam impinges on the sample at grazing incidence (a few degrees from the sample plane) and is forward scattered. The scattering obeys the basic rules of diffraction outlined above and the Ewald construction is frequently used to aid interpretation of diffraction patterns, which are observed on a phosphor screen.

The RHEED experiment is illustrated in Figure 3.13, where we see how the Ewald sphere intersects with the extended reciprocal lattice spots or *rods* (for a surface), giving rise to streaked diffraction spots, Figure 3.13(a). The diffraction images, Figure 3.13(b) for a Si(111) surface along two different directions, show a series of rings of diffracted beams. Each ring, known as a *Laue zone,* arises from a

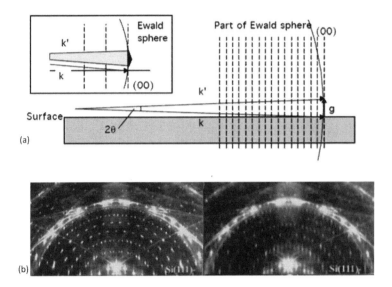

FIGURE 3.13: Experimental set-up for RHEED. (a) Ewald construction showing the intersection of the Ewald sphere and reciprocal lattice rods. (b) RHEED images for the Si(111) surface along two different directions of the crystalline surface.

successuverow of reciprocal lattice rods back from the zeroth order. Note that for a surface only two Miller indices are required. The surface of a crystal may have a special ordering which does not exist in the bulk of the crystal. Any restructuring is called a *surface reconstruction*, and in the case shown in Figure 3.13(b) corresponds to a 7×7 reconstruction. This refers to the change in periodicity in the two directions of the surface with respect to the bulk structure of the same atomic plane, which is (111) in this case. Therefore RHEED, and for that matter LEED, are excellent tools for investigating the structural changes at a crystal surface. Also noted in the diffraction pattern are some broad, bright bands and lines, these are called *Kikuchi* bands or lines and arise from multiple scattering and channelling processes inside the crystal. These are generally indicative of good bulk crystalline order. We note that the grazing incidence provides good surface sensitivity since it prevents the electrons penetrating to more than a few atomic layers into the crystal. Typical energies for electrons beams in RHEED are of the order of 10 - 20 keV, corresponding to wavelengths of around 0.09 - 0.12 Å. One of the principal applications of the RHEED technique is the study of the dynamics of thin film deposition, which is possible due to the experimental geometry. This is usually performed by measuring the intensity of the specularly reflected electron beam as a function of deposition time, where intensity oscillations are indicative of atomic layer type growth.

In a LEED experiment, the electron beam, of energy in the range: 10 - 1000 eV ($\lambda \simeq 0.4 - 3.9$ Å), is incident perpendicularly to the crystal surface. Since the electrons have a much lower kinetic energy, they penetrate much less, thus making them surface sensitive. The geometry of the set-up means that the diffraction pattern is observed via elastically backscattered electrons, as illustrated in Figure 3.14(a). We note that in the experimental set-up a grid is placed in front of the phosphor screen, which has an electrical potential applied to it. This is to energy filter the scattered electrons to ensure only those electrons that are elastically scattered are measured. Inelastic scattering can arise from a number of processes and can give a bright background, which must be eliminated from the LEED pattern. The lower electron energy used in LEED results in a smaller Ewald sphere, Figure 3.14(b), than in RHEED. The diffraction pattern for the same surface, Si(111) with a 7 × 7 reconstruction, as given for the RHEED case above, is shown in Figure 3.14(c). Note the shadow in this image is from the electron gun.

3.6.3 Neutron Diffraction

The neutrons used in diffraction experiments are generally obtained as a by product in a nuclear reactor. Another common source of neutrons is from a so-called *spallation* source. This produces intense neutron beams that arise from the the collision of ("bunches" of) high-energy protons that collide with a target, typically liquid mercury. The high energy neutrons can be slowed down using a moderator. The energy of neutrons appropriate for diffraction from crystals, as given by Equation (3.17), is of the order of about 0.1 eV. This energy is of the same order of energy as thermal neutrons (0.025 eV, which corresponds to a temperature equivalent of 300K, as given by the thermal energy: $k_B T$). The output from a reactor source requires the neutrons to be wavelength selected, which is usually performed using a crystal monochromator. The collimated beam of white neutrons are incident on a large crystal, which

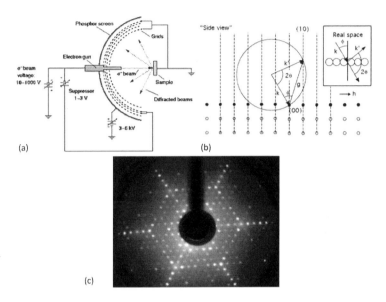

FIGURE 3.14: (a) Experimental geometry for LEED. (b) Ewald construction showing the intersection of the Ewald sphere and reciprocal lattice rods as well as the diffraction condition. (c) LEED image for the Si(111) surface.

is oriented to satisfy a Bragg condition for a specific chosen wavelength. The angular range over which the neutrons are collected will determine the degree of monochromaticity of the experiment. The neutrons interact with the nuclei of the sample, the cross-section for scattering is much smaller than the wavelength and the atomic form factor (called *scattering length* for neutrons) has no angular dependence. The neutron interaction with the nuclei of the sample is such that it momentarily forms a compound after which it is re-emitted. If the nuclei do not possess spin, then the scattering will be coherent, being the same for all nuclei, and the resulting diffraction pattern will be the same as for x-rays. Since neutrons only interact with the atomic nuclei, the penetration depth will be much larger than that of electrons or even x-rays. This is advantageous since less sample preparation will be necessary.

For nuclei that possess spin, I, the neutron scattering will be fundamentally different, since the compound nucleus, upon interaction with the neutrons, will have a spin of $I \pm 1/2$. The scattering may be either coherent or incoherent, where the latter results in an absence of phase correlation and thus produces no interference effects. Since neutrons interact with the nucleus, there is no Z correlation to the scattering strength. However, there are strong differences in scattering between different isotopes of the same element. Another important property of the neutron is that it possesses spin or a magnetic moment and can therefore interact differently with magnetic atoms (i.e., those with a net electron spin) than with to non-magnetic atoms. Such a situation means that there can be scattering of a purely magnetic nature and neutron diffraction is frequently used to assess the long range magnetic order in ferromagnetic, ferrimagnetic and antiferromagnetic media. Details of magnetic order will be discussed in Chapter 10. Such properties makes neutron diffraction an extremely important measurement for magnetic materials, particularly in the case of antiferromagnetic samples. Since antiferromagnets have no net magnetization, they are very difficult to assess by other techniques. Neutron diffraction allows a magnetic signal to be obtained from the periodic condition of the magnetic structure and purely magnetic diffraction peaks can be observed.

3.7 SUMMARY

In this chapter, we have reviewed some of the main principles of crystal structure determination. The reciprocal lattice is an important geometrical construction which has a special symmetry relationship with its real space counterpart. They both form Bravais lattices with related symmetries. The importance of this construction was shown when we considered the phenomenon of diffraction in crystals. In fact, we saw that the reciprocal lattice is a Fourier transform of the real space lattice and aids in the interpretation of diffraction patterns, which is a widely used method for the elucidation of crystalline structures. Later on, in Chapter 7, we will again use the reciprocal lattice as an aid to the description of electronic energy states in periodic structures. The theory of diffraction was shown to be based on wave interference phenomena, where we considered the Bragg and the von Laue approaches. These were shown to be essentially equivalent. The von Laue approach is a more general condition which allows the consideration of all possible diffraction events using the Ewald sphere construction. This is a powerful tool frequently used in the interpretation of diffraction patterns.

The diffraction of waves from a periodic structure allows us to infer crystalline order due to the specific constructive interference effects that occur. This is due to the scattering of waves by the atoms in the crystal, the strength of this scattering depends in most cases on the element involved and has a specific form factor. The structure factor, which can be seen to be a Fourier transform, permits us to assess the form of the reciprocal lattice of a crystal. It has the added advantage of giving the relative intensity of the points in reciprocal space, which is related to the intensity variations observed in diffraction patterns.

We described some of the more important experimental diffraction techniques used. These concerned the use of x-rays, electrons and neutrons as incident radiation. The differences between these techniques are related to the nature of the radiation and how they interact with (and scatter from) the atoms in a crystalline solid. Specific information can be gleaned from each type of experiment.

REFERENCES AND FURTHER READING

Basic Texts

- J. S. Blakemore, *Solid State Physics*, Cambridge University Press, Cambridge (1985)
- H. P. Myers, *Introductory Solid State Physics*, Taylor and Francis, London (1998)

Advanced Texts

- M. A. Wahab, *Solid State Physics: Structure and Properties of Materials*, Alpha Science International Ltd., Harrow (2007)
- N. W. Ashcroft and N. D. Mermin, *Solid State Physics*, Saunders College, Philadelphia (1976)
- G. E. Bacon, *Neutron Diffraction*, Oxford University Press, (1975)

EXERCISES

Q1. Use the reciprocal lattice vectors, expressed in Equations (3.1) or (3.2), to determine the reciprocal lattice for the body-centered tetragonal structure.

Q2. Explain why Equations (3.1) and (3.2) are equivalent.

Q3. Use Equation (3.13) to derive the general interplanar separation for cubic crystals, c.f. Equation (2.11).

Q4. Prove that $\mathbf{a}_1 \cdot (\mathbf{a}_2 \times \mathbf{a}_3) = \mathbf{a}_2 \cdot (\mathbf{a}_3 \times \mathbf{a}_1) = \mathbf{a}_3 \cdot (\mathbf{a}_1 \times \mathbf{a}_2)$. What do these represent?

Q5. Calculate the energies for x-rays, electrons and neutrons if they have a wavelength of 0.75 Å.

Q6. Calculate the Bragg angles for diffraction from Ag, which has fcc structure and a lattice constant of 4.09Å with x-rays (Cu Kα radiation), for the (111) planes. Use $n = 1$.

Q7. How would a a variation in atomic position manifest itself in the Ewald sphere construction? (Note this can happen due to thermal vibrations of the atoms in a crystal, which is a general condition for room temperature diffraction experiments.) What does this imply for the resulting diffraction image?

Q8. Use the structure factor, Equation (3.33), to evaluate the reciprocal lattices for the diamond and the zinc-blende structures.

Q9. In a Laue diffraction pattern from an fcc crystal, with lattice parameter 3.5 Å, determine the distance from the center of the pattern to the reflections which occur from planes with maximum spacing. Consider the x-rays to be produced from a 50 kV tube and the distance between the crystal and the film to be 5.5 cm.

Q10. An antiferromagnetic crystal of CoO is subject to a neutron diffraction experiment with neutrons of energy 0.05 eV. If the lattice parameter for this crystal is 4.2615 Å, evaluate the Bragg angle for a magnetic reflection for an angle of incidence of 20° to the (110) plane.

Q11. Describe the formation of an x-ray diffractogram using the rotating crystal method in terms of the Ewald sphere construction.

Q12. Explain why electrons are used for surface diffraction techniques. Under what conditions can electrons be used for bulk crystal diffraction? Explain.

NOTES

[1] For a plane wave, which we can represent as: $A \exp 2\pi i(\frac{x}{v} - t) = A \exp 2\pi i(\frac{x}{\lambda} - \nu t)$, the wavelength can be written as $\lambda = (v/\nu)$ and angular frequency is defined as $\omega = 2\pi\nu$. We defined the wave-vector as $k = (2\pi/\lambda) = (\omega/v)$. As such we can now write the instantaneous wave amplitude as: $A \exp i(kx - \omega t)$. In three dimensions this is more generally expressed in the form: $A \exp i(\mathbf{k}\cdot\mathbf{r} - \omega t)$. The momentum associated with a quantized wave of wave-vector $\mathbf{k}$ is written as $\mathbf{p} = \hbar\mathbf{k}$.

[2] Moseley's law is generally expressed as the square root of the emitted x-ray frequency, $\sqrt{\nu}$, being proportional to the atomic number, $Z - b$, where b is a constant depending on the target element in question and is related to nuclear screening. Later it was put on a theoretical footing with the Bohr model of the atom.

[3] See for example D. B. Williams and C. B. Carter, *Transmission Electron Microscopy: A Textbook for Materials Science*, Plenum Press, New York, (2009)

CHAPTER 4

IMPERFECTIONS IN CRYSTALLINE ORDER

"Even imperfection itself may have its ideal or perfect state."
—Thomas de Quincey

"If someone is too perfect they won't look good. Imperfection is important."
—Eric Cantona

4.1 INTRODUCTION

Thus far, we have only considered solids as being a perfectly arranged periodic array of atoms. In reality the order in solids is far from perfect. The principal properties of solids does indeed come, for the most part, from this ordered portion of the sample. However, it would be a gross oversight to ignore the effects of imperfections in crystalline order. Of particular importance are mechanical and electronic properties, which are greatly affected by structural disorder. The consideration of perfect order is an important aid to the theoretical modelling of physical properties and conceptualisation of solids. Many would consider this as a first approximation to the understanding of the physical properties of solids. However, as we have stated, we do need to consider the departures from this perfect image of the crystal/solid to gain a fuller insight in to their physical properties.

Structural defects in solids can take a number of forms and generally we classify defects as to their dimensionality. That is to say, whether they are 0, 1, 2 or 3 dimensional defects. In the following sections, we will follow more or less this order in our overview of the defects that can occur in real solids. In the first case, zero-dimensional disorder consists of *point defects*, which pertain to single atomic positions. Following this we have one dimensional disorders, which occur along a crystal direction, and are called *disclocations*. Two-dimensional disorder is a planar defect, such as a *slip plane* and even surfaces in a crystal. Finally, a three dimensional defect will be some form of volume imperfection such as *granular* structure and foreign particle inclusions. Indeed, grains and grain boundaries play an important role in materials and their properties.

It is worth stressing that a perfect solid does not exist in reality. Even the most carefully prepared solid crystals have defects. This inevitable situation must be taken into account in any consideration of a solid. Even lattice vibrations can be considered to be a departure from perfect crystalline order. This will be treated separately in the next chapter.

A complete lack of long range crystalline order can also be considered as a form of three-dimensional defect. Such materials with no crystalline order are called *amorphous* or *non-crystalline* materials. Despite this random structure, we can still make some important conclusions with regards to these materials. Non-crystalline solids will be discussed in the last section of this chapter.

4.2 POINT DEFECTS

4.2.1 Types of Point Defect

The point imperfection is the simplest of structural defects and can be as simple as a one or two atom disturbance to the crystalline order. There are a number of types of point defect, from a simple lattice vacancy, i.e., missing atom, to a substitutional impurity atom, i.e., a foreign atom replacing an atom in a normal lattice site. Atoms can also sometimes position themselves in the spaces between regular lattice points and are called *interstitial sites*. This could be either the atoms normally in the crystal (*self-interstitial*) or impurity atoms (*impurity-interstitial*). In any case, these interstitial atoms can cause some local stresses in the lattice. Smaller atoms can occupy interstitial sites with little or no distortion to the crystal lattice. In Figure 4.1. we show a schematic view of this and other point defects.

More complex point defects consist of combinations of vacancies, substitutions and interstitials. One such combination can be

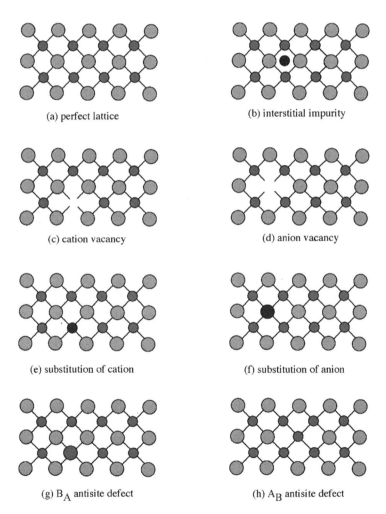

FIGURE 4.1: Schematic diagram of point defects.

generated by moving an atom from its regular lattice position to an interstitial site. Such a combined imperfection in a crystal lattice is called a *Frenkel defect*. This type of disorder can occur in thermal equilibrium and can be generated by bombardment, such as radiation damage. A similar type of disorder is called the *Schottky defect*, which occurs in ionic crystals, where the ions of both types (cation and anion) migrate to the crystal surface leaving a vacancy on both ionic sites and thus maintains charge neutrality of the crystal. We note that the vacancies can be separated in the crystal. These defects are illustrated in Figure 4.2.

Crystals which are polyatomic should have exact proportions to their components, such as NaCl, Fe_2O_3 etc. This is referred to as *stoichiometry* or *chemical stoichiometry*. Antisite defects can occur in which atoms of one type are located in the position of the other component, such defects are illustrated in Figure 4.1(g) and (h). Preferential vacancies can lead to off-stoichiometry in a crystal and can have an effect on the resulting physical properties of the solid if present in significant proportions. Any disorder in the regular structure of compound solids can give rise to *antistructure*, where the components swap lattice positions, while maintaining stoichiometry. Such a situation is likely to occur when the components of the crystal are of similar atomic size. The degree of off-stoichiometry depends on the preparation conditions of crystals and annealing at relatively high temperatures.

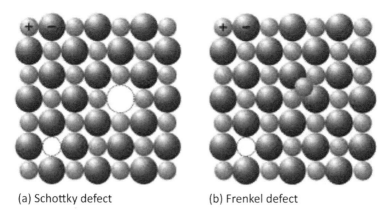

(a) Schottky defect   (b) Frenkel defect

FIGURE 4.2: Schematic diagram Frenkel and Schottky defects.

The existence of point defects can be controlled in a number of ways and we have already mentioned that heating or *annealing* a sample can introduce vacancies. Impurities can also be introduced in a controlled manner at the growth stage by allowing a specific concentration of another atomic species into the melt or in deposition. Such impurities can be welcome, such as in the doping of semiconductors (Chapter 9), which control the charge carrier concentrations. Defects will also affect the electrical resistance of a material by introducing scattering centers into the solid. More defects will essentially mean that the residual resistance of the material increases. (We will further discuss this in Chapter 6, where we consider the electronic properties of solids.)

4.2.2 Thermodynamics of Defect Density

The formation of the point defects discussed above depend on the energy required to move atoms in the crystal lattice. Different types of defect will have different *activation energies*. One of the simplest cases we can consider is that of vacancy formation. This process can occur due to thermal energy even while at thermal equilibrium. We will consider a crystal with N atoms in which there are n vacancies, randomly distributed over the available lattice sites of the crystal. Since we are introducing disorder into the crystal we will increase the entropy, S, in the system. A perfect crystal will have zero configurational entropy, S^{config}. The entropy due to this disordering will thus depend on the possible combinations of our n vacancies over the N sites. Since there are $N!/n!(N-n)!$ ways of doing this, the configurational entropy can be expressed as:

$$S^{con\,fig} = k_B \ln \frac{N!}{n!(N-n)!} \qquad (4.1)$$

here k_B is the Boltzmann constant. Now if the energy to remove an atom is given as, E_v, we can evaluate the change in free energy of the imperfect crystal in the formation of these vacancies as:

$$\Delta F = \Delta U - T\Delta S = nE_v - k_B T \ln \frac{N!}{n!(N-n)!} \qquad (4.2)$$

where T is the temperature of the crystal. The first term is simply the energy required to create the n vacancies. We now follow the minimization procedure, using the first derivative of this energy with respect to the number of vacancies;

$$\frac{d(\Delta F)}{dn} = 0, \quad (4.3)$$

This can be evaluated using the Stirling approximation, ln $X! \simeq X\ln X - X$, which is valid for large X. From this it is a simple matter to show that:

$$\frac{n}{N-n} = e^{-E_v/k_B T} \quad (4.4)$$

Since we have $n \ll N$, we can write:

$$\frac{n}{N} \simeq e^{-E_v/k_B T} \quad (4.5)$$

The form of the above equation is typical for thermally activated processes, where the exponent expresses an activation energy of a process divided by the thermal energy, giving the statistical probability of the process. In the above case, we consider the number of atoms which are removed from their equilibrium positions by random thermal motion. Therefore the important factor is the relation of the energy necessary to remove the atom from its position (which will be related to the number of nearest neighbors and the bond strength) compared to the energy supplied in the form of thermal motion of the atoms. Typically, E_v is of the order of a few eV, thermal energy at room temperature is about 1/40 eV, therefore n/N will be small but not zero at room temperature. Clearly an increase of temperature will then increase the number of defects according to Equation (4.5). Other defect densities can be evaluated in a similar manner using the appropriate energies of formation. In the case of compound defects, such as Schottky and Frenkel defects, the energies of each part of the defect must be taken into account.

4.2.3 Diffusion in Crystals

Vacancies and impurities can move in solids via transport mechanisms activated by concentration gradients and thermal activity. Such processes are referred to as diffusion. In either case, the motion requires a specific energy associated with the atoms or vacancies. The motion of atoms of the lattice is called *self-diffusion*, while that of different atoms is called *impurity diffusion*. Let us consider a crystal in which there is a concentration, c, gradient in the x-direction, such that $dc/dx < 0$. While we have not said what the concentration is, it could be either vacancies or impurity atoms for example. There is no difference in the physical process of diffusion, only in the activation energies involved. If there is a separation of a between atomic planes with concentrations c_1 and c_2, we say that unit areas of these plane will have ac_1 and ac_2 impurities or vacancies. Assuming that all atoms vibrate with the same frequency ν, we can expect that in time δt, an equal number of atoms move to the right and to the left, which in either case would be $\nu \delta t$ atoms. If we say that the probability of an atom jumping to the left or right is P, then the net flux of defects will be:

$$\delta n \simeq \nu \delta t a (c_1 - c_2) P \qquad (4.6)$$

However, due to the concentration gradient we have:

$$c_2 = c_1 + a \frac{dc}{dx} \qquad (4.7)$$

Now we can write:

$$\frac{\delta n}{\delta t} = j \simeq -\nu a^2 P \frac{dc}{dx} \qquad (4.8)$$

where j represents the material current density, which we express as:

$$j = -D \frac{dc}{dx} \qquad (4.9)$$

This equation is known as *Fick's first law* and can be expressed in three-dimensions as:

$$\mathbf{j} = -D \nabla c \qquad (4.10)$$

The constant D is called the coefficient of diffusion and follows an Arrhenius type equation of the form:

$$D = D_0 e^{-E_D/k_B T} \qquad (4.11)$$

cf. Equation (4.5). Here E_D represents the activation energy for diffusion and D_0 is a constant. Typically E_D is of the order of a few eV per atom while D_0 ranges from roughly 0.01 - 3 cm²s⁻¹ in metals. The diffuion mechanism acts to reduce and ultimately eliminate concentration gradients. Therefore we can expect the concentration in any given volume element to change over time. If our crystal maintains the number of vacancies and impurities then we will have:

$$\frac{dc}{dt} = -\frac{dj}{dx} \qquad (4.12)$$

which in three-dimensions is written:

$$\frac{dc}{dt} = -\nabla \cdot \mathbf{j} \qquad (4.13)$$

Inserting Equation (4.10) we have:

$$\frac{dc}{dt} = \nabla \cdot (D\nabla c) = D\nabla^2 c \qquad (4.14)$$

where we assume D to be a constant. Equation (4.14) is known as *Fick's second law*. The study of the diffusion process demands the use of appropriate boundary conditions, which must be considered for any specific case. In a crystal the motion of the impurity or vacancy is subject to random motion or *random walk*. The statistical average distance moved in time t of an impurity can be found from:

$$<x^2> = 2Dt \qquad (4.15)$$

The relevant activation energy for vacancy diffusion must take into account the vacancy and migration energy, from which we write: $E_D = E_v + E_m$. Diffusion is an extremely important physical phenomenon and controls alloy composition and high temperature creep strength in metals. Since diffusion is a thermally activated process, sample temperature is critical and the process will depend on the ratio of activation to thermal energy.

4.2.4 Color Centers

In the above, we noted that certain compound point defects require a charge neutrality, which is maintained by having the same number of vacancies on the cation and anion sites (Schottky), or by having an equal number of vacancies and interstitials of the same constituent (Frenkel). However, it is possible that charge neutrality can be maintain by having a localized electron in the vicinity of a negative ion vacancy. The missing anion leaves a charged vacancy. This acts as a local potential which can attract and bind an electron. The bound electron will have characteristic energy levels, similar to those of an atom. This will create a spectrum of energy levels and excitation between these levels produces a series of optical absorption lines analogous to atomic line spectra. Since these optical transitions usually occur in a *forbidden* optical band, they will stand out from the rest of the crystal producing striking peaks in optical absorption spectra. Such defects are known as *color centers* for this reason. Crystals, such as alkali halides, which are generally transparent in the optical region, when subject to electron or ion bombardment can be visibly seen to have color in the region of radiation damage. This is a direct result of the introduction of vacancies to produce these color centers. Annealing can also produce color center defects. Different crystals will have different characteristic colors, depending on the binding energy and localized electron spectra associated with these point defects.

An electron bound to a single negative ion vacancy is also known as an *F-center*, while two neighboring negative ion vacancies can bind two electrons, producing what is called an *M-center*. An additional negative ion vacancy (i.e., three negative ion vacancy), can bind three electrons and is called an *R-center*. While these centers can be expected to have distinct optical spectra, due to the altered electron states available, in reality they are more difficult to observe since their spectra are not as sharp as the single negative ion vacancy. In fact, the M and R-centers are more strongly coupled to the crystal lattice and can lose energy by producing vibrations in the lattice; this process is called *phonon emission*. We will discuss phonons in the next chapter.

The absence of electrons in solids are known as *holes*. These can act as positive charge carriers in the solid, much in the same way as an electron is a negative charge carrier. We will discuss this subject in more detail when we come to consider the electronic properties of semiconductors (Chapter 9). For now it is sufficient to consider a hole as simply the absence of an electron, being subject to similar properties as the electron and is known as a *quasi-*particle. The *quasi-* here indicates that it isn't a real entity, as is the electron, but acts as one in terms of it physical attributes and how it can be treated. We have introduced this concept since we may expect that holes can be bound, in a similar way to electrons, by positive ion vacancies. However, no such cases have been reported. Hole centers do exist, but are not associated with vacancies as such. The V_K-center is one such entity and is produced by the binding of a hole to two neighboring negative Cl ions (Cl_2^-), producing a spectrum similar to that of Cl_2^-. Another hole or *H*-center results from interstitial of Cl^- ions.

4.3 DISLOCATIONS

Dislocations are a one-dimensional defects and have a very different nature to point defects. They are not the same as a line of vacancies. Line defects can also be curved inside the crystal, but can be associated with specific crystalline directions. The existence of line defects has enormous consequences for the mechanical properties of solids and is a vital component of the description of their behavior under the application of tensile, compressional and shear forces. We distinguish between two types of line defect in a solid as outlined below.

4.3.1 Edge Dislocations

The edge dislocation is defined as the termination of a crystal plane within the solid as is illustrated in Figure 4.3. The deformation of the crystal lattice in the region of the edge dislocation is such that in the region above it the lattice is in compression, while in the

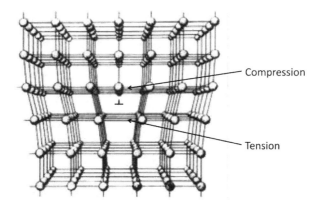

FIGURE 4.3: End on view of an edge dislocation.

region below, where the lattice plane is missing, the crystal lattice is in tension.

4.3.2 Screw Dislocations

In a screw dislocation, see Figure 4.4(a), part of the crystal lattice appears to be displaced in a direction parallel to the dislocation line. The screw and edge defects can combine in a crystal, as shown in Figure 4.4(b). The accommodation of the dislocation is transformed from one to the other as it changes direction in the crystal. The screw dislocation can be revealed by the deposition of material on a crystal. Since the screw dislocation produces a partial atomic step on the crystal surface, the growth of a deposit on this surface will be different at the step region to that on a flat region of the surface. The step is a more stable region for atoms to deposit since it offers more bonding sites to an adatom. This preferential growth at the step is then revealed as an advance of the step edge around the screw dislocation line. This is beautifully demonstrated in Figure 4.5.

4.3.3 The Burgers Vector

The *Burgers vector* is a form of quantification of the line defect and is different for the edge and screw dislocation types and was first introduced by Dutch physicist Jan Burgers. To calculate the Burgers vector, we can make a loop around the defect line (i.e., in the plane perpendicular to the line) in steps of lattice vectors. In

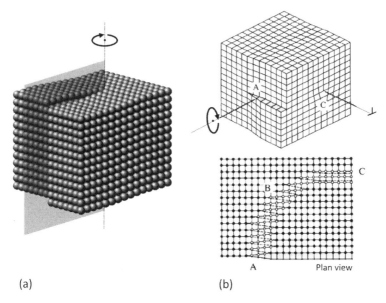

(a) (b)

FIGURE 4.4: (a) The screw dislocation. (b) Transition between pure edge and pure screw dislocations.

FIGURE 4.5: Surface image of a SiC crystal resulting from the existence of a screw dislocation on the surface.

each direction we move, we count the number of lattice spacings to return to the point of origin. The difference gives the magnitude and direction of the Burgers vector, which is designated as **b**. This is illustrated in Figure 4.6 for both types of dislocation. We note that the Burgers vector for an edge dislocation is perpendicular to the direction of the dislocation line, while for the screw dislocation it is parallel. We can further note that in the transformation from one to the other, as shown in Figure 4.4(b), the Burgers vector remains unchanged in direction and magnitude. The Burgers vector is generally given in terms of the lattice constant of the material. Therefore, the magnitude of the Burgers vector may be less than the lattice constant and will depend on the plane in which they occur. For the case illustrated in Figure 4.6, and assuming a simple cubic structure, the Burgers vectors can be written as: $\mathbf{b} = a[0\bar{1}0] = -a[010]$ for the edge dislocation and $\mathbf{b} = a[\bar{1}00] = -a[100]$ for the screw dislocation. Note that in the evaluation of the edge dislocation, we can start at point O, we then move three lattice spacings, in the [001] direction, to position P. Now we move to point M in the [010] direction 5 lattice spacings. Then in the [00$\bar{1}$] direction we move three spacings to point N. We finally close the loop, moving in the [0$\bar{1}$0] until

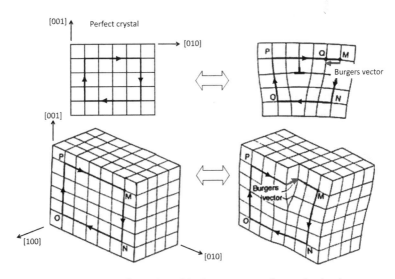

FIGURE 4.6: Illustration of the Burgers vectors for an edge (top) and a screw (bottom) dislocation.

we reach point O again, which is four lattice spacings. Summing the number of lattice spacings in each direction we find the difference of one lattice spacing in the [010] direction, which corresponds to the Burgers vector indicated above for this defect. Clearly for a perfect crystal, there is no residual difference. We can compare the perfect case on the left of Figure 4.6.

4.3.4 Dislocations and Mechanical Properties of Solids

Dislocations are mobile entities, and can be created and activated by the application of external forces to the crystal. In fact, such processes are important in the mechanical properties of solids. In general, the mechanical properties of a material depend on the internal structure of the sample. This is usually anisotropic, meaning that depending on the direction in which the external forces are applied, the response of the material will be different. A rigorous treatment requires the use of tensor analysis, which can resolve how each force acts in the various directions of the crystal. Such treatment is beyond the scope of the current text and the interested reader is referred to the further reading section at the end of this chapter. In the following we shall use a more direct and simplified approach, which will bring out the main ideas and illustrate the important points of discussion. The most important mechanical response that we can consider is how a material reacts to mechanical loading. As with most physical properties, we can measure the response of the material to the specific action of an externally applied force. The classical case would be to load the sample gradually with a mass and see how it deforms. This allows us to characterize materials in terms of their *strength* and *ductility*. In many materials, with the initial response to loading, the material stretches in an *elastic* manner. That is, the material returns to it original form (shape) when the loading force is removed. We generally display this type of behavior in a *stress - strain* curve. We define these quantities as follows: stress is the force per unit area (or cross-section) of the sample, which we can write as:

$$\sigma = \frac{F}{A} \qquad (4.16)$$

Imperfections in Crystalline Order • 109

This is expressed in the units of pressure, the Pascal (Pa). The strain is defined as the extension of the material per unit length, and can be expressed as:

$$\epsilon = \frac{\delta x}{x} \qquad (4.17)$$

where x is the original length of the sample. The stress - strain curve will then parameterize the response of the material to the applied external tensile force, F. By parameterizing we are removing the extrinsic factors of the sample dimensions; area, A and length, x. In Figure 4.7, we show a typical stress - strain curve, such as observed in metals. The initial portion of the curve is linear, corresponding to the elastic response of the material. The gradient of this portion is defined by the *elastic constant* or *Young's modulus* of the material. This can be expressed mathematically using Equations (4.16) and (4.17) as:

$$E = \frac{stress}{strain} = \frac{F/A}{\delta x / x} \qquad (4.18)$$

Typical values for metals lie in the range 2 - 50 GPa. Since the stress is a dimensionless quantity, the units are again that of pressure.

We note that in Figure 4.7, the linear region ends at what is reffered to as the *yield strength* or *yield point*. Beyond this point the material can further extend itself, however, the response is no longer elastic but *plastic*. This means that when we remove the applied force, the material will no longer return to its original size and shape and is said to be *plastically deformed*. It is this plastic deformation that results from the permanent alteration of the internal structure of the material and the introduction of dislocations and other defects in the solid. As we stretch the material in the plastic zone, the material becomes harder, a process called *strain* or *work hardening*. This is seen by the change of slope of the curve. Work hardening is a result of the production of dislocations which become entangled in the material and provides additional resistance to deformation since they impede the motion of other dislocations. The maximum stress that can be attained is called the *ultimate strength* and the material will then undergo rapid deformation which causes a narrowing or *necking* of the sample and quickly ends with the breaking or *fracture* of the material. The degree of necking is important, since as the sample cross-section thins, the stress further accumulates in this region; see Equation (4.16), as A reduces, the stress will increase, and therefore the material snaps very soon afterwards. A material, such as a metal, which has a visible deformation like this is called *ductile*. The opposite behavior, where the material snaps before any visible plastic deformation, is called *brittle fracture*. Chalk and ceramics generally fall into this category. In such materials virtually no plastic variation is observed.

Elastic deformation is due the the stretching of interatomic bonds along the direction of applied stress. It essentially involves all atoms, or interatomic bonds, in the crystal. In plastic deformation, there is a localized structural alteration and involves relatively few atoms, where a *slip* process occurs. Slip is where there is motion of atoms in an atomic plane, and will usually occur between close-packed planes. In Figure 4.8(a), we show the schematic process of slip, while in Figure 4.8(b) a transmission electron micrograph (TEM) of slip planes in a Cd sample. In general the slip planes form an angle, ϕ, with respect to the direction of the applied tensile stress.

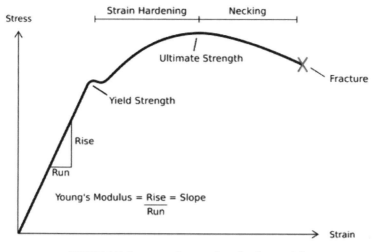

FIGURE 4.7: Stress - strain curve for a ductile material.

Imperfections in Crystalline Order • 111

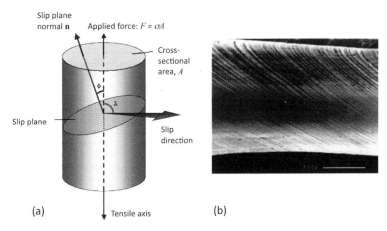

FIGURE 4.8: (a) Resolution of tensile stress in a sample showing a slip plane. (b) TEM image showing slip planes in a Cd crystal. The lines correspond to the slip planes.

Let us consider an applied stress, σ, to a sample of cross-section, A. The spacing between atoms is a with an interatomic separation of d. The applied stress will be resolved into the slip plane, since these will be the weakest points, excluding other defects. This can be considered to be a shear stress, which causes the atomic planes to slip over one another. For small elastic strain, the shear stress will be related to the displacement, x as follows:

$$\tau = \frac{\mu x}{d} \qquad (4.19)$$

where μ represents the shear modulus. Since the atoms in a crystal are periodically arranged, we can make the following approximation:

$$\tau = K \sin\left(\frac{2\pi x}{a}\right) \qquad (4.20)$$

where K is a constant. Equation (4.20) is interpreted as the stress being periodic in x, the displacement, so when the planes move (via slip) from its original position to $x = a$, we essentially return to a condition which is the same as that at the start. For very small displacements, $x \ll a$, we can now write:

$$\tau = \frac{\mu x}{d} = K\left(\frac{2\pi x}{a}\right), \qquad (4.21)$$

from which we obtain the constant of proportionality:

$$K = \frac{\mu a}{2\pi d}. \qquad (4.22)$$

We now express Equation (4.20) as:

$$\tau = \frac{\mu a}{2\pi d}\sin\left(\frac{2\pi x}{a}\right) \qquad (4.23)$$

The critical stress, τ_c, necessary to produce plastic deformation via slip in a perfect crystal can thus be expressed as the maximum amplitude of the sinusoidal wave, i.e., the maximum value of τ:

$$\tau_c = \frac{\mu a}{2\pi d} \simeq \frac{\mu}{2\pi} \qquad (4.24)$$

If an angle of ϕ exists between the slip plane normal and the axis of tensile stress and an angle λ between the slip direction and the tensile stress, the effective stress will be reduced from its maximum value. We can resolve to find that the force acting along the slip direction is $F\cos\lambda$, while the area of the slip plane will be $A/\cos\phi$. The shear stress resolved along the slip plane will now be:

$$\tau = \frac{F\cos\lambda}{A/\cos\phi} = \left(\frac{F}{A}\right)\cos\phi\cos\lambda = \sigma\cos\phi\cos\lambda \qquad (4.25)$$

The critical stress can now be expressed in the form:

$$\tau_c = \sigma_c\cos\phi\cos\lambda \qquad (4.26)$$

In a sample of Al, the value of shear modulus is about 3×10^{10} Nm^{-2}, so according to Equation (4.24) we would expect a valus for the critical shear stress to be roughly 4.8×10^9 Nm^{-2}. However, measured values of τ_c turn out to be $\sim 10^7$ Nm^{-2}, which is around three orders of magnitude smaller than the value expected. This discrepancy can only be explained in terms of the existence of defects in the crystal in the form of dislocations. It will be these dislocations that provide the *weak link* in the structure from which the solid ruptures and in effect no rigid form of slip will take place, i.e., the motion of one atomic plane over an adjacent plane.

The motion of dislocations is provoked by the application of stresses on the crystal lattice, where we can think of the displacement occurring in steps equal to the interplanar spacing, as illustrated in Figure 4.9. This can be viewed as the effect of a shear force on the crystal. If the applied force is sufficient to cause the dislocation to move, the slipped area will grow at the expense of the unslipped region, where the force per unit length can be expressed as:

$$F = \tau b \qquad (4.27)$$

This force will be parallel to the direction of motion of the slip and perpendicular to the dislocation line. The above equation shows that the displacement occurs in portions of the amount, b, the Burgers vector in the slip plane. The process of slip will create a step at the crystal surface, as illustrated for both edge and screw dislocation, in Figure 4.10.

The motion of dislocations can be impeded by the presence of other imperfections in the crystal, where pinning of the dislocation can occur. In such a situation, the dislocation can still move, but will bow out from the pinning sites. We can consider the dislocation pinned between two obstacles a distance, l, apart, such that the an applied stress of τ will give a a normal force of $\tau b l$ on the dislocation line. This force is balanced by the tension, T, in the line, such that we can write:

$$\tau b l = 2T \sin\theta \qquad (4.28)$$

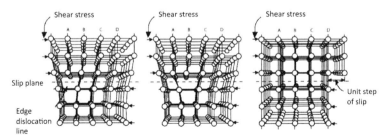

FIGURE 4.9: Edge dislocation movement can occur via the slipping of lower portion of the crystal with respect to the upper portion.

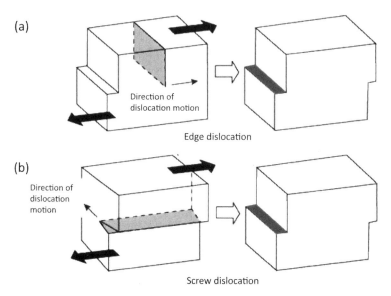

FIGURE 4.10: Creation of steps by the action of slip for (a) edge and (b) screw dislocations.

The line tension, T, is a vector and has a magnitude given by:

$$T = Gb^2 \tag{4.29}$$

where G is the elastic shear modulus, from which we can obtain:

$$\tau = \frac{2Gb \sin \theta}{l} \tag{4.30}$$

Therefore the maximum stress required to cause bowing in the dislocation will be when the line forms a semi-circle, i.e., when $\theta = 90°$, thus we can write:

$$\tau_{max} = \frac{2Gb}{l} \tag{4.31}$$

There are other types of motion for dislocations. For example, when an edge dislocation moves in a direction which is perpendicular to its slip plane, the resulting motion is called *climb*. Such motion must be accompanied by the generation or annihilation of vacancies or interstitials, depending on whether the motion is in a positive or negative sense. This is because the number density of defects (interstitials or vacancies) must be maintained in accord with the temperature of the sample. Another form of motion is *cross-slip*. This is where a change of slip plane occurs in the movement of a screw dislocation, as shown in Figure 4.11.

4.3.5 Dislocation Energy

The presence of dislocation implies additional energy in the crystal and as such they are thermodynamically unstable. We can consider the energy associated with the screw dislocation by evaluating the shear strain around the screw line at a radius, r, as shown in Figure 4.12. Consider a screw dislocation of length, l, Burgers vector, b, along the direction of its axis. The shear strain, γ, of a thin annular section, of radius r and thickness, dr is:

$$\gamma = \frac{b}{2\pi r} \tag{4.32}$$

The elastic energy per unit volume, dE/dV, of the annular portion can be expressed as:

$$\frac{dE}{dV} = \frac{1}{2}\tau\gamma = \frac{1}{2}G\gamma^2 = \frac{G}{2}\left(\frac{b}{2\pi r}\right)^2 \tag{4.33}$$

Given that the volume of the annular ring is: $dV = 2\pi r l dr$, the energy per unit length of the shell is:

$$dE = \frac{lGb^2}{4\pi}\left(\frac{dr}{r}\right) \tag{4.34}$$

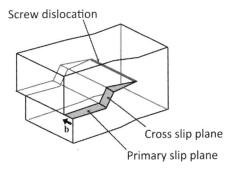

FIGURE 4.11: The change of slip direction in a screw dislocation is the cross-slip.

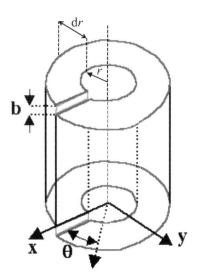

FIGURE 4.12: Consideration of the elastic strain energy associated with a screw dislocation.

From this we can write the total elastic energy per unit length of the screw dislocation as:

$$\epsilon = \frac{E}{l} = \int_{r_0}^{R} \frac{Gb^2}{4\pi} \frac{dr}{r} = \frac{Gb^2}{4\pi} \ln\left(\frac{R}{r_0}\right) \quad (4.35)$$

where R and r_0 are chosen as the upper and lower limits of r. Since the energy is relatively insensitive to the ratio R/r, we generally express the energy of the screw dislocation as:

$$E \simeq lGb^2 \quad (4.36)$$

The corresponding result for an edge dislocation is:

$$E = \frac{1}{1-\nu} \frac{lGb^2}{4\pi} \ln\left(\frac{R}{r_0}\right) \simeq \frac{lGb^2}{1-\nu} \quad (4.37)$$

where ν is called the Poisson ratio and expresses the contraction or transversal strain to the extension or axial strain of the material. Metals have values typically of the order of $\nu \simeq 1/3$. Since the energy of dislocations are additional energies in crystals, they will be minimized for smaller b values and this will usually confine the dislocations to close-packed directions.

4.3.6 Interactions between Dislocations

We have already seen that dislocations are characterized by their Burgers vector. These can be negative or positive since they are real vectors. For the case of edge dislocations, a positive (negative) Burgers vector would correspond to the cases where the extra halfplane is above (below) the dislocation line. It is a simple matter to understand that edge dislocations with opposite sign would attract one another while those of the same sign repel. In the former case, the attraction ultimately leads to the annihilation of the dislocations if they lie on the same slip plane, while the repulsion of same sign dislocations avoids increasing the local tensions in the lattice. If the dislocation lines lie on different atomic planes they can produce a line of vacancies or a line of interstitials if they overlap. In certain cases they can give rise to fractional Burgers vectors, called a partial dislocation.

We previously illustrated the edge dislocation as a straight line. This, however, is not a necessary condition and the line can have steps in them, called *jogs*, such as when two dislocations cross; see Figure 4.13(a). The transmission electron microscope (TEM) permits the visualization of dislocations as dark lines. The micrograph in

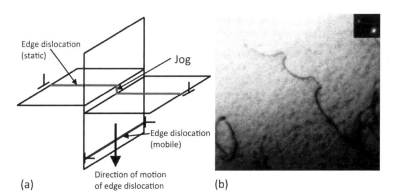

FIGURE 4.13: The crossing of two edge dislocations can introduce jogs in a dislocation, as illustrated in (a). In (b) we see a TEM micrograph of a dislocation with jogs along it.

Figure 4.13(b) shows a dislocation with jogs. Jogs can play an important role in the mechanical properties of the solid since they can act as an anchor point. Dislocations that cannot move are called *sessile*.

It is important to note that straight dislocations, while mobile, can be pinned by jogs or connections to sessile dislocations. The build-up of dislocations in a solid is important in the work-hardening process and implies that the working (application of stress) of a solid creates more dislocations. Frank and Read suggested the process by which this can occur. Consider a dislocation line stretched between two anchor points in a solid. If tension is applied to the solid the dislocation line can bow out. Continuing to apply this tension will cause the bowing to increase to the point where the line will eventually bow out backwards and meet up to form a closed loop. At this point we have a dislocation loop plus a dislocation line between the original two anchor points. Therefore we have created a new dislocation in the solid. Such a process is known as a *Frank - Read source* and is illustrated in Figure 4.14. Continuing this process can allow further dislocations to be produced, giving a series of dislocation lines and multiple slip on the same plane.

Since the motion along slip planes requires the least energy, they will be the preferred directions of dislocation motion. The combination of a slip direction and slip plane is called a *slip system* and depends on the lattice type. For example, in the fcc lattice, the slip planes are $\{111\}$, while the slip direction is $<1\bar{1}0>$. Care must be taken to define the plane and the particular set of directions that exist in that plane. There are $4 \times 3 = 12$ slip systems in the fcc lattice.

Dislocations can build up in networks and there must be a conservation of the Burgers vector when dislocations meet at a node. Therefore we can write:

$$\mathbf{b} = \mathbf{b}_1 + \mathbf{b}_2 \qquad (4.38)$$

where the dislocation of Burgers vector, $\mathbf{b}$, dissociates into two different dislocations, $\mathbf{b}_1$ and $\mathbf{b}_2$. This is called a *dislocation reaction*. Using the fcc example, we can consider the slip plane (111), see Figure 4.15, where from the hard sphere model we can see that the Burgers vector $\mathbf{b} = (a/2)[10\bar{1}]$. However, given the contour of

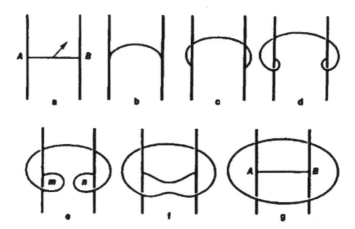

FIGURE 4.14: The Frank - Read source: A straight dislocation line anchored at two points is bent by the application of stress causing it to bow and eventually meet up to form a closed loop and a new dislocation line between the two pinning points, A and B.

the spheres, it is possible to see that the slip is more likely to occur in two steps; displacement in the $[11\bar{2}]$ direction followed by the displacement in the $[2\bar{1}\bar{1}]$ direction. This composite motion will have a zig-zag form, but means that the lattice plane doesn't have to go directly over an atom and slips in the grooves between them. We can write out the Burgers vectors as follows:

$$\mathbf{b} = \frac{a}{2}[10\bar{1}] \tag{4.39}$$

$$\mathbf{b}_1 = \frac{1}{3}\left(a[10\bar{1}] - \frac{a}{2}[1\bar{1}0]\right) = \frac{a}{6}[11\bar{2}] \tag{4.40}$$

$$\mathbf{b}_2 = \frac{a}{6}[2\bar{1}\bar{1}] \tag{4.41}$$

From Equation (4.38) we can write:

$$\frac{a}{2}[10\bar{1}] = \frac{a}{6}[11\bar{2}] + \frac{a}{6}[2\bar{1}\bar{1}] \tag{4.42}$$

Here **b** is said to be a perfect dislocation, while $\mathbf{b}_1$ and $\mathbf{b}_2$ are called *partial dislocations*. To check whether such a process is likely to occur we can consider the energies associated with the dislocation

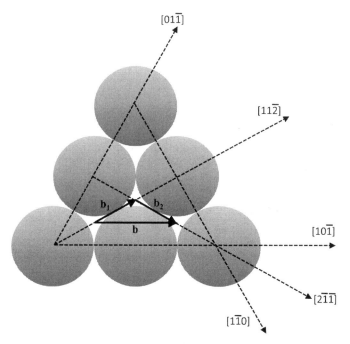

FIGURE 4.15: The (111) plane showing the various principal crystalline directions. The Burgers vector **b** for the perfect dislocation is indicated along with the partials, **b₁** and **b₂**, into which it can dissociate.

and the partials, since they are proportional to b^2. From this we find the following:

$$b^2 = \frac{a^2}{2}; b_1^2 = b_2^2 = \frac{a^2}{6}; b_1^2 + b_2^2 = \frac{a^2}{3} \quad (4.43)$$

From this we note that $b^2 < b_1^2 + b_2^2$ and is thus energetically favorable for a split into the two partials. The two partial dislocations may mutually repel to create a region of partial slip called a *stacking fault*. We note that the fcc structure usually has a stacking sequence which is given as ABCABC..., which corresponds to the positions of the the atoms in successive atomic plane, as shown in Figure 2.12. Now supposing that the slip occurs in plane A, then the passage of the partial dislocation $\mathbf{b_1}$ causes the crystal above the slip plane to shift one partial step, such that plane B becomes plane C etc. Thus we can see why this partially slipped region is referred to as a stacking fault. Since the stacking fault disturbs the the lattice periodicity, it will have a definite surface energy associated with it and this will determine whether the stacking fault can propagate in the crystal. The energy criterion for this can be expressed as:

$$b^2 < b_1^2 + b_2^2 + \Delta \qquad (4.44)$$

where Δ represents the contribution from the stacking fault energy. Therefore the stacking fault can hinder the motion of dislocations along their slip directions.

It is clear that the motion of dislocations requires work to be done on the lattice. Therefore we have an internal resistance to dislocation motion called *lattice* or *internal friction* and is also known as the *Peierls - Nabarro force*. We previously mentioned that the intersection of dislocations causes this internal friction to increase leading to work hardening. Precipitates in the material can also act as pinning sites for dislocations and leads to a process called *precipitation hardening*.

4.4 PLANAR DEFECTS

As the name would suggest, these types of defects occur on surfaces and can be a few atomic layers thick. The main planar defects are: grain boundaries, tilt boundaries, twin interfaces and stacking faults. We have already discussed the latter since it was also relevant to our discussion on dislocation motion. While grains are indeed three-dimensional objects, the grain boundary can be thought of as a surface.

4.4.1 Grain Boundaries

In polycrystalline materials the solid is made up of regions of crystalline order with different orientations of the crystalline axes. Such regions of crystalline material are called *grains*. The region (usually disordered) between the grains is called the *grain boundary*. In Figure 4.16, we show a schematic illustration of grains and grain boundaries. Since the physical properties of a material is strongly dependent on the local environment and the number of nearest neighbors, the existence of grains can have a significant effect on the overall properties of a polycrystalline solid. Grain size is a crucial factor, since it will

122 • Solid State Physics

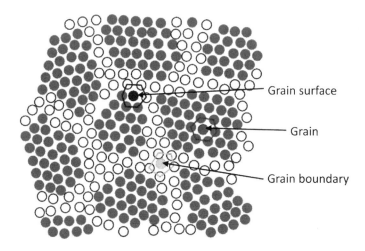

FIGURE 4.16: Crystalline grains are surrounded by grain boundaries, which separate the zones of periodic order in the grains. Atoms can find themselves in distinct positions, such as inside the grains, at the surface of the grains and within the grain boundary itself.

determine the proportion of atoms which are in regions that differ from the normal bulk crystal; i.e., those on grain surfaces and within the grain boundaries themselves. Polycrystallinity can, for example, significantly alter the electrical resistance of a metal. The mechanical properties will also be strongly dependent on grain structure.

4.4.2 Tilt Boundaries

A tilt boundary is a form of grain boundary in which there is a small angle between the crystalline axes of one grain and the adjacent grain. In Figure 4.17, we shown a schematic illustration of a tilt boundary, which can be viewed as an interface with regular arrays of edge dislocations. Since the edge dislocations have the same sign, if they are regularly spaced it is possible to evaluate the angle of tilt between the two adjacent grains. This can be expressed as:

$$\frac{b}{h} = \tan\theta \qquad (4.45)$$

where b is the magnitude of the Burgers vector of the edge dislocations and h is the distance between the edge dislocations.

A twist boundary is similar to the tilt boundary, where the edge dislocations are replaced by screw dislocations, whose axis is

Imperfections in Crystalline Order • **123**

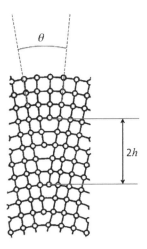

FIGURE 4.17: The tilt boundary consists of regularly spaced edge dislocations of the same sign.

perpendicular to the grain boundary. In this case one crystal grain will be rotated with respect to the other.

4.4.3 Twin Boundaries

The *twin boundary* is a very specific interface between grains. Essentially it is formed when the misorientation angles are the same in both grains and are then joined such that the plane between then appears to be a plane of reflection. This situation is illustrated in Figure 4.18. Twinning can occur during the growth of a crystal since

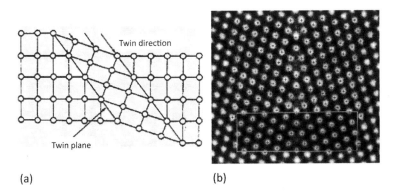

FIGURE 4.18: The twin boundary appears to be a plane of reflection for the crystal and corresponds to a specific plane, which is equivalent for the normal and twinned portion of the sample. A HRTEM (high resolution TEM) image shows a real example of twinning.

the twin plane represents the same energy for both orientations of the crystal in the untwinned and the twinned regions. Plastic deformation can result from the introduction of twinning in a crystal. The occurrence of twinning is quite common in metals with the bcc or hcp structure.

4.5 NON-CRYSTALLINE MATERIALS

So far we have considered solids to be the result of the regular arrangement of atoms in a crystal with the presence of some structural disorders, as we have discussed above. It is possible, however, for solids to be formed under specific conditions which results in no long range ordering of the atoms in the solid. Such materials are called *non-crystalline* or *amorphous*. Common examples are glasses and plastics. Typically amorphous solids are produced by rapidly quenching the material from a melt. The rapid cooling is so fast that atoms do not have the time to find the position of lowest energy or equilibrium, which would occur for a slowly cooled melt. Annealing of amorphous materials can lead to the crystallization of the solid in question, though it is more likely to result in a polycrystalline solid. In Figure 4.19, we show the difference between an ordered solids and a non-crystalline one. In a diffraction pattern, we see that while the ordered structure

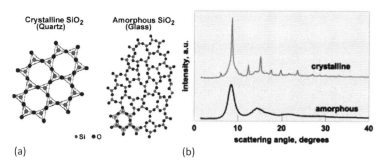

FIGURE 4.19: Illustration of ordered SiO_2 and glassy SiO_2. In the former, the atoms have definite positions and a regular periodic structure. For the glassy state, the atoms have no long range regularity, though the interatomic separations are fairly constant. On the right hand side we illustrate the form of x-ray diffraction patterns for a crystalline material and an amorphous one. In the latter we note a lack of sharp pronounced peaks, which are replaced by broad maxima, known as an amorphous halo.

has well-defined peaks corresponding to the diffraction (constructive interference) conditions being satisfied along specific atomic planes, in the amorphous sample, the diffraction spectrum consists of broad peaks. These broad maxima are called *amorphous halos* and correspond to the average interatomic spacings in the structure.

Amorphous materials can have significantly different properties from their crystalline and polycrystalline counterparts and have attracted much interest in solid state physics and materials science. While glasses tend to be quite brittle in nature, amorphous metals can be quite strong and flexible. Also electronic and magnetic properties can vary appreciably from their normal equilibrium structures. In magnetic materials, for example, there has been much interest in the partial devitrification from the amorphous state brought about by specific annealing. This allows the formation of magnetic nanocrystals in an amorphous magnetic matrix. The combination of the two magnetic phases gives rise to extremely soft magnetic materials, with coercive field as low as 0.01 Oe. Such properties do not exist in the same materials with a crystalline structure. It is in fact the random nature of the material which suppresses magnetic anisotropies which produces such low coercive fields. Amorphous Si and SiO_2 has also attracted much attention in semiconducting materials, which again, due to the lack of atomic periodicity gives rise to altered properties and offer new applications. Amorphous Si appears to be a good option in the manufacture of solar cells.

The formation of glasses is a well developed area of study with many types of glass available. The main principle of glass formation is the rapid cooling through the melting temperature, T_m, to the glass formation temperature, T_g. A change in slope of the specific volume versus temperature is observed at this point. A slow cooling would however be accompanied by a sharp drop in the specific volume at the melting point, which arises from the crystallization of the melt.

4.6 SUMMARY

While the principal properties of a solid derive from its periodic atomic or crystalline structure, imperfections occur frequently.

Crystal imperfections in a solid can have profound consequences for material properties. For example, the residual electrical resistance of a solid depends directly on the concentrations of defects in the material. The form of defects can be point like, linear, planar and three-dimensional. In this chapter, we have described in detail these different types of defect. Point defects, for example, come in a variety of forms; they can be a simple missing atom, or vacancy, to an atom in a non-regular position called an interstitial. Impurity atoms can occupy regular or interstitial sites to further complicate the situation. The disorder of binary systems also constitutes another point defect. The concentration of vacancies and other point defects can be calculated in terms of their activation energies of formation, generally having an Arrhenius type dependence, as is common for thermally driven processes. This was also illustrated for the diffusion process, by which defects can move in a solid. Concentration gradients also provide a driving force in diffusion. In addition to the interruption of the periodicity in a crystal, point defects can have specific properties which can attract charge carries. The transition between the eigenstates of these localized charge carriers can give rise to the emission of radiation, and hence such structures are referred to as color center.

Linear defects come in the form of deformations of the lattice structure and are designated as edge and screw dislocations. The former can be considered as a missing "halfplane", while the latter are formed by a displacement of part of the crystal along a line, which appears as a twist in the lattice plane. Dislocations are characterized by their Burgers vector, which gives its direction and magnitude. The edge dislocation has its Burgers vector perpendicular the the defect line, while the screw dislocation has a Burgers vector parallel to the line defect. The existence of dislocations has a strong influence on the mechanical properties of solids, where interactions between dislocations lines can cause a resistance to dislocation motion making the material more resistant to applied forces. Such a process is called work hardening. The shape of the stress - strain curve gives a good idea of the mechanical properties of a solid. The linear portion is due to elastic deformation, having a slope equal to the Young's modulus of the material. Plastic deformation occurs after the yield point is passes and the material suffers permanent damage by the introduction of dislocations into the solid. Motion of dislocations takes place according to the slip system (slip-plane and slip direction) of the crystal. The movement of a dislocation can be dissociated into partial dislocations, where they are energetically favorable.

Planar defects occur in solids also in a variety of forms. The most common are the region between crystalline portions of a polycrystalline solid. These are called grain boundaries and can also have a significant effect on the physical properties of a solid. This is because the properties depend crucially on the local environment of atoms, and when the grains are relatively small, a large proportion of atoms can lie at the surface of grains or within the grain boundaries themselves. These can have very different symmetries to atoms within a crystalline grain. Tilt, twist and twin boundaries are also planar defects, which are fairly common. A further linear defect is the stacking fault, in which there is a disorder in the normal stacking sequence of atomic planes. This situation can occur in atomic planes with more than one equivalent minimum energy position.

Three-dimensional defects can occur in solids, such as inclusions of foreign atom aggregates. We also outlined the situation where no long range order exists in amorphous solids. These can be produced by rapidly cooling or quenching a material from its molten state. The speed of cooling will determine the degree of disorder. Amorphous materials can have significantly different physical properties to their crystalline counterparts and has generated much interest among material scientists always on the search for new physical properties and applications.

REFERENCES AND FURTHER READING

Basic Texts

- J. S. Blakemore, *Solid State Physics*, Cambridge University Press, Cambridge (1985)

- H. P. Myers, *Introductory Solid State Physics*, Taylor and Francis, London (1998)

Advanced Texts

- M. A. Wahab, *Solid State Physics: Structure and Properties of Materials*, Alpha Science International Ltd., Harrow (2007)
- N. W. Ashcroft and N. D. Mermin, *Solid State Physics*, Saunders College, Philadelphia (1976)
- D. Hull and D. J. Bacon, *Introduction to Dislocations*, Butterworths - Heinemann (Elsevier), 5th Edition, Oxford, (2011)
- J. F. Nye, *Physical Properties of Crystals: Their Representation by Tensors and Matrices*, Oxford University Press (1985)

EXERCISES

Q1. Derive Equation (4.4)

Q2. Use the same procedure to show the corresponding density of Frenkel defects in a solid.

Q3. Determine the density of vacancies in Cu at room temperature and just below the melting temperature of 1356 K. Copper has an energy of vacancy formation of 1 eV with an fcc structure and lattice parameter of 3.61 Å.

Q4. The density of Schottky defects in a specific sample of sodium chloride is 5×10^{11} m^{-3} at room temperature. If the interatomic separation is 2.82 Å, determine the average energy required to create a single Schottky defect in this crystal.

Q5. Consider the case of 1D diffusion in a medium bounded by planes at $x = 0$ and $x = l$. In a steady-state solution the concentrations at these positions will remain constant, with constant diffusion coefficient, D. Show that this leads to:

$$\frac{d^2 c}{dx^2} = 0 \qquad (4.46)$$

Find the form of the solution to this equation and show that the steady-state flux can be expressed as:

$$J = D\frac{c_1 - c_2}{l} \qquad (4.47)$$

Q6. If the ratio of diffusion rate of Ag in Si at temperatures of 1350 °C and 1100 °C is 8 in a specific doping process, calculate the activation energy.

Q7. Calculate the diffusion coefficient of Cu in Ag at 550 °C, assuming that the energy involved is 121 kJmol^{-1} and $D_0 = 0.25 \times 10^{-4}$ m^{2}s^{-1}. Evaluate the approximate distance of penetration of Cu atoms after 1 hour.

Q8. A color center in an NaF crystal has a blue tinge. Estimate the effective charge of the negative ion vacancy using the Bohr model. Make any assumptions required.

Q9. Illustrate graphically and analytically that the first two dislocations add to give the third dislocation in the following reaction:

$$\tfrac{a}{6}[21\bar{1}] + \tfrac{a}{6}[121] = \tfrac{a}{2}[110]$$

Determine whether this reaction is energetically favorable.

Q10. What is the spacing between edge dislocations in a tilt boundary in fcc Ni if the tilt angle is 2°? N.B. Ni has a lattice parameter of 3.52 Å.

Q11. A hexagonal crystal may twin on $\{10\bar{1}2\}$ in a $<10\bar{1}1>$ direction. Make a sketch of this twin.

CHAPTER 5

LATTICE VIBRATIONS

"If you want to find the secrets of the universe, think in terms of energy, frequency and vibration."

—Nikola Tesla

5.1 INTRODUCTION

We have considered solids, up to now, as an assembly of atoms in fixed positions on the sites of a Bravais lattice. In reality atoms are not fixed, but have a certain kinetic energy from which they can move either to another lattice site, as we saw in the previous chapter, or can undergo elastic vibrations. We will assume, in this latter case, that the atoms have a mean equilibrium position which corresponds to the Bravais lattice site, about which this vibrations takes place. We also assume that the amplitude of vibration is small with respect to the lattice spacing. The driving force for the motion of the atoms/ions in the solid is from thermal energy. So when we come to describe this situation in more detail, we have made the assumptions which are necessary for a simple treatment, called the *harmonic approximation*. Thus we say that atoms vibrate about the position of the Bravais lattice sites, though at any one instance, the atoms in a solid are not fixed at those points. The average deviation of the atom from its corresponding lattice site will depend on the thermal energy available. In fact, we equate thermal energy as lattice vibrations, which occur as a series of superposed sound waves with a frequency spectrum determined by the elastic properties of the crystal.

The discrete vibration of the lattice is called a *phonon*, in analogy to the photon, the quantum energy for electromagnetic radiation. The analogy extends further than just a name, and phonons have much in common with photons. Both, in addition to having discrete energies, can be created and destroyed in collisions, and they are described by Bose - Einstein statistics, being *bosons*. The phonon, as we shall see, is an extremely important particle (or quasi-particle), having fundamental roles in thermal properties of solids as well as being a form of energy transfer between, for example, electrons and the crystal lattice.

In this chapter, we will outline the fundamental theory of lattice vibrations in one-dimension for a monatomic and a diatomic assembly or chain. We will further discuss some of the main properties of phonons and how we can describe the thermal properties of materials. Of course this treatment is also pertinent to the propagation of waves in a periodic lattice. The limits we have set ourselves is that of linearity, within the harmonic theory of solids. Beyond this limit, non-linear effects or *anharmonicity* become important. Such effects are relevant for the consideration of the interactions between photons and phonons.

5.2 VIBRATIONAL MODES OF A MONATOMIC LATTICE

5.2.1 One-Dimensional Chain

We will consider a chain of identical atoms of mass, m, which are separated by a distance, a. In this way we can consider the one-dimensional Bravais lattice as having vectors: $\mathbf{R} = n\mathbf{a}$, where n is an integer. In addition to the harmonic approximation indicated above, it is usual to invoke the *adiabatic approximation*, which states that the ion cores move independently of the electrons, being much heavier. Furthermore, we consider that the atoms are held together via interatomic bonds which can be viewed as being held together by a sequence of springs. The simple model we are constructing even considers that the springs obey Hooke's law. It can be noted that while this approach may seem over simplified, actually we could consider the same approach for a three-dimensional crystal in which a perturbation of a longitudinal vibration is propagated in a crystal, where instead of individual atoms, we consider atomic planes, where each plane will have mass, $M = Nm$, where we consider there to be N atoms per plane. Let $u(na)$ be the displacement along the line from the equilibrium position, of the atom that oscillates about position na, where we have assumed that the atomic chain commences at $x = 0$, see Figure 5.1. In terms of our monatomic chain, the force on the nth atom will be, given in terms of Hooke's law:

$$F_n = K(u_{n+1} - u_n) - K(u_n - u_{n-1}) = K(u_{n+1} + u_{n-1} - 2u_n) \quad (5.1)$$

where we only consider the interaction between neighboring atoms. The equation of motion for the nth atom can be written using Newton's second law:

$$F_n = m \frac{d^2 u_n}{dt^2} \quad (5.2)$$

Now since the atoms have harmonic motion we can express the displacement as a plane wave of the form:

$$u_n = A e^{i[kna - \omega t]} \quad (5.3)$$

In this equation A represents the amplitude, k is the wave-vector, na is the position of the nth atom, $\omega = 2\pi \nu$ with ν being the frequency

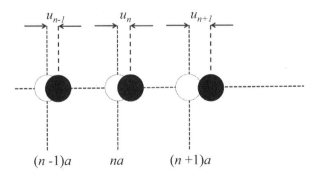

FIGURE 5.1: Schematic diagram showing the atomic positions on a monatomic chain.

of oscillation of the atom. We can now take the second derivative of Equation (5.3):

$$\frac{d^2 u_n}{dt^2} = -\omega^2 A e^{i[kna-\omega t]} = -\omega^2 u_n \qquad (5.4)$$

Combining the above equations into Equation (5.2), we obtain:

$$-m\omega^2 u_n = K(u_{n+1} + u_{n-1} - 2u_n)$$
$$= K(e^{ika} + e^{-ika} - 2)u_n \qquad (5.5)$$

We have used the fact that: $u_{n\pm 1} = u_n e^{\pm ika}$, which can be easily demonstrated. Eliminating the displacement, u_n, and recognizing; $e^{ika} + e^{-ika} = 2\cos(ka)$, we find:

$$-m\omega^2 = 2K[\cos(ka) - 1] = -2K[1 - \cos(ka)] \qquad (5.6)$$

This allows us to express the angular frequency, $\omega = \omega(k)$, as:

$$\omega(k) = \left\{ \left(\frac{2K}{m}\right)[1 - \cos(ka)] \right\}^{1/2}$$
$$= 2\sqrt{\frac{K}{m}} \left| \sin\left(\frac{ka}{2}\right) \right| \qquad (5.7)$$

where we have made use of the half-angle formula for the sine function. This solution, called the *dispersion relation*, gives the frequency of vibration of the atoms/ions as a function of the wave-vector, k, for the one-dimensional lattice. Equation (5.7) can be written as:

$$\omega(k) = \omega_m \left| \sin\left(\frac{ka}{2}\right) \right| \qquad (5.8)$$

where $\omega_m = \sqrt{4K/m}$ is the maximum value that the frequency can take. The dispersion relation, shown in Figure 5.2, is a convenient way of presenting the frequency spectrum since the phase velocity, (ω/k), and group velocity, $(\partial \omega/\partial k) = \nabla_k[\omega(k)]$, are directly related to this curve[1]. We note from Figure 5.2 that the frequency is periodic in k, with a periodicity of $2\pi/a$. Since we have introduced the phase

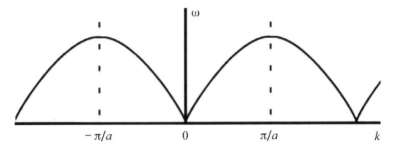

FIGURE 5.2: Schematic diagram of the dispersion relation for a monatomic chain.

and group velocities and using Equation (5.7), we can now express these as follows:

Phase velocity:

$$v_p = \frac{\omega(k)}{k} = \frac{2}{k}\sqrt{\frac{K}{m}}\left|\sin\left(\frac{ka}{2}\right)\right| = a\sqrt{\frac{K}{m}}\left[\frac{\sin(ka/2)}{(ka/2)}\right] \quad (5.9)$$

Group velocity:

$$v_g = \frac{\partial \omega(k)}{\partial k} = a\sqrt{\frac{K}{m}}\cos\left(\frac{ka}{2}\right) \quad (5.10)$$

In the region of long wavelengths or small k, which corresponds to the acoustic wave regime, we can approximate $ka \ll 1$, for which we can write $\sin(ka/2) \simeq (ka/2)$. In this case we have $v_p \simeq a\sqrt{K/m} \simeq v_g$. This is called the non-dispersive regime, and corresponds to the linear portion of the dispersion relation, near the origin. This simple treatment has already given much insight into the propagation of a disturbance in a chain of atoms and has much relevance for a normal 3D solid. For example sound, or elastic, waves in a solid propagate with a velocity given by $v = \sqrt{E/\rho}$, so a knowledge of the Young's modulus, E, and density, ρ, of a solid is enough to estimate the speed of sound in a solid. For a 3D solid we can write that the density is: $\rho = m/a^3$ and $E = K/a$. Substituting these into the previous expression we obtain the same relationship; $v = a\sqrt{K/m}$, as indicated above. For example, copper has a density of 8960 kg m³ and a Young's modulus of 13.4 × 10¹⁰ Pa This gives a speed of sound of 3870 ms⁻¹.

The low frequency regime is generally valid for $\nu \leq 10^{12}$ Hz, which includes the normal frequencies of acoustic waves. For the Cu example above, experimental results at ultrasonic frequencies (> 20 kHz), give a good agreement with the theory. Such sound waves will have wavelengths of the centimeter range and thus correspond to hundreds of millions of interatomic spacings. Even at the frequencies in the GHz range, the wavelength is around 0.4 μm, this corresponds to roughly 1500 atomic spacings.

From the form of the dispersion relation, we see that the maximum permitted frequency occurs at $|k| = \pi/a$. Now what will this mean in physical terms for the oscillations of the atoms? Firstly, this will correspond to the case where the wavelength is equal to two atomic spacings ($\lambda = 2a$). This must correspond to a natural limit for the phonons, so we can limit our attention to the region of k for which we have: $|k| \leq \pi/a$, setting the limit to a region of $2\pi/a$, which corresponds to the periodicity of Equation (5.7) and interestingly corresponds to the primitive reciprocal lattice vector for the monatomic chain of lattice parameter a. As we mentioned previously, for low frequencies there is no velocity dispersion, however, for larger values of ω this is no longer the case and significant velocity dispersion is observed. Low frequencies equate to long wavelengths ($\lambda \gg a$), with the lower limit being $\omega = 0$, which corresponds to $k = 0$ and an infinite wavelength, implying that the atoms are at rest (or moving rigidly with no relative displacement). As the frequency increases, the wavelength decreases until we reach our upper limit, $\omega_m = 2\sqrt{K/m}$. The vibrational states or modes are a property of the lattice and not the individual atoms and the dispersion relation expresses this property of the crystal lattice. These changes are accompanied by the variations of the phase and group velocities. We have already noted that for low frequencies $v_g \simeq v_p$, however, as the frequency increases and dispersion emerges, then $v_g < v_p$. At the maximum frequency the group velocity falls to zero, while the phase velocity at this point is about 64% of its value at zero frequency[2]. Something very important is happening here. We have seen that it makes no sense to consider wavelengths below 2 atomic spacings, but what is happening at this limit? Since the frequency spectrum is only repeated outside the reciprocal lattice unit cell, we can say that all of the frequency spectrum is thus contained within it. In fact, the limit of the reciprocal lattice unit cell for $\mathbf{k} = \pm(\pi/2)$, corresponds to the Wigner-Seitz cell, mentioned in Chapter 2, for our 1D lattice. The area within this region is called the *first Brillouin zone*. The second Brillouin zone occupies the same area on either side of the first Brillouin zone. We will mention Brillouin zones again for other structures when we discuss electronic states and band structures in Chapter 7. Any set of atomic displacements consistent with Equation (5.3) for a value of $|k| > (\pi/a)$ are equally valid with any wave-vector $\mathbf{k}' = \mathbf{k} + \mathbf{G}$, where $\mathbf{G}$ is a reciprocal lattice vector. Which is precisely what we said about all of the spectrum being represented in the first Brillouin zone; the primitive unit cell of the reciprocal lattice. Now returning to the condition $|k| = (\pi/a)$; this is equivalent to saying $\lambda = 2a$, since we can only propagate our waves along the direction of the chain, we can see that this is actually the Bragg condition, where we have:

$$\lambda = 2d \sin\theta = 2d \sin(\pi/a) = 2a \quad (5.11)$$

here we have set θ to $\pi/2$, since the direction of propagation of the wave (phonon) is perpendicular to the (imagined) planes and $d = a$. Thus we conclude that the phonons are diffracted at the point where their wavelength corresponds to a Bragg condition; i.e., at the Brillouin zone boundary. The limit value of the wave-vector corresponding to the Brillouin zone boundary is a *standing wave* rather than a travelling wave, and the Bragg condition is through a 180° angle, such that it interferes with the incoming wave. This type of condition will also occur in two and three-dimensional lattices when the wave-vector reaches the Brillouin zone boundary. In such cases, the wave must have a zero component of its group velocity in the direction perpendicular to the zone boundary line of plane in k-space.

So far we have only discussed the possibility of the atoms being displaced in the direction of the chain, i.e., longitudinal vibrations. However, the linear lattice can also support motion in the lateral directions; these are called *transversal modes*. The forces acting on transversal modes are different from those of the longitudinal modes and are thus independent, and both sets give rise to different branches on the dispersion curve. Actually, since there is more than one transversal direction, these modes can be degenerate. This will become more evident when we discuss the vibrational modes in three-dimensional lattices in the following section.

5.2.2 Extension to Three-Dimensions

Much of the physics discussed above remains valid when we consider two and three-dimensional lattices. While we will not here treat the 3D case mathematically, we will illustrate some of the features of normal modes of vibrations in 3D lattices. The types of vibrational modes can be both longitudinal and transversal, and only in special cases will planes move in unison, as can occur in cubic crystals along special crystalline directions. For cubic crystals it is possible for there to be purely longitudinal or purely transverse modes in [100], [110] or [111] directions. In general, most waves in three-dimensions will be a mixture of longitudinal and transverse. We note that while in 1D the transverse modes are degenerate, in 3D this will not necessarily be the case and in most circumstances will not be so. Therefore, we may expect to see three branches of the dispersion relation, one for longitudinal modes and two for transversal modes. In certain directions the transverse vibrations will be degenerate.

Experimentally, the phonon spectrum can be measured using inelastic scattering of thermal neutrons, since they have the appropriate energies for transfer to the crystal lattice. Electrons will not have much interaction with neutrons of these energies[3], while there can be appreciable changes in magnitude and direction of the neutrons with phonons. In the scattering process, the neutrons impart some of their kinetic energy to the lattice in the form of phonons. Given that the initial energy, E, of the neutrons is known and their subsequent energy (after collision), E', can be measured, we can determine the phonon energy as the difference. We note that the neutron kinetic energy can be expressed as:

$$E = \frac{\hbar k_n^2}{2m_n} \tag{5.12}$$

and

$$\mathbf{k}_n = \frac{m_n \mathbf{v}_n}{\hbar} \tag{5.13}$$

being the wave-vector of the neutron of velocity v_n and mass, $m_n = 1836.65\, m_e = 1.675 \times 10^{-27}$ kg. The energy and wave-vector of the scattered neutron have similar expressions. The conservation of energy can be expressed as:

$$E - E' = \pm \hbar \omega \tag{5.14}$$

While the conservation of momentum takes the form:

$$\mathbf{k}_n - \mathbf{k}'_n = \pm \mathbf{k} + \mathbf{G} \tag{5.15}$$

where $\mathbf{k}$ is the phonon wave-vector generated in the collision and $\mathbf{G}$ is a reciprocal lattice vector. It is customary to write the *crystal momentum* of a phonon as $\hbar \mathbf{k}$. The change in momentum of the neutron is the momentum gain by the crystal in the form of the phonon to within an additive reciprocal lattice vector. The additive reciprocal lattice vector can be ignored when we consider the energy conservation, which is a periodic function of the reciprocal lattice:

$$\omega(\mathbf{k} \pm \mathbf{G}) = \omega(\mathbf{k}) \tag{5.16}$$

In a neutron scattering experiment the energy and momentum of the incident neutron will be known. The energy of the scattered neutrons will thus reveal the phonon spectrum:

$$\omega(\mathbf{k}) = \frac{E - E'}{\hbar} \tag{5.17}$$

In general the measurements are taken as a function of the orientation of the crystal, so that we can fully construct the dispersion relation for the material. In Figure 5.3, we show the dispersion relation for fcc Pd[4], which is expressed as the variation of the phonon frequency with the wave-vector. In the notation given, this refers to the directions in the (real and reciprocal) lattice, where the points of high symmetry are expressed as greek capital letters (Γ, K, X, etc.).

5.2.3 Number of Modes: Density of States

Since we have access to the complete dispersive behavior of the phonons in the reciprocal lattice unit cell or Brillouin zone, we should be able to evaluate the vibrational frequency spectrum. That

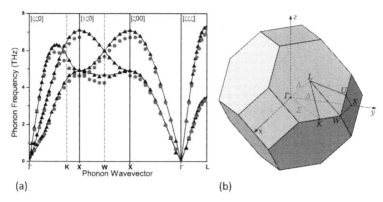

FIGURE 5.3: (a) Phonon dispersion for fcc Pd. The symbols on the x-axis represent the positions of the Brillouin zone boundary in different orientations of the crystal lattice. The experimental data are shown in red dots, while the theoretical calculation is shown black. (b) The first Brillouin zone for the fcc structure. The points of high symmetry are shown and illustrate how the phonon spectrum is constructed with respect to the orientation of the crystal. (Figure 5.3(a) published with permission: D. A Stewart, "Ab initio investigation of phonon dispersion and anomalies in palladium", New J. Phys., **10**, 043025 (2008); ©IOP Publishing Ltd and Deutsche Physikalische Gesellschaft. Published under a CC BY-NC-SA licence.)

is the number of normal modes of vibration as a function of frequency or energy. Since our crystal has finite dimensions, the number of modes of vibration will also be finite in size due to the discrete nature of the crystal lattice. We will start by considering the one-dimensional monatomic chain with a total of $(N+1)$ atoms, giving the chain a length of Na. If we assume that the atoms at the end of the chain are fixed ($u_1 = u_{N+1} = 0$), then longitudinal or transversal vibrations can exist for which there are integral numbers of half-wavelengths along the length of the chain. Therefore, we can write the allowed wave-vectors as:

$$k_\nu = \frac{\nu\pi}{Na} = \frac{\nu\pi}{L}; \nu = 1, 2, 3, \ldots, N \qquad (5.18)$$

where we have written that the length of the chain is, $L = Na$. Since the chain is long, the spacing between the allowed modes of vibration will be very small. It is usual therefore to discuss the number of states or modes of vibration between k_ν and $k_\nu + dk_\nu$. A simple differentiation of Equation (5.18) allows us to give this as $(Na/\pi)dk_\nu$. Now the number of states per unit length of the chain, in an interval dk_ν, will be:

$$g(k)dk = \left(\frac{1}{\pi}\right)dk, \quad k \leq (\pi/a)$$
$$= 0, \quad k > (\pi/a) \qquad (5.19)$$

where $g(k)$ is referred to as the density of states with respect to k. (Note that we have dropped the subscript ν for simplicity.) We note that in this 1D case, $g(k)$ doesn't have a k-dependence, which is not the case for higher dimensions. Despite the fact that we imposed the fixed boundary conditions on our evaluation of the density of states, actually, this has no limitation on the actual density of states for free motion of the end atoms in the chain. The only important point is our assumption of large, N, the number of atoms in the chain. Often it is useful to express the density of states in terms of the frequency, ω, or energy, E, of the phonons. To convert we need to look at the relationship between the wave-vector and the other parameters. For example, the density of states as a function of frequency can be obtained as follows:

$$g(\omega) = g(k)\frac{dk}{d\omega} = \left(\frac{1}{\pi}\right)\frac{dk}{d\omega} \qquad (5.20)$$

Using Equation (5.8) we can write:

$$\frac{d\omega}{dk} = \frac{a\omega_m}{2}\cos\left(\frac{ka}{2}\right)$$
$$= \frac{a}{2}(\omega_m^2 - \omega^2)^{1/2} \qquad (5.21)$$

Substituting into Equation (5.20) we obtain the frequency density of states as:

$$g(\omega) = \left(\frac{2}{\pi a}\right)(\omega_m^2 - \omega^2)^{-1/2}$$
$$= \left(\frac{2}{\omega_m \pi a}\right)\left[1 - \left(\frac{\omega}{\omega_m}\right)^2\right]^{-1/2}$$
$$= \left(\frac{1}{\pi v_p^0}\right)\left[1 - \left(\frac{\omega}{\omega_m}\right)^2\right]^{-1/2} \qquad (5.22)$$

We see that $g(\omega)$ does have a frequency dependence. In terms of the number of allowed modes in reciprocal space, the size of the lattice is important. Which makes sense since there are more modes available if we increase the size of the crystal. For our 1D k-space, the spacing between modes is $2\pi/L$.

We can extend our discussion for the 3D case, where now the k-space available per mode is now $(2\pi)^3/L_1L_2L_3$, where the L_i are the lengths of the crystal along the three Cartesian axes. The volume of k-space will be a spherical shell centered on the origin, with a radius of $k \equiv |\mathbf{k}|$ and a thickness of dk. This volume will be $4\pi k^2 dk$ and the corresponding density of states per unit volume can be expressed as:

$$g_{3D}(k)dk = \frac{1}{\Omega_{crystal}}\left[\frac{4\pi k^2 dk}{(2\pi)^3/L_1L_2L_3}\right]$$
$$= \frac{k^2 dk}{2\pi^2} \qquad (5.23)$$

where $\Omega_{crystal}$ represents the crystal volume and is equal to $L_1L_2L_3$. It should be evident that the calculation for a 3D lattice is significantly more complex than for the 1D case, though for the low frequency (non-dispersive) limit, it can be shown that $g(\omega)$ varies as ω^2. The

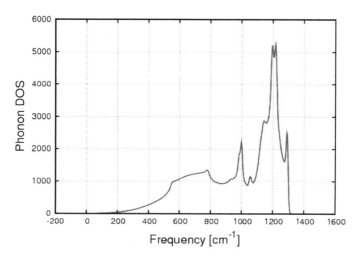

FIGURE 5.4: Phonon density of states as a function of frequency. The different peaks correspond to the transverse and longitudinal vibrational modes.

total number of vibrational states for a 1D lattice was found to be equal to N, the number of atoms. In a three-dimensional solid this is $3N$. Of these, one-third (N) are longitudinal modes and two-thirds $(2N)$ are transverse modes. In Figure 5.3, the transverse (T) and longitudinal (L) modes are indicated for a Pd fcc lattice. In Figure 5.4, we show a phonon density of states as a function of frequency for a fcc structure. The sharp peaks are common in such curves and arise from abrupt changes in slope, which occur due to the group velocity of the phonons vanishing at these points. Such critical points are known as *van Hove singularities*. These also occur in the electronic spectra of solids for similar reasons and can be of considerable use in understanding electronic band structures, which we will discuss in Chapter 7.

5.3 VIBRATIONAL MODES OF A DIATOMIC 1D LATTICE

We will return to a 1D chain of atoms, which can be considered to be a one-dimensional lattice with a basis. Here we will consider a chain of alternating atoms of masses, m_1 and m_2, where for argument sake we can write $m_1 > m_2$ and separated by a distance a. We will follow a similar analysis to that given above for the monatomic case. Since the two atoms in our diatomic chain have different masses, we can expect them to have different displacement amplitudes, such that:

$$u_{2n} = Ae^{i[2kna-\omega t]} \quad (5.24)$$

$$u_{2n+1} = Be^{i[k(2n+1)a-\omega t]} \quad (5.25)$$

We now write the equations of motion for the two atoms as:

$$-m_1\omega^2 u_{2n} = m_1 \frac{d^2 u_{2n}}{dt^2} = K(u_{2n+1} + u_{2n-1} - 2u_{2n}) \quad (5.26)$$

$$-m_2\omega^2 u_{2n+1} = m_2 \frac{d^2 u_{2n+1}}{dt^2} = K(u_{2n+2} + u_{2n} - 2u_{2n+1}) \quad (5.27)$$

Using the relevant relations given in Equations (5.24) and (5.25) and extending for u_{2n+2} and u_{2n-1}, we obtain:

$$(2K - m_1\omega^2) A = KB (e^{ika} + e^{-ika}) \quad (5.28)$$

$$(2K - m_2\omega^2) B = KA (e^{ika} + e^{-ika}) \quad (5.29)$$

We re-arrange these as:

$$(2K - m_1\omega^2) A = 2KB\cos(ka) \quad (5.30)$$

$$(2K - m_2\omega^2) B = 2KA\cos(ka) \quad (5.31)$$

Eliminating the constants A and B, we obtain:

$$(2K - m_1\omega^2)(2K - m_2\omega^2) = 4K^2\cos(ka) \quad (5.32)$$

This dispersion relation is quadratic in ω^2, which can be solved as:

$$\omega_\pm^2(\mathbf{k}) = K\left(\frac{1}{m_1} + \frac{1}{m_2}\right) \pm K\left[\left(\frac{1}{m_1} + \frac{1}{m_2}\right)^2 - \frac{4\sin^2(ka)}{m_1 m_2}\right]^{1/2} \quad (5.33)$$

The solution has two distinct parts which are highly significant and is shown graphically in Figure 5.5.

The solution we see in Equation (5.33) looks rather more complex than the solution for the monatomic case. However, if we put $m_1 = m_2 = m$, we find that the solution reduces to:

$$\omega_\pm^2(\mathbf{k}) = \frac{2K}{m}[1 \pm \cos(ka)] \quad (5.34)$$

This situation is shown by the green line, where $\alpha = m_2/m_1 = 1$. The negative solution is identical to that given in Equation (5.7). We will return to the second solution shortly. The negative solution to Equation (5.33) corresponds to the lower branch called the *acoustic branch*. The representation given in Figure 5.5 is restricted to the first Brillouin zone ($|k| \leq \pi/2a$). We note that since we have kept the atomic separation at a distance, a, the periodicity of the diatomic chain is $2a$. This means that the upper portion appears to be folded

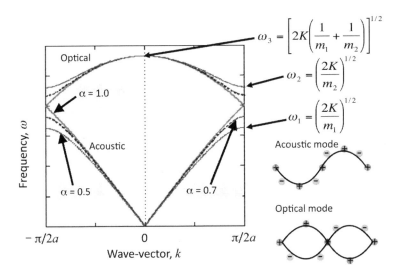

FIGURE 5.5: Dispersion relation for a diatomic 1D lattice for longitudinal wave propagation. This representation shows just the first Brillouin zone, which contains all the necessary frequency behavior. Here $\alpha = m_2/m_1$. The upper branches correspond to the optical modes, while the lower branches correspond to the acoustic modes. A schematic representation of the acoustic and optical modes are shown in the lower left of the figure.

back at the Brillouin zone boundaries. Actually, closer inspection will reveal that this upper branch, called the *optical branch*, arises from the positive solution and allows us to represent all frequencies within the Brillouin zone. From Figure 5.5 we note that there are three critical points indicated by, ω_1, ω_2 and ω_3. The first of these corresponds to the frequency of the acoustic branch at the Brillouin zone boundary, the second to the solution of the optical branch at the Brillouin zone boundary and the last solution corresponds to the optical branch for $k = 0$. It is useful to express our solution in terms of these frequencies:

$$\omega_\pm^2(\mathbf{k}) = \frac{\omega_3^2}{2} \pm \left[\frac{\omega_3^4}{4} - \omega_1^2 \omega_2^2 \sin^2(ka) \right]^{1/2} \quad (5.35)$$

Now we can express the acoustic (A) and optical (O) branches of our longitudinal vibrations for the diatomic chain as:

$$\omega_-^2(\mathbf{k}) = \omega_A^2(\mathbf{k}) = \frac{\omega_3^2}{2} - \left[\frac{\omega_3^4}{4} - \omega_1^2 \omega_2^2 \sin^2(ka) \right]^{1/2} \quad (5.36)$$

$$\omega_{+}^{2}(\mathbf{k})=\omega_{O}^{2}(\mathbf{k})=\frac{\omega_{3}^{2}}{2}+\left[\frac{\omega_{3}^{4}}{4}-\omega_{1}^{2}\omega_{2}^{2}\sin^{2}(ka)\right]^{1/2} \qquad (5.37)$$

The solution for $k = 0$ gives: $\omega_A^2(k) = 0$ and $\omega_O^2(k) = \omega_3^2$, while at $k = \pi/2a$, the Brillouin zone boundary, it is a simple matter to show $\omega_A^2(k) = \omega_1^2$ and $\omega_O^2(k) = \omega_2^2$. (N.B. if you want to try this, use Equation (5.33) instead of (5.36) and (5.37)[5].) Of great significance is the difference in the optical and acoustic frequencies at the zone boundary. Effectively we have a frequency region in which no modes of vibration can exist. The larger the difference in the two masses, the greater this frequency gap will be. The forbidden region will disappear when the masses of the atoms in the basis are equal.

Further insight can be gained in the consideration of the amplitudes of vibration. From Equations (5.30) and (5.31), we can express the ratio of amplitudes of the two atoms in the basis as:

$$\frac{A}{B} = \frac{2K\cos(ka)}{2K - m_1\omega^2} = \frac{2K - m_2\omega^2}{2K\cos(ka)} \qquad (5.38)$$

For long wavelength (low frequency) acoustic waves, the amplitude ratio is practically unity and all atoms move in the same way. We can evaluate the velocity in this region by considering:

$$\lim_{k\to 0}(\omega_A) \to 0 \qquad (5.39)$$

Under this condition we obtain:

$$\begin{aligned}\lim_{k\to 0}(\omega_A^2) &= K\left(\frac{1}{m_1}+\frac{1}{m_2}\right) - K\left[\left(\frac{1}{m_1}+\frac{1}{m_2}\right)^2 - \frac{4k^2a^2}{m_1m_2}\right]^{1/2} \\ &= K\left(\frac{1}{m_1}+\frac{1}{m_2}\right)\left[1-\left(1-\frac{2m_1m_2k^2a^2}{(m_1+m_2)^2}\right)\right] \\ &= K\left(\frac{m_1+m_2}{m_1m_2}\right)\frac{2m_1m_2k^2a^2}{(m_1+m_2)^2} \\ &= K\frac{2k^2a^2}{(m_1+m_2)} \qquad (5.40)\end{aligned}$$

In this limit we can thus write:

$$\omega_A(k) = \sqrt{\frac{2K}{m_1 + m_2}} ka \qquad (5.41)$$

From which we can express the phase and group velocities in this region as:

$$v_p^0 = \sqrt{\frac{2K}{m_1 + m_2}} a \qquad (5.42)$$

$$v_g^0 = \sqrt{\frac{2K}{m_1 + m_2}} a = v_p^0 \qquad (5.43)$$

As the value of k increases from zero, the frequency initially increases linearly. The ratio (A/B) will also initially increase from unity in a linear fashion. Further increase of k will be followed by further increases of frequency and (A/B), but not linearly. The limiting case is when we reach $\omega_A = \omega_1 = \sqrt{2K/m_1}$, where $\mathbf{k} = \pm(\pi/2a)$. At this point the phase and group velocities will be:

$$v_p = \sqrt{\frac{8Ka^2}{\pi^2 m_1}} \qquad (5.44)$$

$$v_g = \frac{\partial \omega_A}{\partial k} = 0 \qquad (5.45)$$

Let us now turn our attention to the optical branch, ω_O, which is so named since these modes can be excited by light of an appropriate frequency. (In fact, the techniques of Brillouin light scattering and Raman spectroscopy are based on the scattering of light by phonons or other quantized excitations in solids.) In these modes the solutions are given by the positive sign in Equations (5.33) and (5.35). For the long wavelength limit, where $\mathbf{k} \to 0$, we have the frequency maximum; $\omega_O \to \omega_3$. At this point we have the phase and group velocities:

$$v_p = \frac{\omega}{k} \to \infty \qquad (5.46)$$

$$v_g = \frac{\partial \omega_O}{\partial k} \to 0 \qquad (5.47)$$

We note that the group velocity tends to zero where the gradient of the dispersion relation goes to zero. Substituting $k = 0$ and $\omega = \omega_3$ in Equation (5.38), we find the amplitude ratio takes the form:

$$\frac{A}{B} = -\frac{m_2}{m_1} \qquad (5.48)$$

Therefore, we conclude for the long wavelength limit, the optical vibrations have neighboring atoms out of phase, such that the center of mass remains unmoved. The wave becomes a standing mode, since v_g goes to zero. The (A/B) ratio remains negative throughout the entire optical branch, and as we reach the Brillouin zone boundary at $\mathbf{k} = \pm(\pi/2a)$, we find $\omega \to \omega_2$, its short wavelength limit ($\lambda = 4a$). Again the phase velocity at this limit diverges and the group velocity tends to zero. We note that at the Brillouin zone boundary, both optical and acoustic branches become standing wave modes, corresponding to the Bragg diffraction effect through an angle of 180°. The mass ratio (m_2/m_1) will determine the size of the frequency (band) gap between the two branches at this point, as illustrated in Figure 5.5. If $m_1 \gg m_2$ the optical modes will tend to a straight line since $\omega_3 \to \omega_2$, and the gap will be wide. At this point we can ask, what happens at frequencies in the region of the gap and why they do not excite lattice vibrations? Since the solutions are limited to extremal values by the sine function, the only way that we can extend the solution into the forbidden region is by making the wave-vector imaginary; $k = \pm(\pi/2a + i\alpha)$. In this case we obtain a decaying exponential term in our solution, which is an attenuation factor. This factor becomes large for the forbidden regions and effectively damps all normal modes of vibration.

As with the monatomic case, the vibrational modes can be both longitudinal and transverse, which adds further branches to the phonon spectrum, one longitudinal and two transversal. Vibrational modes in solids are frequently labelled; LO, LA, TO and TA, for the various combinations of longitudinal optic (LO) etc. Transverse

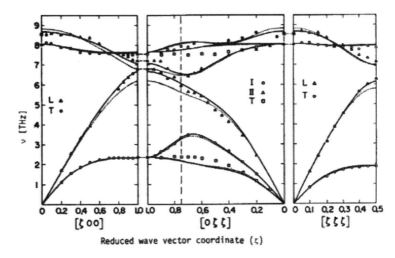

FIGURE 5.6: The phonon dispersion curve for GaAs, which is a diatomic material. The solid lines are theoretical models and points correspond to experimental data. The vertical dashed line in the [0 ζ ζ] direction represents the zone boundary. In this direction, points labeled I, II refer to modes whose polarization vectors are parallel to the (0$\bar{1}$1) mirror plane. Other modes are either strictly longitudinal (L) or transverse (T). (Reprinted figure with permission from: J. L. T. Waugh and G. Dolling, Phys. Rev. **132**, 2410, (1963). Copyright 1963 by the American Physical Society.)

modes can be degenerate in certain cases, as for the monatomic case. In Figure 5.6, we show an example of a 3D phonon spectrum of a diatomic material (GaAs), which shows the various branches we have discussed.

5.4 THERMAL PROPERTIES OF SOLIDS

Since we associate heat in a crystal lattice with the vibrations of its atoms, the study of the vibrational modes of atoms in crystals should provide much practical insight to the thermal properties of solids. It is worth reviewing some of the basic principles of thermal properties of solids here since it is related to lattice modes, in other words, the vibrational motion of atoms in a solid is directly related to the heat content of the system.

5.4.1 Classical Specific Heat: Dulong and Petit's Law

The specific heat of a system is defined as the heat energy that must be supplied in order to raise its temperature. We can express this as:

$$C_v = \left(\frac{\partial U}{\partial T}\right)_v \quad (5.49)$$

The subscript, v, indicates that this process takes place at constant volume. The motion of an atom in a solid can be expressed in classical terms as: $E = \frac{mv^2}{2} + \frac{Kx^2}{2} = (m/2)[v^2 + \omega^2 x^2]$, being the sum of kinetic and potential energies. Averaging over a Boltzmann (classical) distribution is made as follows:

$$<E> = \frac{\int\int E e^{-E/k_B T} dv dx}{\int\int e^{-E/k_B T} dv dx} \quad (5.50)$$

with the integration over the limits of velocity and position. Using the above energy, it can be shown that $<E> = k_B T$. For an assembly of N atoms in a solid with three degrees of freedom will give a lattice energy of $U = 3Nk_B T$, which, using Equation (5.49) gives a constant value of the specific heat of $3N_A k_B = 3R = 24.94$ Jmol^{-1}K^{-1}.[6] This is the classical result known as the *Dulong and Petit law*, dating from 1819. It was only superseded in the early 20th century with the advent of early quantum theory and the Einstein model, by which time it was evident that the classical theory was severely deficient.

5.4.2 Einstein's Model

The Einstein model is based on the Planck quantum hypothesis (1901). In this, Einstein assumed that each atom of the solid vibrates about its equilibrium position with an angular frequency, ω and all atoms have the same frequency, vibrating independently from the other atoms of the solid. The quantum mechanical theory of the harmonic oscillator gives an energy spectrum of:

$$E = \left(n + \frac{1}{2}\right)\hbar\omega \quad (5.51)$$

where $n = 0, 1, 2, \ldots$ The zero point energy ($n = 0$) has no significant role in the specific heat since we are only interested in the change of internal energy due to the increase of temperature, as such it makes no difference if we write: $E_n = n\hbar\omega$. The occupancy or probability of this energy state can be expressed as:

$$f(E_n) = \frac{e^{-n\hbar\omega/k_BT}}{\sum_{n'=0}^{\infty} e^{-n'\hbar\omega/k_BT}} \tag{5.52}$$

then the total energy of the solid can be written:

$$U = \frac{3N\sum_{n=0}^{\infty} n\hbar\omega e^{-n\hbar\omega/k_BT}}{\sum_{n=0}^{\infty} e^{-n\hbar\omega/k_BT}} \tag{5.53}$$

Using the following:

$$\frac{d}{dx}\left(\ln \sum_{n=0}^{\infty} e^{-nx}\right) = \frac{\sum_{n=0}^{\infty} ne^{-nx}}{\sum_{n=0}^{\infty} e^{-nx}} = \frac{d}{dx}\ln(1-e^{-x})^{-1} \tag{5.54}$$

we can write Equation (5.53) as:

$$U = \frac{3N\hbar\omega}{e^{\hbar\omega/k_BT} - 1} \tag{5.55}$$

The factor $(e^{\hbar\omega/k_BT} - 1)^{-1}$ corresponds to the average phonon occupancy for the mode of frequency ω, at temperature T. Indeed it is a special case of the so-called *Bose - Einstein distribution function*, which applies to particles called *bosons*, of which phonons and photons are examples. In his model Einstein assumed that a solid with N atoms will have N modes of vibration, with all the atoms having the same frequency, ω_E. Following Equation (5.49) we evaluate the heat specific heat capacity as:

$$C_v = 3Nk_B \left(\frac{\hbar\omega_E}{k_BT}\right)^2 \frac{e^{\hbar\omega_E/k_BT}}{(e^{\hbar\omega_E/k_BT} - 1)^2} \tag{5.56}$$

It is customary to define the Einstein temperature as: $\Theta_E = \hbar\omega_E/k_B$, such that the above equation takes the form:

$$C_v = 3R\left(\frac{\Theta_E}{T}\right)^2 \frac{e^{\Theta_E/T}}{(e^{\Theta_E/T} - 1)^2} \tag{5.57}$$

In the high temperature range, $\hbar\omega_E \ll k_B T$, we can write $e^{\hbar\omega_E/k_B T} \simeq 1 + \hbar\omega_E/k_B T$, or in terms of the Einstein temperature: $\Theta_E \ll T$ and $e^{\Theta_E/T} \simeq 1 + \Theta_E/T$. It is evident in this limit that $C_v \to 3R$, and we recover the Dulong - Petit law. At the other extreme, where $\hbar\omega_E \gg k_B T$ and $e^{\hbar\omega_E/k_B T} \gg 1$, we find:

$$C_v = 3R \left(\frac{\Theta_E}{T}\right)^2 e^{-\Theta_E/T} \tag{5.58}$$

This has a dominant exponential variation at low temperatures and while being a reasonable approximation is physically unrealistic in expecting all vibrational modes to have the same frequency. Since the atoms are coupled, they do not act independently.

5.4.3 The Debye Model

The Debye model overcomes the shortfall of the Einstein approach, which while following from his example in maintaining the energy and phonon occupancy relations, allowed the angular frequency of a vibrational mode to depend on the wave-vector, k. The angular frequency must have a maximum value or cut-off frequency, ω_m, such that the total number of modes is finite and can be expressed as:

$$3N = \int_0^{\omega_m} g(\omega)\,d\omega \tag{5.59}$$

Where we have considered a 3D solid. The cut-off frequency will affect the total vibrational energy, which will now take the form:

$$U = \int_0^{\omega_m} \frac{(\hbar\omega)g(\omega)d\omega}{(e^{\hbar\omega/k_B T} - 1)} \tag{5.60}$$

To progress we need to evaluate the density of states, $g(\omega)$, which in fact is the principal correction Debye made to Einstein's approach. To solve this problem, Debye suggested that the phase velocity, $v_p = (\omega/k)$, be chosen as the speed of sound in the solid for all modes of vibration. This requires that an upper limit of the frequency, called the *Debye frequency*, ω_D, be imposed in the integral, Equation (5.60). This is necessary to limit the total number of modes to $3N$. While this cut-off is artificially imposed, it works at low temperatures since there are very few high frequency modes in this range. A *Debye characteristic temperature* is also defined, much in the same way as the Einstein temperature, and is expressed as: $\Theta_D = \hbar\omega_D/k_B$. The density of states for a three-dimensional solid was given in Equation (5.23), which when adjusted to include the one longitudinal and two transverse modes can be expressed as:

$$g(\omega) = \frac{\omega^2}{2\pi^2}\left(\frac{1}{v_L^3} + \frac{2}{v_T^3}\right), \quad \omega \ll (v_L/a) \quad (5.61)$$

which refers to the low temperature acoustic limit. If transverse and longitudinal waves are transmitted at the same velocity we obtain:

$$g(\omega) \simeq \frac{3\omega^2}{2\pi^2 v^3}, \quad 0 < \omega < \omega_D \quad (5.62)$$

In Figure 5.7, we illustrate the form of the density of states for the Debye model and is compared to the case for an fcc metal.

This allows us to express the vibrational energy per unit volume of the crystal as:

$$U = \left(\frac{3\hbar}{2\pi^2 v^3}\right)\int_0^{\omega_D} \frac{\omega^3 d\omega}{(e^{\hbar\omega/k_B T} - 1)} \quad (5.63)$$

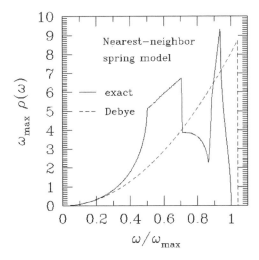

FIGURE 5.7: Phonon density of states in the Debye model and that for an fcc metal. While there is good agreement for low frequencies, discrepancies emerge for the higher temperature regime.

If we put $x = \hbar\omega/k_B T$, we can simplify the above expression to:

$$U = \left(\frac{9Nk_B T^4}{V\Theta_D^3}\right) \int_0^{(\Theta_D/T)} \frac{x^3 dx}{e^x - 1} \quad (5.64)$$

where we have taken into account that the integration over the density of states gives, $3N/V = \omega_D^3/2\pi^2 v^3$. Now we can evaluate the specific heat as:

$$C_v = \left(\frac{\partial U}{\partial T}\right)_v = \left(\frac{9Nk_B T^3}{V\Theta_D^3}\right) \int_0^{(\Theta_D/T)} \frac{x^4 e^x dx}{(e^x - 1)^2} \quad (5.65)$$

We can express the extremes of temperature analytically, but a full range evaluation must be done numerically. The value of Θ_D is obtained as a fitting parameter.

In Figure 5.8, we show the comparison of the Debye and Einstein models of the specific heat of a solid as a function of reduced temperature. We note that both the Einstein and Debye models tend to the Dulong - Petit limit at high temperatures, though there is some discrepancy at low temperatures, with the Debye model having a good agreement with experiment over the whole temperature range.

In the high temperature limit, the integral reduces to $\frac{1}{3}(\Theta_D/T)^3$, and we recover the Dulong - Petit result of $C_v = 3N_A k_B$. For low

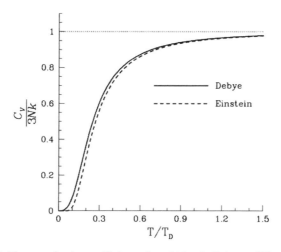

FIGURE 5.8: The curve for the specific heat of a solid for the Debye and Einstein models. The red line corresponds to the Dulong - Petit law.

temperatures $(T < \Theta_D/10)$, the integral in Equation (5.65) can be taken to range up to infinity, giving the integral a value of $(\pi^4/15)$, from which we obtain:

$$C_v \simeq \frac{12\pi^4 N k_B}{5V}\left(\frac{T}{\Theta_D}\right)^3 \quad (5.66)$$

This exhibits the cubic dependence observed experimentally. In the case of metals, the specific heat has an additional contribution at low temperatures and can be shown to have a variation of the form:

$$C_v = AT + BT^3 = C_{el} + C_{latt} \quad (5.67)$$

where the first term is attributed to the specific heat of the free electron gas in a conductor and the second term is the Debye lattice or phonon contribution. (This will be elaborated upon in the following chapter, see Section 6.8.) The Debye characteristic temperature is a material constant, which is about 100 K in Pb and as high as 732 K in LiF. Clearly the value of Θ_D will depend on the elastic constant of the material and its density through the speed of sound in the solid.

It is evident from Figure 5.7 that the density of states (DOS) is more complex than the Debye picture. Corrections can be made by considering that the wave-vectors be limited to $k_D = \omega_D/v$ and not just being limited to the first Brillouin zone. The structure of the real DOS arises from the consideration that the velocity of propagation differs for the longitudinal and two transverse modes.

5.5 ANHARMONIC EFFECTS

In the preceding sections, we have relied on the fact that the amplitude of the atomic oscillations about their equilibrium positions are small and that we are only concerned with a linear harmonic response. While the first approximation is quite reasonable in a majority of cases for temperatures well below the melting temperature of a solid, the second assumption can prove too limiting in reality. Deviations from the latter lead to what are termed *anharmonic effects*. Such effects arise due to higher order terms in the potential, which we previously assumed to obey Hooke's law, see Equation (5.1). However, the true potential has a form more akin to that illustrated in Figure 1.6, see Chapter 1. We note that for very small oscillations about the equilibrium position Hookes law is an acceptable approximation, but if the vibrations extend beyond a very small amplitude we need to consider higher order terms. Expanding the potential energy in terms of the deviation from the equilibrium position, gives a relation of the form:

$$V(u_n) = U_0 + Ku_n^2 - K'u_n^3 - K''u_n^4 \qquad (5.68)$$

where the coefficients are positive. Another form of the anharmonic potential is given by the Morse potential, which again has an asymmetric variation. The general form of anharmonic potential is shown in Figure 5.9. One of the consequences of the asymmetry in the potential is that the potential energy changes with interatomic separation. Therefore for higher temperatures more phonons will be excited and anharmonicity effects will be more pronounced. Another consequence is that while the harmonic potential has evenly spaced modes, as given by Equation (5.51), the anharmonic potential means that the modes are note regularly spaced and for higher

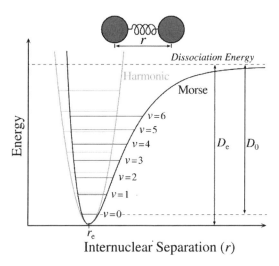

FIGURE 5.9: The asymmetric potential between atoms in a crystal gives rise to non-linear effects called anharmonicity.

order vibrational modes, the energy can be significantly different from that given by the harmonic approximation.

5.5.1 Thermal Expansion

An important effect of anharmonicity is thermal expansion. The expansion of the crystal lattice with temperature parallels that of the heat capacity. It is always important to bear in mind that when considering the thermal behavior of solids, the energy content of the solid resides in the phonons and not in the individual atoms. That is, we cannot impart energy to a single atom since its vibrations will cause the neighboring atoms to vibrate. Therefore thermal energy is manifest as a collective excitation of the solid in the form of a phonon. We can make a crude estimation of the thermal expansion as follows: Firstly, the classical value of the mean square displacement has the form:

$$<u^2> = \frac{1}{2}\left(\frac{k_B T}{K}\right) \qquad (5.69)$$

The mean displacement is non-zero due to the cubic term of Equation (5.68). This can be assessed by taking $\partial V/\partial u = 0$ and neglecting terms higher than cubic and takes the form:

$$<u> = \frac{3K'<u^2>}{2K} = \frac{3K'k_B T}{4K^2} \qquad (5.70)$$

For high temperatures this is linear in T, while at low temperatures it varies as T^4 in the same way $<u^2>$ does. The change in volume of the crystal will also affect the frequency of the vibrations, which we can express in the form:

$$\frac{\Delta \omega}{\omega} = -\gamma \frac{\Delta V}{V} \qquad (5.71)$$

where γ is called the *Grüneisen constant*. This then relates the changes in phonon frequency with those of the lattice volume, and can be alternatively expressed as:

$$\gamma = -\frac{V}{\omega}\frac{d\omega}{dV} = -\frac{d(\ln \omega)}{d(\ln V)} \qquad (5.72)$$

In practice, the Grüneisen constant has a value close to unity. The coefficient of thermal expansion can be expressed in terms of the Grüneisen constant as:

$$\alpha = \frac{\gamma C_v}{3E} \qquad (5.73)$$

where E here refers to the Young's modulus of the solid. Since this is relatively independent of temperature, the thermal expansion coefficient should tend to a constant value at large temperatures, while as $T \to 0$ it will vary as T^3. In metals at low temperature, there will be a further contribution to the thermal expansion due to the electronic specific heat, as noted above.

5.5.2 Thermal Conduction

In an ideal crystal within the harmonic approximation, two phonons can pass through one another without interacting. They would be reflected at the crystal surface and continue to move through the solid. In real materials, defects and anharmonicity cause phonons to interact, be scattered and decay. Phonons behave much like particles in a gas which collide and are in thermal equilibrium at a given temperature. In non-conducting materials we might expect that the heat be transported through the crystal with the speed of sound. However, in reality the phonons interact and have a *mean free path*[8], which leads to a more diffusive heat conduction, which is a much slower process. Of course metallic materials can conduct heat via electrons, which can dominate thermal conductivity.

We will now consider the thermal conduction due to a temperature gradient. We recall that the phonon concentration at temperature, T, which are excited to a vibrational mode of wave-vector, k, and angular frequency, $\omega(k)$, is expressed as:

$$<n_{\mathbf{k}}> = \frac{1}{e^{\hbar \omega(\mathbf{k})/k_B T} - 1} \qquad (5.74)$$

At thermal equilibrium, when no temperature gradient exists, we can write: $<n_{\mathbf{k}}> = <n_{-\mathbf{k}}>$, i.e there is a an equal flow of phonons on any opposite pair of directions, and as such there will be no heat flow. If we establish a temperature gradient, ∇T, then the thremal conductivity will be determined by this gradient and a rate of energy flow, per unit area, will be produced of the form:

$$\mathbf{Q} = -\kappa_l \nabla T \tag{5.75}$$

where κ_l is the thermal conductivity. The kinetic theory of gases can be used to assess the thermal conductivity, if we assume that phonons behave as gas particles, where the phonon velocity is constant (which is true below the Debye energy). This gives the following expression for the thermal conductivity:

$$\kappa_l = \frac{1}{3} C_v v \Lambda \tag{5.76}$$

where Λ is the mean free path of the phonons and v their velocity. The value of Λ can vary quite strongly with temperature, being as low as a few atomic spacings near the melting point and as large as 1 mm at very low temperatures. The variation of the thermal conductivity as a function of temperature is shown in Figure 5.10[9], which from Equation (5.76) depends on the product of the specific heat and the mean free path of the phonons.

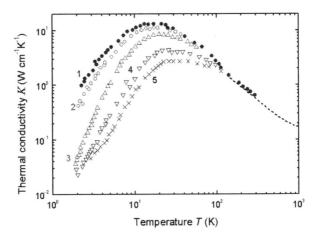

FIGURE 5.10: The variation of the thermal conductivity with temperature for a p-Ge sample with doping concentrations: 1) $10^3 cm^{-3}$, 2) $10^{15} cm^{-3}$, 3) $2.3 \times 10^{16} cm^{-3}$, 4) $2.0 \times 10^{18} cm^{-3}$, 5) $10^{19} cm^{-3}$.

5.5.3 Umklapp Processes

When two phonons collide they can produce a third phonon. The scattering process is however dependent on the size of the anharmonic terms in the potential, Equation (5.68). In our discussion of the inelastic scattering of neutrons to produce phonons in a crystal lattice we mentioned that this occurs via the conservation of energy and momentum. The latter is more strictly the crystal momentum, $\hbar\mathbf{k}$, which is physically different from the ordinary sense we have of the momentum of a moving particle for example. The wave-vector here corresponds to that of the phonon. Now, for phonons with a wave-vector corresponding to the Brillouin zone boundary, the resulting vibration must be a standing wave which has zero momentum. In any case real momentum will be transferred in the inelastic scattering of a neutron. The details of the transfer of momentum depends on the circumstances of the transition itself, as will be discussed below.

For what is called a *normal process* (or N - process), the crystal momentum is conserved when two phonons collide to form a third. Such a process can be expressed as:

$$\hbar\mathbf{k}_1 + \hbar\mathbf{k}_2 = \hbar\mathbf{k}_3 \tag{5.77}$$

$$\hbar\omega_1 + \hbar\omega_2 = \hbar\omega_3 \tag{5.78}$$

which correspond to the conservation of momentum and energy, respectively. However, the thermalization process for phonons can also occur, which satisfy the conservation of energy, as given above, but with the following vector relation:

$$\mathbf{k}_1 + \mathbf{k}_2 = \mathbf{k}_3 + \mathbf{G} \tag{5.79}$$

where $\mathbf{G}$ is a reciprocal lattice vector. This process, pointed out by Peierls[10], is possible since $\mathbf{k}_3 + \mathbf{G}$ is indistinguishable from $\mathbf{k}_3$, as we have pointed out in previous discussions on the reciprocal lattice. It must therefore be a valid conservation law. Such a scattering event is termed an *Umklapp process* (or U - process)[11]. The N and U - processes are schematically illustrated in Figure 5.11.

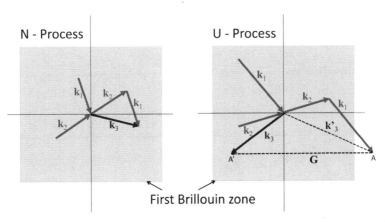

FIGURE 5.11: The N and U processes can be understood in terms of the wave-vectors and the first Brillouin zone. In the former, all vectors lie within the first Brillouin zone, while in the latter, a reciprocal lattice vector is used to define the scattering process to bring it back to within the first Brillouin zone. We note that points A and A' are equivalent.

It is a curious aspect of the Umklapp process that it apparently "destroys" momentum and changes the direction of energy flow. The U - process thus can provide thermal resistance to phonon flow and thermalize the phonon distribution.

In the high temperature regime, the mean free path of U - processes vary as T^{-1}. When we consider low temperatures, we note that the mean free path increases and the wave- vectors will get smaller. Now since the U-process requires that $\mathbf{k}_1 + \mathbf{k}_2$ extends beyond the first Brillouin zone, these processes will reduce in probability and eventually be "frozen out". The limiting case will obviously be when $\mathbf{k}_1 = \mathbf{k}_2 = \mathbf{G}/2$, which will just take two aligned phonons to the edge of the first Brillouin zone. This can be of critical importance for the low temperature thermal conductivity, since only N - processes can occur and the total phonon wave-vector will be conserved, Equation (5.77). This means, in theory, that for an infinite crystal the thermal conductivity for an insulator should be infinite in the absence of the Umklapp process. This is similar to *Knudsen* or *Molecular flow* in gases or *ballistic transport*, which occur when the mean free path between collisions is greater than the dimensions of the supporting medium. In reality this doesn't occur, since there will still be some small probability of momentum destroying U - processes. Also we need to consider the scattering of phonons by defects, which will cause the mean free path of the phonons to be reduced. This will be particularly important in polycrystalline materials, since the mean free path will always be limited by the average grain size and not just by the dimensions of the sample.

5.6 SUMMARY

A simple monatomic one-dimensional model of lattice vibrations allowed us to determine the dispersion relation for the quantized modes of excitation in the form of phonons. We note that there is an upper limit on the frequency of oscillation, which is determined by the elastic constants and density of the material. The dispersion relation provides a wealth of information on the physical processes involved in the production of phonons. Of particular interest is the limiting case where the wave-vector of the phonon reaches the edge of the first Brillouin zone. Here we saw that this corresponds to the lower limit of the phonon wavelength, the speed of propagation of the phonons drops to zero and the phonon corresponds to a standing wave. We extended this view to consider three-dimensional crystals, where the changing periodicity with direction in the crystal means that the 3D dispersion relation is more complex in structure, but is essentially related to the period and extremal lengths of the wave-vector to the Brillouin zone boundary.

We further developed the phonon dispersion relation for diatomic crystals. We note here that due to the mass difference of the component atoms, the dispersion relation has upper and lower sections called the optic and acoustic branches. There is generally a frequency or energy region in which no phonons can exists called a forbidden zone. It exists for a similar reason to the maximum frequency value in the monatomic case, and corresponds to a Bragg condition, which produces standing wave modes.

The thermal properties of solids are intimately related to the vibrations of atoms in a solid. In general we consider low amplitude harmonic oscillations, which serve to describe some of the main features of thermal effects. It is essential to understand that the thermal energy exists as a collective excitation in the solid in the form of phonons. We reviewed the consequences of the Einstein model, which considers all atoms to vibrate at some fixed frequency. This provides a simple model in which the variation of the specific heat varies exponentially at low temperature and saturates to the classical (Dulong - Petit) value at high temperatures. This low temperature behavior does not agree completely with experiment and Debye corrected the Einstein model by allowing a distribution of frequencies, described by the density of states for phonons, up to a cut-off point, called the Debye frequency. This gives the experimentally observed low temperature behavior. An additional contribution to the specific heat occurs in metals due to free electrons in the solid, which can also transport heat in collisions with the lattice.

Finally, we described anharmonic effects, which occur in real systems and accounts for the non-symmetric potential. This has a number of rather important consequences. It gives rise to thermal expansion effects in solids and allows us to account for the thermal conductivity of materials. Defect scattering and Umklapp processes were also described and take up important roles in the thermal conductivity of real solids.

REFERENCES AND FURTHER READING

Basic Texts

- J. S. Blakemore, *Solid State Physics*, Cambridge University Press, Cambridge (1985)
- H. P. Myers, *Introductory Solid State Physics*, Taylor and Francis, London (1998)

Advanced Texts

- M. A. Wahab, *Solid State Physics: Structure and Properties of Materials*, Alpha Science International Ltd., Harrow (2007)
- N. W. Ashcroft and N. D. Mermin, *Solid State Physics*, Saunders College, Philadelphia (1976)

- J. M. Ziman, *Electrons and Phonons*, Oxford University Press, Oxford (1960)
- M. T. Dove, *Lattice Dynamics*, Cambridge University Press, Cambridge (2009)

EXERCISES

Q1. Use the Taylor expansions of u_{n+1} and u_{n-1} about the position $u_n(x)$ in the equation of motion (Equations (5.1) and (5.2)) to show that the atomic vibrations adhere to the general wave equation:

$$\frac{\partial^2 u_n(x)}{\partial t^2} = v^2 \frac{\partial^2 u_n(x)}{\partial x^2} \qquad (5.80)$$

where $v = \sqrt{\dfrac{K}{m}} a$ is the velocity of sound in a solid medium.

N.B. $u_{n+1}(x) = u_n(x+a) = u_n(x) + a\dfrac{\partial u_n(x)}{\partial x} + \dfrac{a^2}{2}\dfrac{\partial^2 u_n(x)}{\partial x^2}$.

Q2. Find the energy density of states of phonons for a 1D monatomic chain.

Q3. Show the mean square of the displacement for an atom in a solid, in the harmonic approximation can be expressed as:

$$<u^2> = \frac{1}{2}\left(\frac{k_B T}{K}\right) \qquad (5.81)$$

Q4. Find the heat capacity for a 2D solid using the Debye model.

Q5. Show that the restoring forces on transverse vibrations are smaller than for longitudinal ones.

Hint: Use a simple elastic spring to argument your reasoning.

Q6. Since heat and sound are propagated via phonons, explain why they travel at such different velocities.

Q7. Calculate the dispersion relation for a diatomic chain of atoms with masses m and $2m$.

Q8. Find the dispersion relation for a linear chain of atoms which have alternating spring constants of K_1 and K_2. Compare this with our solution in Equation (5.33).

Q9. Show that for a linear chain of atoms, the group velocity for phonons goes to zero at the Brillouin zone boundaries and tend to the phase velocity at the origin of k-space.

Q10. The Debye temperature for diamond is 2230 K. Calculate the maximum vibrational frequency and the heat capacity at 10 K.

Q11. Estimate the Debye temperature of for copper, which has an atomic weight of 63.55, a density of 8.96 g cm^{-3} and a speed of sound of 3880 ms^{-1}.

NOTES

[1] The group velocity is the velocity with which the waves move, while the phase velocity corresponds to the rate at which the phase of the wave propagates through the crystal. The quantities can have very different values and one can even be negative with respect to the other

[2] Incidently, we can evaluate the maximum frequency based on typical values for metals. Now $\omega_m = 2\sqrt{K/m} = (2/\pi)v_p^0 k \simeq 2v_p^0/a$. where v_p^0 is the long wavelength limit of the phase velocity. Putting in some typical values $v_p^0 = 5000\,\text{ms}^{-1}$; and $a = 2\,\text{Å}$, we obtain $\omega_m \simeq 10^{14}$ rads^{-1}, which corresponds to $v_m \simeq 10^{13}$ Hz, i.e., ~ 10 THz.

[3] There can be interactions where electrons and phonons couple, though these are generally observed to be small deviations from the normal behavior in some metals and are known as *Kohn anomalies*.

[4] D. A Stewart, New J. Phys., **10**, 043025 (2008)

[5] Another way of doing this is to recognize $\omega_3^2 = (\omega_1^2 + \omega_2^2)$

[6] We have used the Avogadro constant $N_A = 6.022 \times 10^{23}$ mol^{-1} and the gas constant, $R = N_A k_B = 8.314$ JK^{-1}mol^{-1}.

[7] Keeping the energy as $E = \left(n + \frac{1}{2}\right)\hbar\omega$ will add a contribution to the distribution function which would disappear upon differentiation with respect to temperature when we calculate the specific heat.

[8] The mean free path is the average distance travelled by a particle without suffering a scattering event.

[9] Data from J. A. Carruthers *et al.* Proc. Roy. Soc., **238**, 502, (1957).

[10] R. Peierls *Quantum Theory of Solids*, Oxford University Press (1955).

[11] The word *umklapp* derives from the German *umklappen*, which means to turn over

CHAPTER 6

FREE ELECTRONS IN METALS

"Everybody gets so much information all day long that they lose their common sense"

—Gertrude Stein

"For a successful technology, reality must take precedence over public relations, for Nature cannot be fooled."

—Richard P. Feynman

6.1 INTRODUCTION

Metals are relatively common materials and serve many functions in modern society. They are known to be generally strong materials, but are overwhelmingly known for their electrical properties, being excellent conductors. In fact, we generally use metallic behavior as a synonym for conducting. Such properties have been know for many centuries, but the explanation has emerged relatively recently. We have already introduced metals in the first chapter when we discussed bonding mechanisms, where we mentioned the existence of free electrons in a solid. We now need to restrict this comment a little, by which we mean that the electrons are bound to the solid, but are delocalized from the individual ions which reside near the equilibrium positions, which define the crystal lattice of the metal. The electrons are liberated from the outer electron shells of the metal atoms when the solid is formed. These electrons are principally responsible for the high electrical and thermal conductivities observed in metallic substances. It is fairly common to talk of the free *electron gas* in metals that are responsible for the conductive properties. As we will see in the development of the theory of electrons in solids, "free" in this context is rather generous, and we will need to make some adaptations when we consider how charge carriers, such as electrons move and exist in the periodic potential of a crystalline material. The electrons are also confined to the solid by a potential barrier, called the *work function* of the metal. This can be measured experimentally through the *photoelectric effect*.

We have discussed in some detail the periodic nature of crystalline solids and soon we will consider the periodic nature of the potential of the ion cores, which are situated at the lattice sites of crystalline solids. Indeed, the periodic potential has some profound effects on the energy and momentum of the electrons in solids. We will leave the detailed discussion of this to the next chapter when we discuss some of the basic models of the *band theory* of solids.

In this chapter, we will introduce some of the basic concepts of the physical description of electrical conduction and metallic behavior. The early theory of metals was based on the works of Drude and Lorentz, who made some basic assumptions, which are summarized as follows:

i) Metals consist of positive ion cores with valence electrons, which are free to move between the ions as if they form a gas; ii) The ion cores are held together in the crystal by electrostatic forces between the positive ions and the negatively charged electron gas; iii) The Coulomb repulsion expected between the electrons is considered to be negligible; iv) The potential due to the positive ions is assumed to be uniform, such that electrons move in a uniform way throughout the crystal; v) The electrons can collide with the positive ions at any given temperature and their velocities can be determined from the Maxwell-Boltzmann distribution law.

The results of the Drude theory were surprisingly good given some of the broad sweeping assumptions that are made, managing to explain such properties as the electrical and thermal conductivities, thermionic emission, as well as thermoelectric and galvanomagnetic effects. Of course there are serious shortcomings of the theory, which mainly concern those aspects that depend on the internal structure of the solid. Importantly it could say nothing as to why some materials are conductors while others are insulators.

With the advent of quantum theory in the early 20th century, Sommerfeld, Pauli, Fermi and many other began to construct a much more solid basis to the theory of metals. In particular, the establishment of quantum statistics and the band theory of solids aided our understanding to the more sophisticated level that it holds today. In this chapter, we will go through some of the early developments, as they are instructive and will assist further developments in subsequent chapters.

6.2 METALLIC BEHAVIOR

Before we consider some of the early theories of metals, it is useful to review some of the basic properties of these materials. The following list gives some of the most important properties of metals:

1. Under normal conditions of temperature, metals obey Ohm's law, which is most commonly expressed as:

$$V = IR \qquad (6.1)$$

where V is the applied voltage, I the current flowing in the metal and R the electrical resistance. This form of Ohm's law is, however, better expressed as:

$$\mathbf{J} = \sigma \mathbf{E} = \frac{\mathbf{E}}{\rho_e} \qquad (6.2)$$

which removes any geometric factors inherent in Equation (6.1). Here σ is called the conductivity (expressed in units of $\Omega^{-1}\text{m}^{-1}$), being the inverse of the resistivity, ρ_e (Ωm). These are constants of proportionality between the current density, $\mathbf{J}$, (units: Am^{-2}), and the electric field, $\mathbf{E}$ (units: Vm^{-1}).

2. Metals are good conductors electricity, with conductivities typically in the range: 10^6 to $10^8 \, \Omega^{-1}\text{m}^{-1}$.

3. Metals have large electronic thermal conductivities, κ_e. The relation between good thermal and electrical conductivities was noted by Wiedemann and Franz (1853), who noted that (κ_e/σ) in metals is fairly consistent at a given temperature. This is referred to as the *Wiedemann - Franz law*. Later Lorenz (1881) observed that the quantity $(\kappa_e/\sigma T)$ is virtually temperature independent and varies very little among metals. The numerical value of $L \equiv (\kappa_e/\sigma T)$ is known as the *Lorenz number*.

4. At low temperatures, the conductivity reaches a plateau, which depends on the the level of impurities and defects in the solid. The electrical resistivity $(1/\sigma)$, reaches a residual value, have a behavior well described by the *Matthiessen rule*, which can be expressed as:

$$\rho_e(T) = \rho_e^0(T) + \rho_e^{defect} + \rho_e^{impurity} \tag{6.3}$$

or

$$\frac{1}{\sigma(T)} = \frac{1}{\sigma^0(T)} + \frac{1}{\sigma^{defect}} + \frac{1}{\sigma^{impurity}} \tag{6.4}$$

We note that the defect and impurity contributions are temperature independent and are additive at any given temperature. The residual value of the conductivity or resistivity will therefore be almost entirely due to these contributions.

5. In ferromagnetic metals, the electrical resistance depends on an applied magnetic field. This is called *magnetoresistance*. See Chapter 10.

6. Many metals become superconducting at very low temperatures. See Chapter 11.

7. Free electrons add a small contribution to the specific heat of a metal. The electrons also add a small paramagnetic susceptibility, which is temperature independent.

8. There exist complex galvano-thermo-magnetic effects in metals, which can produce small additional electrical currents in the solid.

9. There can exist, in pure single crystal metals, complex oscillatory behavior in the presence of strong magnetic fields.

6.3 THE MAXWELL - BOLTZMANN VELOCITY DISTRIBUTION

While we know that the classical treatment is not an accurate description of metals, it does provide some insights and was at the root of some of the earlier models of metallic behavior. The Maxwell - Boltzmann velocity distribution relates the number of electrons per unit volume with velocities between v and a small increment to $v + dv$, which is expressed mathematically as:

$$f_{MB}(\mathbf{v}) = n \left(\frac{m}{2\pi k_B T}\right) e^{-mv^2/2k_B T} \tag{6.5}$$

where $n = N/V$ is the number density of electrons. The quantity f_{MB} is an equilibrium distribution function, and allows us to evaluate the probability function for the speeds in the same range:

$$P(\mathbf{v})dv = \left(\frac{2m}{\pi k_B T}\right)^{3/2} v^2 e^{-mv^2/2k_B T} dv \tag{6.6}$$

The average kinetic energy of the electrons can be evaluated as:

$$\frac{1}{2}m<v^2> = \frac{3}{2}k_B T \tag{6.7}$$

From which we can write the root mean square (rms) velocity as:

$$v_{rms} = \sqrt{<v^2>} = \sqrt{\frac{3k_B T}{m}} \tag{6.8}$$

This gives a value of 1.17×10^5 ms^{-1} at room temperature.

In our subsequent discussion of the Drude theory of metallic conductivity, it is useful to consider the concept of the *mean free path* and *relaxation time* of the electrons in a solid. The former refers to the average distance that the electron moves between elastic collisions with the positive ion cores and is related to the *collision cross-section*, which gives the probability of a collision event. This can be expressed in the form:

$$\lambda_{mfp} = \frac{1}{\sigma_c n_i} \simeq \frac{1}{\pi r_c^2 n_i} \qquad (6.9)$$

where n_i is the number density of ions and $\sigma_c \simeq \pi r_c^2$ is the scattering cross-section. If we know the mean velocity of the electrons, which we can take as the thermal velocity, $v_{th} \simeq v_{rms}$, we can then define the mean free time between collisions or relaxation time as:

$$\tau_m = \frac{\lambda_{mfp}}{v_{th}} \qquad (6.10)$$

Of course we have not taken into account the effect of an applied electric field on the velocity of the electron, to produce the *drift velocity* of the electron, but this has only a small effect for small to moderate fields.

6.4   THE DRUDE THEORY

The Drude model dates from 1900 and expresses a classical view of the electron as a free gas particle inside the metal. At any instant in time and in the absence of an applied electric field we can expect the electrons to have a velocity distribution given by Equation (6.5) and to have a random orientation. As such the net flow of electrons will be zero, giving a current density also of zero. If we apply a potential to the metal an electric force will act on the electronic charges, which is expressed as:

$$\mathbf{F}_e = -e\mathbf{E} \qquad (6.11)$$

where e is the electronic charge (which has a numerical value of -1.602×10^{-19} C). The current density in the presence of an electric field has the form of Equation (6.2). However, macroscopically, we can write this as, $\mathbf{J} = -\rho_c \mathbf{v}_d$, where $\mathbf{v}_d$ is the drift velocity of the electrons due to the applied electric field and ρ_c is the charge density, which we can express as: $\rho_c = -n_c e$, with n_c being the number density of the charge carriers. Since the instantaneous velocity of the electrons will still have a random nature due to thermal motion, it is more customary to express the current density in the form:

$$\mathbf{J} = -e\sum_{i=1}^{n}\mathbf{v}_i = -n_c e \mathbf{v}_d \tag{6.12}$$

with $\mathbf{v}_i$ being the velocity of the ith electron. From this equation we see that the drift velocity expresses the number average of the invdividual electrons in the metal. In the absence of electric field we have $\mathbf{E} = 0$ and $\sum_i \mathbf{v}_i = 0$ and no net current is apparent. From Equation (6.11) it is a simple matter to show that the drift velocity of an electron in the presence of an electric field can be expressed as:

$$\mathbf{v}_d = -\frac{e\mathbf{E}\tau}{m} \tag{6.13}$$

where τ is the average time the electron has travelled since its last collision. Combining Equations (6.12) and (6.13) we obtain:

$$\mathbf{J} = \frac{n_c e^2 \tau}{m}\mathbf{E} \tag{6.14}$$

so a comparison with Equation (6.2) yields the conductivity as:

$$\sigma = \frac{n_c e^2 \tau}{m} \tag{6.15}$$

where we have assumed the same value of τ for all electrons. The value of τ is usually equated with τ_m given in Equation (6.10). Another quantity of interest is the *mobility* of charge carriers, with symbol μ_m, which is related to the conductivity as:

$$\sigma = n_c e \mu_m \tag{6.16}$$

such that we can write:

$$\mu_m = \frac{e\tau_m}{m} \tag{6.17}$$

Furthermore, we can incorporate the mean free path into the conductivity in the following way:

$$\sigma = \frac{n_c e^2 \tau}{m} = \frac{ne^2 \lambda_{mfp}}{m v_{th}} = \frac{ne^2 \lambda_{mfp}}{\sqrt{3mk_B T}} \tag{6.18}$$

The variation of conductivity with $1/\sqrt{T}$ in the above goes against empirical findings and lead Drude to argue that the collision of electrons with the fixed ion cores must also have a similar dependence to obtain the expected $1/T$ dependence. As a point of interest, the typical values for the mean free path of electrons in metals is of the order of around 100 nm, where we consider the typical relaxation time of 10^{-14} s and thermal velocities for the electrons at room temperature are around 10^7 ms^{-1}. The Drude model uses kinetic theory and this can also be used to show that the electronic contribution to the thermal conductivity can be expressed as:

$$\kappa_e = \frac{2}{3}\tau_m v^2 C_{kin} \qquad (6.19)$$

where C_{kin} is the kinetic or classical specific heat for an electron gas, c.f. Equation (5.76), which should have a value of $3k_B n_c/2$. We now obtain:

$$\kappa_e = \tau_m v^2 k_B n_c = \frac{3n_c k_B^2 T \tau_m}{m} \qquad (6.20)$$

where we have used the thermal velocity. If we substitute in for the conductivity, Equation (6.15), we can express the Lorenz number as:

$$L \equiv \left(\frac{\kappa_e}{\sigma T}\right) = 3\left(\frac{k_B}{e}\right)^2 \qquad (6.21)$$

which has a value of 2.23×10^{-8} V^{2}K^{-2}. This is in rather good agreement with experimentally found values for many metals and was a major success of the Drude theory since it also agreed with the Wiedemann - Franz law. Despite this early success, subsequent work by H. A. Lorentz (not to be confused with Lorenz) in 1905 along with the Hall effect and magnetoresistance, provided evidence that the Drude model had a number of weaknesses. We will return to these topics at a later stage (Chapter 8).

6.5 FERMI - DIRAC STATISTICS OF AN ELECTRON GAS

By the mid to late 1920s it was known that electrons were particles which have a spin of $\frac{1}{2}$ and obeyed the laws of quantum mechanics. Also they adhere to the Pauli exclusion principle, which states that no two electrons (or *fermions*) can possess the same set of quantum numbers, see Chapter 1. In 1926 both Fermi and Dirac had pointed out, independently, that a perfect gas of electrons would behave significantly differently to a perfect classical gas, which can be described by the Maxwell - Boltzmann distribution. One of the main consequences of the Pauli exclusion principle is that at a temperature of absolute zero, all electrons must be in their ground state, but cannot all occupy the same lowest energy state. So if we have a system of N electrons, at $T = 0$ K, the lowest N states will be filled, with all higher states being vacant. In a metal, which we are assuming to be our container for the electrons, we assume that the Coulomb interaction between the electrons is negligible and that the potential due to the ion cores is uniform. As such, the Fermi - Dirac description is that of a perfect non-interacting electron gas.

As we have described the electron energy levels above, at $T = 0$ K all states will be occupied up to a certain energy. We say that the probability of occupation below this energy is unity, while above this energy the probability of occupation is zero. This probability function is called the *Fermi - Dirac distribution function* and is denoted by $f_{FD}(E, T)$. The upper limit to the energy at zero temperature is called the *Fermi energy* or *Fermi level*, E_F. Our situation at absolute zero can thus be more concisely expressed as:

$$f_{FD}(E,0) = \begin{cases} 1, & E < E_F; \\ 0, & E > E_F. \end{cases} \qquad (6.22)$$

The functional form for the Fermi - Dirac distribution can be written for any temperature as:

$$f_{FD}(E,T) = \frac{1}{e^{(E-\mu)/k_B T} + 1} \qquad (6.23)$$

The function $f_{FD}(E, T)$ is an equilibrium distribution. We have used the *chemical potential*, μ, in the above equation to distinguish it from the Fermi energy, where we note that:

$$\lim_{T \to 0} \mu = E_F \qquad (6.24)$$

At higher temperatures they diverge, though they remain fairly close in metals up to room temperature and many authors do not make this distinction. The functional variation of the Fermi - Dirac distribution is illustrated in Figure 6.1. It is noted that the distribution at 0 K is a step function and spreads out as the temperature increases. It will be seen that the curve will alway pass through the point of $f_{FD}(E,T) = \frac{1}{2}$. Summing over all states of the particles in the system, we must find:

$$\sum_n f_{FD}^{(n)}(E,T) = N \qquad (6.25)$$

In the classical limit, $f_{FD}(E, T)$ must behave as the function $f_{MB}(E, T)$, which in terms of energy can be expressed as:

$$f_{MB}(E,T) = Ce^{-E/k_B T} \qquad (6.26)$$

and is essentially the same as given in Equation (6.5). The classical limit will be reached for the condition: $f_{MB}(E,T) \ll 1$. This means we require: $(E - \mu)/k_B T \gg 1$, for all E. In this case we have:

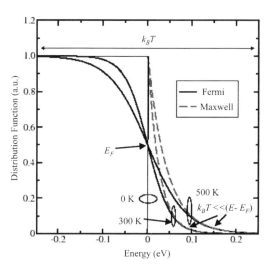

FIGURE 6.1: The Fermi - Dirac distribution function for temperatures of 0, 300 and 500 K. The energy indicated corresponds to $E - \mu$. For comparison the Maxwell - Boltzmann distributions for 300 and 500 K are also shown.

$$f_{FD}(E,T) = \frac{1}{e^{(E-\mu)/k_BT}} \simeq e^{-(E-\mu)/k_BT} = e^{\mu/k_BT} e^{-E/k_BT} \quad (6.27)$$

which, since μ is a constant, has the required form of $f_{MB}(E, T)$. The Maxwell - Boltzmann distribution is compared with that of the Fermi - Dirac function in Figure 6.1.

6.6 THE SOMMERFELD MODEL

As we mentioned earlier, in 1905 Lorentz found some inconsistencies in the Drude model. It was clear that something was missing from the theory. With the development of quantum mechanics in the 1920s, these inconsistencies were becoming more evident. In 1928, Sommerfeld used the Fermi - Dirac statistics to evaluate the electronic properties of solids. He considered the non-interacting electron gas with a uniform positive charge (potential), to maintain charge neutrality. Effectively this meant that the electron was considered to be moving inside an empty box. We note that the effect of non-interacting means that the electron is essentially alone and the only effect of the other electrons in the gas is to reduce the overall positive potential inside the box. While still grossly simplified (and we shall look at this problem in more complex ways later), it provided a much improved model to that of Drude.

We will consider a cube of metal of side, L, and an electron to be confined within it. The Schrödinger equation, in the steady - state, can be written as:

$$\left\{-\frac{\hbar}{2m}\nabla^2 + U\right\}\psi(\mathbf{r}) = E\psi(\mathbf{r}) \quad (6.28)$$

where U is the uniform potential inside the box (metal). Appropriate boundary conditions must apply to ensure that the wave-functions disappear at the limits of the material. Using a general wave-function of the form: $\psi(\mathbf{r}) = Ce^{i\mathbf{k}\cdot\mathbf{r}}$, where C is a constant, we obtain:

$$E = \frac{\hbar^2}{2m}(k_x^2 + k_y^2 + k_z^2) = \frac{\hbar^2 k^2}{2m} \quad (6.29)$$

The components of the wave-vectors conform to:

$$k_i = \frac{2\pi}{L} n_i, \text{ where } i = x, y, z \qquad (6.30)$$

since we have chosen a volume, which is a cube of side, L. The n_i are real integers. This means we can write Equation (6.29) as:

$$E = \frac{2\pi^2 \hbar^2}{mL^2} \left(n_x^2 + n_y^2 + n_z^2 \right) \qquad (6.31)$$

The normalization constant C can be found in the usual way, by taking $\int_{-\infty}^{\infty} \psi(\mathbf{r})\psi(\mathbf{r})* dV = 1$. This will be left as an exercise for the student, see Exercise 6.3. It will be noticed that the form of our wave-function doesn't seem to satisfy the condition that it should go to zero at the boundaries of the metal. This problem is overcome by using the *periodic boundary condition*, which ensures that at the limits of the solid the wave-function in each direction is equal, which we can express as: $\psi(x = L) = \psi(x = 0)$ etc. This may seem at first a little unconvincing. However, this actually turns out to not be a problem, since we can consider the electron to be a wave packet described by a set of states of the form $e^{i\mathbf{k}\cdot\mathbf{r}}$, which move through the conductor. It responds to an applied field much like the classical electron we spoke of earlier. Importantly, it will travel only a finite distance through the solid, about a mean free path length, before undergoing a scattering event. So if the mean free path is much smaller than the size of the sample, then it will spend most of its time far from the surface anyway and boundary conditions are not of interest to it. In the final chapter we will see some of the consequences of considering very small samples, when we look at nanomaterials.

6.7　THE DENSITY OF STATES

We have already discussed densities of states for the case of phonons and here we shall do the same for electrons, with much the same arguments. Using the form of the components of the wave-vector given in Equation (6.30), we can evaluate the number of possible values these can take. Maintaining the same notation, we can ask then how many values of k_i there are in a range dk_i. We use the expression (6.30) to obtain:

$$dn_i = \frac{L}{2\pi} dk_i \qquad (6.32)$$

In terms of the volume of k-space, we can think of the volume $d^3\mathbf{k} = dk_x dk_y dk_z$, such that the volume element can support:

$$dn_x dn_y dn_z = \left(\frac{L}{2\pi}\right)^3 dk_x dk_y dk_z = \left(\frac{L}{2\pi}\right)^3 d^3\mathbf{k} \qquad (6.33)$$

states. We can consider the n_i to be quantum numbers, each of which can support two states (corresponding to the two spin states, $\pm\frac{1}{2}$). The total number of states that are available in the volume element of k-space, $d^3\mathbf{k}$, will be:

$$dN_s = 2 dn_x dn_y dn_z = \frac{V}{4\pi^3} d^3\mathbf{k} \qquad (6.34)$$

For our free electron model we obtained an energy, which was given in Equation (6.29). We can therefore states that all electrons of energy, E, can be represented as states that lie on the surface of a sphere of radius $k = \sqrt{2mE}/\hbar$ in k-space. Now states with energy in the range E to $E + dE$ will have corresponding wavevectors k to $k + dk$ and exist inside the shell defined by spheres of these radii. This corresponds to a region of k-space of volume: $4\pi k^2 dk$. We now define the number of states per unit volume, i.e., the density of states, between energies E and $E + dE$ as:

$$g(E)dE = \frac{dN_s}{V} = \frac{4\pi k^2 dk}{4\pi^3} = \frac{k^2 dk}{\pi^2} = \frac{1}{2\pi^2}\left(\frac{2m}{\hbar^2}\right)^{3/2} E^{1/2} dE \qquad (6.35)$$

We therefore find a variation of the density of states with $\sqrt{E}$. Of course not all states will be occupied. There is a difference between this and the phonon density of states. This arises from the difference in the statistics, where phonons obey the Bose - Einstein relation, see Equation (5.74), and can take any energy (i.e., do not obey the Pauli exclusion principle) with no limit to the number of states. In the case of fermions (particles with spin of integral numbers of one-half, such as electrons), there is a finite limit to the number of available states, with occupancy also limited to one per allowed state, as expressed by the four quantum numbers. The total electron density can thus be expressed as:

$$n = \int_0^\infty f_{FD}(E,T)g(E)dE = \frac{1}{2\pi^2}\left(\frac{2m}{\hbar^2}\right)^{3/2}\int_0^\infty \frac{E^{1/2}dE}{e^{(E-\mu)/k_BT}+1} \quad (6.36)$$

Since we know that at $T = 0$ K, all states are filled up to the Fermi energy, E_F, we can write:

$$n = \frac{1}{2\pi^2}\left(\frac{2m}{\hbar^2}\right)^{3/2}\int_0^{E_F} E^{1/2}\,dE = \frac{1}{3\pi^2}\left(\frac{2mE_F}{\hbar^2}\right)^{3/2} \quad (6.37)$$

which has a numerical value of $n = 4.54\times 10^{27} E_F^{3/2} m^{-3}$, where E_F is expressed in energy units of eV[1]. Metals typically contain electron densities of the order of around 10^{28} m^{-3}, so the value of E_F is of the order of a few eV. Alternatively we can write:

$$E_F = \frac{(3n\pi^2)^{2/3}\hbar^2}{2m} \quad (6.38)$$

The *Fermi wave-vector* can be introduced as: $k_F = \sqrt{2mE_F}/\hbar$, such that from the above we obtain:

$$k_F = (3n\pi^2)^{1/3} \quad (6.39)$$

So in k-space, all filled states lie inside a sphere of radius, k_F. The surface of this sphere is called the *Fermi surface*. In Figure 6.2, we illustrate the density of states, $g(E)$, the Fermi - Dirac function and the density of occupied states.

It is worth noting that we can also define other Fermi parameters based on the Fermi energy. The *Fermi temperature* is expressed as: $T_F = E_F/k_B$, while the *Fermi velocity* is written as: $v_F = \hbar k_F/m = \sqrt{2E_F/m}$.

We noted earlier that the Fermi energy, which we defined as the half-occupation level given by the Fermi - Dirac distribution, is temperature dependent and is different from the chemical potential. As the temperature increases some of the electrons can gain energy to occupy levels above the Fermi level, thus leaving

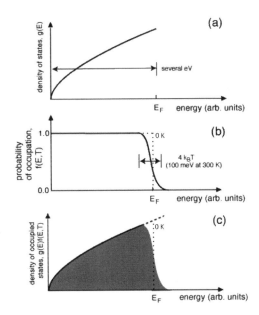

FIGURE 6.2: (a) The density of states function, corresponding to Equation (6.35). (b) The Fermi - Dirac distribution function, or probability of occupation. (c) Combining the functions $f_{FD}(E)g(E)$ gives the density of occupies states.

some states below E_F unoccupied. This gives the broadening in the Fermi - Dirac distribution. Now since the function, $g(E)$ increases with energy, the half-occupancy level must decrease as the thermal broadening increases. The evaluation of (Fermi - Dirac) integrals of the form of (6.36) at non-zero temperatures can be quite complex, expansions can be made or one can numerically determine their values. This is a worthwhile venture since it can yield the relation between the chemical potential and the Fermi energy, which as we have note is a function of temperature. Sommerfeld made an expansion by noting that integrals of the form: $\int_{-\infty}^{\infty} h(E)f(E)dE$, only differ from their zero temperature value, $\int_{-\infty}^{E_F} h(E)dE$, by the form of $h(E)$ in the region of $E = \mu$. As long as $h(E)$ doesn't vary rapidly within the order of $k_B T$ around $E = \mu$, then a Taylor expansion of $h(E)$ can be made around $E = \mu$:

$$h(E) = \sum_{n=0}^{\infty} \left(\frac{d^n}{dE^n} h(E) \right)_{E=\mu} \frac{(E-\mu)^n}{n!} \qquad (6.40)$$

We can now write:

$$\int_{-\infty}^{\infty} h(E) f_{FD}(E) dE = \int_{-\infty}^{E_F} h(E) dE + \sum_{n=0}^{\infty} (k_B T)^{2n} a_n \left(\frac{d^{2n-1}}{dE^{2n-1}} h(E) \right)_{E=\mu} \quad (6.41)$$

where the a_n are dimensionless constants of the order of unity. The above equation is known as a *Sommerfeld expansion*. For our purposes, the expansion up to first order is sufficient[2]. From Equation (6.36) we can write, applying the Sommerfeld expansion:

$$n = \int_0^\mu g(E) dE + \frac{\pi^2}{6} (k_B T)^2 \left[\frac{d}{dE} g(E) \right]_{E_F} + \ldots \quad (6.42)$$

Now we need to correct for the fact that μ differs from E_F, which can be done as follows:

$$\int_0^\mu g(E) dE = \int_0^{E_F} g(E) dE + (\mu - E_F) g(E_F) \quad (6.43)$$

This gives:

$$n \simeq \int_0^{E_F} g(E) dE + (\mu - E_F) g(E_F) + \frac{\pi^2}{6} (k_B T)^2 \left[\frac{d}{dE} g(E) \right]_{E_F} \quad (6.44)$$

The first term is nothing more than n, so we now find:

$$\mu = E_F - \frac{\pi^2}{6} (k_B T)^2 \left[\frac{d}{dE} g(E) \right]_{E_F} \frac{1}{g(E_F)} \quad (6.45)$$

Using Equation (6.35), we obtain:

$$\mu = E_F \left[1 - \frac{1}{3} \left(\frac{\pi k_B T}{2 E_F} \right)^2 \right] \quad (6.46)$$

This gives a very small correction at room temperature, and justifies the use of E_F instead of μ in a majority of cases.

6.8 SPECIFIC HEAT OF AN ELECTRON GAS

As we have already indicated above, Arnold Sommerfeld considered the use of the Fermi - Dirac rather than the Maxwell - Boltzmann distribution. He was aware that at normal temperature, the scattering of electrons changes their energies, but only to within a small window of energy states available to them near the Fermi level, since other states are occupied. This corresponds to the white area below E_F being equal to the grey area above E_F in Figure 6.2(c). Sommerfeld therefore concluded that only electron states near the Fermi level control the principal properties of metals.

The total energy of the electron gas can be expressed as:

$$U = \int_0^\infty E f_{FD}(E,T) g(E) dE = \frac{1}{2\pi^2} \left(\frac{2m}{\hbar^2}\right)^{3/2} \int_0^\infty \frac{E^{3/2} dE}{e^{(E-\mu)/k_B T} + 1} \quad (6.47)$$

This again contains a complex Fermi integral and we will have to use the Sommerfeld expansion again. Now we have:

$$U = \int_0^\mu E g(E) dE + \frac{(\pi k_B T)^2}{6}\left[\mu \frac{d}{dE}g(\mu) + g(\mu)\right] + \ldots$$

$$= \int_0^{E_F} E g(E) dE + E_F \left\{(\mu - E_F) g(E_F) + \frac{(\pi k_B T)^2}{6}\left[\frac{d}{dE}g(E)\right]_{E_F}\right\}$$

$$+ \frac{(\pi k_B T)^2}{6} g(E_F) \quad (6.48)$$

We saw earlier that the term in curly brackets cancels, see (6.44), and the first term is the ground state energy. Therefore we obtain:

$$U = U_0 + \frac{(\pi k_B T)^2}{6} g(E_F) \quad (6.49)$$

It is now a simple matter to obtain the free electron contribution to the specific heat:

$$C_{el} = \left(\frac{\partial U}{\partial T}\right)_n = \frac{(\pi k_B)^2}{3} g(E_F) T \quad (6.50)$$

Using the expression for $g(E_F)$ and n, we can write:

$$C_{el} = \frac{\pi^2 k_B T}{2 E_F} n k_B \quad (6.51)$$

We see that the Fermi - Dirac statistics give a different result to the classical value of $3 n k_B / 2$ and is said to depress the specific heat

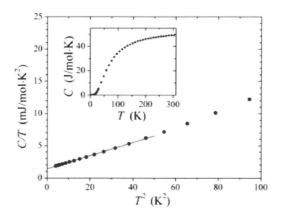

FIGURE 6.3: Variation of C_v/T against T^2 for the PdGa alloy at low temperature. (Figure published with permission: M. Klanjsek et al., "PdGa intermetallic hydrogenation catalyst: an NMR and physical property study", J. Phys.: Condens. Matter, **24**, 085703 (2012); ©IOP Publishing Ltd and Deutsche Physikalische Gesellschaft. Published under a CC BY-NC-SA licence.)

by a factor of $(\pi^2 k_B T/3 E_F)$. An electron gas at $k_B T \ll E_F$ is called a *degenerate electron gas*. Importantly, the Sommerfeld result is confirmed by experiment, where the electronic contribution to the specific heat has a linear T dependence, though at room temperature it is swamped by the phonon or lattice contribution. At low temperature this is more evident however. Where, as we have already seen in the previous chapter, we have a total specific heat of:

$$C_v = \gamma T + AT^3 \qquad (6.52)$$

or

$$\frac{C_v}{T} = \gamma + AT^2 \qquad (6.53)$$

This relationship is illustrated in Figure 6.3 for PdGa[3].

6.9 PAULI PARAMAGNETISM

In 1925, the electron spin was discovered by G. E. Goudsmit and S. A. Uhlenbeck[4]. This lead Pauli to make the point a year later that the electron's magnetic moment should align with an applied magnetic field, **B**. Such an alignment of spin moments due to an applied magnetic field is referred to as *paramagnetism* and the paramagnetic effect of an electron gas, as in metals, is called the *Pauli paramagnetism*. The moment on the electron has a value of one *Bohr magneton*, which is defined as:

$$\mu_B = \frac{e\hbar}{2m} \tag{6.54}$$

and has a numerical value of 9.274×10^{-24} JT^{-1}, which has a direction anti-parallel to the spin angular momentum. The spin of an electron was introduced in Chapter 1 as a quantum number and can take a value of $\pm\frac{1}{2}$. So when a magnetic field is applied the spins align with the field. Electrons that are paired with opposite spins cancel out their magnetic effects, but the alignment of unpaired electrons leads to paramagnetic effects in a number of solids. We may therefore expect that metals, with all their free electrons should have a significant alignment in a magnetic field to produce a magnetic moment or induced magnetization, **M**. If we consider that a magnetic field will align the electron spins, then we can expect there to be a disparity between the number of electrons of each type of spin. The resulting magnetization can then be estimated from:

$$M = -\mu_B[n_\uparrow(E) - n_\downarrow(E)] \tag{6.55}$$

Such an imbalance of the electronic density of states is illustrated in Figure 6.4. In this figure, we separate the spin-up $(+\frac{1}{2})$ and spin-down $(-\frac{1}{2})$ components of the density of states. When a magnetic field, **B**, is applied each set of states is shifted with respect to the origin by $\pm\mu_B B$, as illustrated in Figure 6.4(b). Thermalization of the free electrons will allow that the chemical potential is equalize for both spin-up and spin-down electrons, Figure 6.4(c). The result is that the occupation of the spin states anti-parallel to the applied field will be greater than that parallel and the resulting magnetization will be positive as given by Equation (6.55). This will mean that to first order, the effect should be temperature independent, and we can evaluate the result at absolute zero.

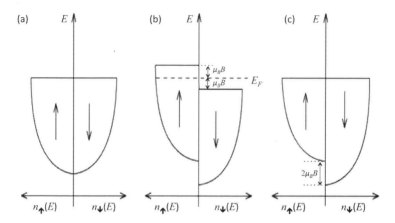

FIGURE 6.4: Density of states for up and down spin directions for electrons with: (a) equal occupation of the two spin states without an applied magnetic field. (b) The density of states are shifted by an amount equal to $\mu_B B$ in opposite directions, under the application of field, B. (c) After thermalization the occupancy becomes unequal and results in Pauli paramagnetism.

We can express the paramagnetic response of the electron gas as the magnetic susceptibility, $\chi_m = \mathbf{M}/\mathbf{H}$, where $\mathbf{H} = \mathbf{B}/\mu_0$. The net magnetic moment of the electrons can thus be evaluated as the imbalance of the spin-up and spin-down contributions:

$$M = \chi_m H = \frac{\mu_B}{2} \int_{-\infty}^{\infty} f_{FD}(E)[g(E+\mu_B B) - g(E-\mu_B B)] dE \quad (6.56)$$

Since we are evaluating at absolute zero, we can write this as:

$$M = \frac{\mu_B}{4\pi^2} \left(\frac{2m}{\hbar^2}\right)^{3/2} \left[\int_0^{E_F+\mu_B B} E^{1/2} dE - \int_0^{E_F-\mu_B B} E^{1/2} dE\right]$$

$$= \frac{\mu_B}{6\pi^2} \left(\frac{2m}{\hbar^2}\right)^{3/2} \left[(E_F + \mu_B B)^{3/2} - (E_F - \mu_B B)^{3/2}\right] \quad (6.57)$$

In the case where $\mu_B B \ll E_F$, which is almost always the case, we can write:

$$M = \mu_B^2 B \frac{1}{2\pi^2} \left(\frac{2m}{\hbar^2}\right)^{3/2} E_F^{1/2} = \mu_B^2 B g(E_F) \quad (6.58)$$

From which we can express the Pauli paramagnetic susceptibility as:

$$\chi_m = \frac{M}{H} = \mu_0 \mu_B^2 g(E_F) \qquad (6.59)$$

We can now see that the magnetic effect of the free electrons lies within an energy of $\mu_B B$ of the Fermi energy, as expected from the Sommerfeld theory and as Pauli predicted, the magnetic susceptibility is degenerated from the classical value. The degree of this degeneracy can be expressed as:

$$\chi_m = \frac{3n\mu_0 \mu_B^2}{2E_F} \qquad (6.60)$$

We will look at the general theory of paramagnetism later in Chapter 10, which can be expressed in terms of the *Curie - Weiss law*, having a temperature dependence in agreement with experiment. The difference here is that we are discussing the degenerate free electron gas, whereas the Curie - Weiss law is derived from the unpaired (non-degenerate) electrons which reside on the ions of the solid. In our discussion above, we have neglected to consider the translational motion which a magnetic field has on the free electron. This problem was considered by L. Landau in 1930, we will address this problem later in Chapter 8, when we consider transport phenomena, see Section 8.8. For now we will just mention that Landau was able to show that the the magnetic field restricts the electron motion in directions perpendicular to the field and produces a susceptibility contribution of $-n\mu_0 \mu_B^2 / 2E_F$, which a third of the Pauli paramagnetism. Since the Landau response is negative, it is a *diamagnetic* effect and the net susceptibility of the free electrons will be:

$$\chi'_m = \frac{n\mu_0 \mu_B^2}{E_F} \qquad (6.61)$$

Which is a temperature independent response and is in good agreement with experiment for many simple metals and alkali elements. We will also discuss diamagnetism, which is a feature of a majority of material solids, in the chapter on Magnetic Materials and Phenomena, see Section 10.3.

6.10 HIGH FREQUENCY RESPONSE AND OPTICAL PROPERTIES

We have already considered the electrical conductivity of metals in the presence of a static applied electric field, which conforms to a shift, $\Delta \mathbf{k}$, of the the Fermi sphere in k-space: $\hbar \Delta \mathbf{k} = -e\mathbf{E}\tau$, see Figure 6.5. The displacement of the Fermi sphere, $\Delta \mathbf{k}$, is small compared to the Fermi vector, and corresponds to the drift velocity of the electrons; $\mathbf{v}_d = -e\tau\mathbf{E}$. This will give rise to a small redistribution of the occupied electron states in the vicinity of the Fermi level; an increase of occupied states near $-k_F$, with a corresponding decrease near k_F. This imbalance causes a flow of charge or current. Removing the applied electric field, the drift of charge will decay back to zero via inelastic scattering processes, and the sphere of occupied states returns to its position at the origin of k-space. The time taken to return back to the stationary state is related to the relaxation time, τ.

We will now consider the effect of an alternating electric field of the form:

$$\mathbf{E} = \mathbf{E}_0 e^{-i\omega t} \tag{6.62}$$

The equation of motion can be expressed as:

$$m\frac{d\mathbf{v}}{dt} + \frac{m\mathbf{v}}{\tau} = -e\mathbf{E} \tag{6.63}$$

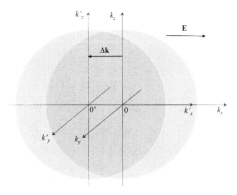

FIGURE 6.5: Displacement of the Fermi sphere from its origin due to the application of an electric field, **E**. The shift is in the opposite direction to that of the applied electric field.

where we note that the second term on the right hand side is due to the *relaxation time* of the electrons. From the time varying electric field, we expect a time varying drift velocity:

$$\mathbf{v}_d = \mathbf{v}_d^0 e^{-i\omega t} \qquad (6.64)$$

From which we obtain the drift velocity as:

$$\mathbf{v}_d = \frac{-e\mathbf{E}\tau}{m(1-i\omega\tau)} \qquad (6.65)$$

Writing the current density as $\mathbf{J}(\omega) = \sigma(\omega)\mathbf{E}(\omega)$, leads to the frequency dependent expression for the conductivity:

$$\sigma(\omega) = \frac{ne^2\tau}{m(1-i\omega\tau)} = \frac{\sigma_0}{(1-i\omega\tau)} \qquad (6.66)$$

where σ_0 is the static or dc conductivity, which will take the form of Equation (6.15). The frequency response of a metal is dependent on the value of the relaxation time, which for most metals is of the order of 10^{-14} s at room temperature. This means, in practice, that the conductivity doesn't change significantly until we are considering rather high frequencies.

Since electromagnetic radiation has an associated electric field, we can consider what the effect of incident radiation is on a metal. In doing so we should remember that under normal circumstances of the incidence of radiation, current is neither entering nor leaving the sample with the electrons responding by moving back and forth in the sample[5]. Also, these induced currents are transverse to the wave-vector of the radiation. For wavelengths greater than the mean free path of electrons[6], we can consider that the response of a metal to be governed by Maxwell's equations. Using the Maxwell version of Ampère's law, we write:

$$\nabla \times \mathbf{H} = \mathbf{J} + \varepsilon_0 \frac{\partial \mathbf{E}}{\partial t}$$
$$= \sigma(\omega)\mathbf{E} - i\omega\varepsilon_0 \mathbf{E} \qquad (6.67)$$

Using the Faraday law:

$$\nabla \times \mathbf{E} = -\mu_0 \frac{\partial \mathbf{H}}{\partial t} \tag{6.68}$$

we can write:

$$\begin{aligned}
\nabla \times (\nabla \times \mathbf{E}) = -\nabla^2 \mathbf{E} &= -\mu_0 \frac{\partial (\nabla \times \mathbf{H})}{\partial t} \\
&= i\omega \mu_0 (\nabla \times \mathbf{H}) \\
&= \omega^2 \mu_0 \varepsilon_0 \left[1 + i\frac{\sigma(\omega)}{\omega \varepsilon_0}\right] \mathbf{E}
\end{aligned} \tag{6.69}$$

This has the form of a wave equation with a complex dielectric constant of the form:

$$\begin{aligned}
\varepsilon(\omega) &= 1 + i\frac{\sigma(\omega)}{\omega \varepsilon_0} \\
&= 1 - \frac{\omega_p^2 \tau^2}{1 + \omega^2 \tau^2} + i\frac{\omega_p^2 \tau}{\omega(1 + \omega^2 \tau^2)}
\end{aligned} \tag{6.70}$$

where we have introduced the *plasma frequency*, $\omega_p = \sqrt{ne^2/\varepsilon_0 m}$, and used the frequency dependent conductivity, Equation (6.66). The plasma frequency corresponds to a charge density oscillation in the electron gas. A quantum of plasma oscillation, called a *plasmon*, has an energy of $\hbar \omega_p$. For metals, the plasma frequency is of the order of 10^{15} Hz, extending into the ultraviolet region of the electromagnetic spectrum.[7] We have explicitly expressed the form of the complex dielectric constant so that it is easy to see the real and imaginary components.

For very low frequencies, $\omega \tau \ll 1$, we recover the dc conductivity, $\sigma(\omega) \simeq \sigma_0$ and the effective dielectric constant tends to $i\omega_p^2 \tau/\omega = i\sigma_0/\varepsilon_0 \omega$. This corresponds to the classical *skin depth* region, where the electromagnetic wave entering the metal is attenuated over a distance:

$$\delta = \sqrt{\frac{2}{\mu_0 \sigma_0 \omega}} \tag{6.71}$$

The refractive index of a medium is related to the ratio of the speed of light in vacuum to the speed of light in the medium; $\mathcal{N} = c/v = \sqrt{\varepsilon(\omega)}$. The refractive index is also a complex number

($\mathcal{N} = n_0 + i\kappa$), where the imaginary part is related to the absorption of electromagnetic radiation.

At high frequencies, $\omega\tau \gg 1$, we have the conductivity being dominated by the imaginary part: $\sigma(\omega) \simeq \sigma_0/i\omega\tau$ and the effective dielectric constant can be expressed as:

$$\varepsilon(\omega) \simeq 1 - \left(\frac{\omega_p}{\omega}\right)^2 \qquad (6.72)$$

When $\omega > \omega_p, \varepsilon > 0$ and the refractive index will be real; $\mathcal{N}^2 = \varepsilon$ as given in Equation (6.72). However, for $\omega < \omega_p, \varepsilon < 0$ and the refractive index will be completely imaginary. The implications of this is that for high frequencies ($\omega > \omega_p$), there will be no absorption and the metal behaves as a dielectric with a poor reflectivity. However, when we are in the low frequency regime ($\omega < \omega_p$), we have $\mathcal{N} = i\kappa$, and the wave is attenuated as it passes through the electron gas[8]. We can express the reflectivity at the interface of a medium, one side of which is the vacuum and the other the metal, as:

$$\mathcal{R} = \left|\frac{1-\mathcal{N}}{1+\mathcal{N}}\right|^2 \qquad (6.73)$$

where we have assumed normal incidence. So for $\mathcal{N} = i\kappa$, in the low frequency regime, we have have $\mathcal{R} = 1$ and we have a perfect reflective medium. The plasma frequency therefore marks a dividing line between the optically absorbing and transmitting regions of the free electron gas. Hence we can understand why metals have a mirrored appearance, where we note that the plasma frequency for metals is above the frequency range for visible light. For the high frequency region the reflectivity falls off rapidly and the metal becomes transparent.

The Drude theory gives a reasonable description of metals in the low frequency regime up to around the plasma frequency. This is also dependent on temperature, since for low temperatures, the mean free path of the electrons will be long, and in the above we have assumed that the electric field is uniform. At low temperatures, the breakdown of the Drude theory in metals can occur in the microwave frequency range. Optical properties of non-conducting solids will be discussed in Chapter 12.

6.11 SUMMARY

Metallic behavior is mainly characterized by the high electrical and thermal conductivities that is empirically found in these materials. The former is well described in terms of Ohm's law. The classical theory of electronic conduction in metals was developed by Drude at the turn of the 20th century and served to describe some of the principal properties of electrical conduction in metals and adheres broadly to the Wiedemann - Franz law. However, deficiencies in the model emerged in terms of the temperature dependent behavior and the specific heat contribution of the free electrons. With the progress of the quantum description of materials, the Fermi - Dirac statistics was found to be the correct form for electrons and allowed many improvements in the description of metallic behavior to be made. Much of this was due to the work of Sommerfeld and Lorentz. Indeed, the the Fermi - Dirac statistics allowed the electronic states to be evaluated and in conjunction with the Pauli exclusion principle, Sommerfeld concluded that the majority of the experimentally observed behavior is due to the electron distribution in the region of the Fermi energy of the metal.

The specific heat of the free electron gas was determined and found to agree very well with observation. Furthermore, Pauli was able to describe the paramagnetic response of the electron gas when an applied magnetic field aligns the electron spins in the region of the Fermi level. A further diamagnetic effect of the ions was proposed by Landau, being one-third of the Pauli paramagnetism, and gives a net paramagnetic response for metals in agreement with measurement.

A consideration of the variation of the electrical conductivity and dielectric constants of metals permits us to understand some of the basic high frequency and optical properties of metals. Below a frequency defined by the excitation of a plasma oscillation in the electron gas, the metals appear highly reflective and are reasonably well understood in terms of the Drude theory. However, at high frequencies, metals become transparent and the Drude model is inadequate.

There are still some topics which we have not yet described with regards to electrical properties, such as the Hall effect and magnetoresistance. These will be considered in Chapter 8, when we deal with transport phenomena. However, before this it is important to consider the effect of the crystal lattice in more detail, which is the subject of the next chapter.

REFERENCES AND FURTHER READING

Basic Texts

- J. S. Blakemore, *Solid State Physics*, Cambridge University Press, Cambridge (1985)

- R. G. Chambers, *Electrons in Metals and Semiconductors*, Chapman and Hall, London (1990)

- H. P. Myers, *Introductory Solid State Physics*, Taylor and Francis, London (1998)

Advanced Texts

- M. A. Wahab, *Solid State Physics: Structure and Properties of Materials*, Alpha Science International Ltd., Harrow (2007)

- N. W. Ashcroft and N. D. Mermin, *Solid State Physics*, Saunders College, Philadelphia (1976)

- M. T. Dove, *Lattice Dynamics*, Cambridge University Press, Cambridge (2009)

EXERCISES

Q1. Prove Equation (6.13). Also find an expression for the average distance travelled by the electrons in the direction of an applied electric field.

Q2. Show that the Fermi - Dirac function must always pass through the point $f_{FD}(E,T) = \tfrac{1}{2}$ at the Fermi level for all T.

Q3. Find the form of the normalization constant for the wave-function: $\psi(\mathbf{r}) = Ce^{i\mathbf{k}\cdot\mathbf{r}}$.

Q4. Evaluate the ground state energy and lowest excited state energy for an electron in a cubic box of side 2 Å. What temperature would be sufficient to excite and electron to this state? What is the degeneracy of this first excited state?

Q5. Consider an electron in a sample of silver, which has a Fermi energy of 5.49 eV. Calculate the corresponding Fermi velocity and temperature. At what temperature can we expect to have a 10% probability of finding an electron with an energy which is 1% above the Fermi level?

Q6. Deduce fully Equation (6.46).

Q7. Calculate the temperature at which the electronic and lattice contributions to the specific heat capacities are equal for Au and Pb. These have Fermi energies of 5.53 eV and 9.47 eV, respectively.

Q8. What are the paramagnetic susceptibilities of Cu and Au? Also evaluate their mean free paths at room temperature. Use the following data: $n_{Cu} = 8.47 \times 10^{22} \, \text{cm}^{-3}$, $n_{Au} = 5.90 \times 10^{22} \, \text{cm}^{-3}$, $E_F^{Cu} = 7.0 \, \text{eV}$, $E_F^{Au} = 5.53 \, \text{eV}$, $\tau_{Cu} = 2.7 \times 10^{-14} \, \text{s}$, and $\tau_{Au} = 3.0 \times 10^{-14} \, \text{s}$.

Q9. Demonstrate the validity of Equation (6.59).

Q10. Calculate the plasma frequency for copper, which has an electron density of $8.47 \times 10^{22} \, \text{cm}^{-3}$. What is the energy of this quantum excitation in eV? Also evaluate the skin depth of copper at this frequency.

Q11. Prove Equation (6.70).

NOTES

[1] Note that 1 eV = 1.602×10^{-19} J.
[2] For the interested reader, several authors have treated this problem in more detail. See for example: N. W. Ashcroft and N. D. Mermin, *Solid State Physics*, Saunders College, Philadelphia (1976) and R. Kim and M. Lundstrom, *Notes on Fermi - Dirac Integrals*, (2008); *https://www.nanohub.org/resources/5475/*, (arXiv: 0811.0116v4).

[3] Data from M. Klanjsek *et al.*, J. Phys.: Condens. Matter, **24**, 085703 (2012).
[4] Actually, four years earlier A. H. Compton had also postulated the spin of the electron to explain experimental data on the magnetic properties of the electron. However, Goudsmit and Uhlenbeck used the spin of the electron to explain atomic spectra and despite initial criticism from Pauli, it caught on.
[5] In the following analysis, we are ignoring the effect of the time varying magnetic field component of the electromagnetic wave since its effects are negligible in normal metals compared to that of the electric field.
[6] This removes complications due to locally varying fields which differ over the trajectory of the electron in between collisions with the ions of the solid. This condition will be satisfied for metals exposed to visible light.
[7] Plasmon frequencies in low dimensional structures, such as nanoparticles, are directly related to the size of the entity which confines the electrons. Such considerations has given rise to a field of study called *nanoplasmonics*
[8] Note that since the wave-vector can be written in terms of the refractive index: $\mathbf{k} = m\mathbf{v}/\hbar = (mc\mathcal{N}/\hbar)\hat{\mathbf{n}} = (imc\kappa/\hbar)\hat{\mathbf{n}}$, the electric field in the medium will become: $\mathbf{E} = \mathbf{E}_0 e^{i(\mathbf{k}\cdot\mathbf{r}-\omega t)} = \mathbf{E}_0 e^{i\mathbf{k}\cdot\mathbf{r}} e^{-i\omega t} = \mathbf{E}_0 e^{i(mc\mathcal{N}/\hbar)\hat{\mathbf{n}}\cdot\mathbf{r}} e^{-i\omega t} = \mathbf{E}_0 e^{-(mc\kappa/\hbar)\hat{\mathbf{n}}\cdot\mathbf{r}} e^{-i\omega t}$, which we note will decay with distance.

CHAPTER 7

BAND THEORIES OF SOLIDS

"Everything happens to everybody sooner or later if there is time enough."
—George Bernard Shaw

"Bad times have a scientific value. These are occasions a good learner would not miss."
—Ralph Waldo Emerson

7.1 INTRODUCTION

In the previous chapter, we considered the quantum theory for a free electron gas to explain the electronic properties of metals. In doing so we assumed that the atoms of the metal release a certain number of electrons, which are then free to roam around the solid. Furthermore, we did not consider the effect of having the positive ions in their lattice positions and how this would affect the motion of the electrons. Indeed, the positive potential due to these ions was assumed to be uniform throughout the metal. Scattering of electrons by the ions was thus considered to be elastic, with a certain mean free distance, λ_{mfp}, between the average collision event, or a mean time between collisions, τ. In reality this model is oversimplified and has nothing to say about the electronic properties of materials with less charge carriers available. For instance, we know that semiconducting materials have extremely important electronic properties and are the working materials for a large majority of electronic devices. The use of band theories applies to all types of material, whether conductor, insulator or semiconductor. The evaluation of the electronic band structure of materials is one of the central problems in modern solid state physics.

In general, the way we take into account the crystalline structure is by permitting the ionic potential to vary within the lattice, such that the periodicity of the potential matches that of the position of the *fixed* ions in the solid. Band theory takes this into account by imposing the periodic potential within the Schrödinger equation to determine the allowed states of the electron within the crystal. A majority of the models that describe the electronic states in a solid do this by:

1. Choosing a wave-function, $\psi(\mathbf{r})$, to describe the eigenstates of the electron in a perfectly periodic potential, such that the eigenenergies, E_n, of the electron can be determined.

2. The choice of the form of the periodic potential, $V(\mathbf{r})$, is important as it will perturb the electron and change the eigenstates for suitable solutions of the Schrödinger equation.

3. It is customary to use a one-electron approximation, which significantly simplifies the form of the Schrödinger equation. The eigenstates obtained in this manner must be in agreement with the Fermi - Dirac statistics for occupancy. This approach is compatible with self-consistent techniques, as introduced by Hartree and Fock in the late 1920s, early 1930s. These remove the need for the dynamic interactions between electrons by averaging over the occupied electron states.

In using the Fermi - Dirac statistics, which effectively takes care of the Pauli exclusion principle that operates on the electrons, we assume that the electrons should not interact. Taking into account the electron - electron scattering is a very complex problem and requires the use of so-called *many-body* techniques. Such an approach is beyond the scope of the current text. In this chapter, we will introduce some of the more simple techniques for evaluating the energy states of electrons in solids, which take into account the lattice structure (periodicity). We will start by introducing some basic concepts which are necessary for the development of these models. As we have mentioned, the models differ essentially in the way we represent the periodic potential and how we set up the electron wave-function. Of the techniques discussed in this chapter, the principal difference is in the former, where we can control the strength and shape of the periodic potential. Of particular interest are the extremes, where we use a weak and a strong potential. These correspond to the *free* and *nearly free electron* models at the weak end of the scale to the *tight-binding* method, which adopts an almost atomic like description of the electron states.

Despite the seemingly simplistic approach we are taking, there is much insight that can be derived from these models, and as we shall see they approximate quite well to the electronic properties of real solids.

7.2 THE PERIODIC POTENTIAL

In describing the periodic potential, it is important that we use the lattice periodicity, so that the correct electron states will be established once we solve the Schrödinger equation. This is generally done as follows: The periodic potential will follow the symmetry of the lattice by using the Bravais lattice vectors, **R**, in the periodic potential such that we satisfy the condition:

$$V(\mathbf{r}) = V(\mathbf{r} + \mathbf{R}) \tag{7.1}$$

where, **r**, is any vector within the crystal. Equation (7.1) says that if we make any displacement from a position, **r**, with a Bravais lattice vector, **R**, then the potential at the new point, $\mathbf{r} + \mathbf{R}$, is indistinguishable from the original position. A schematic illustration of the periodic potential in one-dimension is shown in Figure 7.1. Evidently the scale of the periodicity conforms to the lattice structure and is of the order of Ångstroms.

Once we impose the periodic form of the potential, we can write the one-electron Schrödinger equation as:

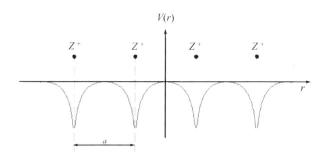

FIGURE 7.1: One-dimensional periodic variation of the potential of the ions in a solid. We note that the condition given in Equation (7.1) is satisfied.

$$\mathcal{H}\psi(\mathbf{r}) = \left[-\frac{\hbar^2}{2m}\nabla^2 + V(\mathbf{r})\right]\psi(\mathbf{r}) = E\psi(\mathbf{r}) \quad (7.2)$$

where of course the potential must be consistent with Equation (7.1). We can compare this with the free electron Schrödinger Equation, (6.28), where the potential is constant or zero and offers the simplest solution to the electron behavior in a solid.

In writing Equation (7.2), we have made a number of assumptions. Apart from the non-interacting electrons, we have also used the adiabatic approximation (see Chapter 5), where we consider the ions to be fixed as they are much heavier than the electrons. Each of the electrons in the solid are then described by Equation (7.2). The wave-functions of these electrons must also conform to the periodic potential and not to the plane wave solutions we considered in the previous chapter. The electrons are referred to as *Bloch electrons* and the corresponding wave-functions as *Bloch functions*.

Since the Fourier series serve to represent periodic functions, we can therefore represent the periodic potential, as given in Equation (7.1), as:

$$V(\mathbf{r}) = \sum_{\mathbf{G}} V_{\mathbf{G}} e^{i\mathbf{G}\cdot\mathbf{r}} \quad (7.3)$$

Where the vectors $\mathbf{G}$ belong to the reciprocal lattice[1] of the crystal and $V_{\mathbf{G}}$ are the Fourier coefficients. We note that the coefficient, V_0, will represent the uniform background potential, which we can choose such that; $V_0 \equiv 0$.

7.3 THE BLOCH THEOREM AND FUNCTIONS

Felix Bloch considered the effect of the periodic potential on the electron in a solid in 1928. The theorem which carries his name has some very important consequences[2]. It is based on the electrostatic potential of the ion cores which are ideally located in the Bravais lattice positions, with no defects or phonons, such that the potential has the translational periodicity of the lattice. The potential due to the other electrons is assumed to have the same average value in each unit cell and thus also has the same periodicity of the crystal lattice. In this way, Bloch attempts to account for the electron-electron interactions. The eigenstates of the electrons in this periodic structure, as described by Equations (7.1) and (7.2), form plane waves with a multiplying factor, which is a function containing the periodicity of the Bravais lattice. The wave- or *Bloch-function* has the form:

$$\psi_{nk}(\mathbf{r}) = e^{i\mathbf{k}\cdot\mathbf{r}} u_{nk}(\mathbf{r}) \tag{7.4}$$

The function, $u_{nk}(\mathbf{r})$, is subject to the periodic condition:

$$u_{nk}(\mathbf{r}) = u_{nk}(\mathbf{r}+\mathbf{R}) \tag{7.5}$$

where the vector, $\mathbf{R}$, is any Bravais lattice vector. A consequence of Equations (7.4) and (7.5) is that we can write:

$$\begin{aligned}\psi_{nk}(\mathbf{r}+\mathbf{R}) &= e^{i\mathbf{k}\cdot(\mathbf{r}+\mathbf{R})} u_{nk}(\mathbf{r}+\mathbf{R}) \\ &= e^{i\mathbf{k}\cdot\mathbf{r}} e^{i\mathbf{k}\cdot\mathbf{R}} u_{nk}(\mathbf{r}) \\ &= e^{i\mathbf{k}\cdot\mathbf{R}} \psi_{nk}(\mathbf{r})\end{aligned} \tag{7.6}$$

This shows that the effect of the addition of a translation vector, $\mathbf{R}$, of the real lattice is to introduce a phase factor in the Bloch function.

Once again we can make use of the Fourier representation for the periodic function. This will involve the Fourier coefficients, C_{nk}, where we will express the wave-vector in terms of a reciprocal lattice vector plus a vector in reciprocal space; say, $\mathbf{k} = \mathbf{q} - \mathbf{G}$. We can now write the Bloch function in the form:

$$\psi_{nq}(\mathbf{r}) = \sum_{\mathbf{G}} C_{n\mathbf{q}-\mathbf{G}} e^{i(\mathbf{q}-\mathbf{G})\cdot\mathbf{r}}$$
$$= e^{i\mathbf{q}\cdot\mathbf{r}} \sum_{\mathbf{G}} C_{n\mathbf{q}-\mathbf{G}} e^{-i\mathbf{G}\cdot\mathbf{r}} = e^{i\mathbf{q}\cdot\mathbf{r}} u_{nq}(\mathbf{r}) \quad (7.7)$$

which is identical to Equation (7.4).

We need to make sure the correct boundary conditions are applied to the wave-functions. This will inevitably impose some form of restriction on the solutions of our eigenvalue Equation, (7.2), and thus on the allowed states, **k**. The most appropriate form of boundary condition is one in which the periodic nature of the solid is taken into account. These are generally expressed as the *periodic* or *Born-von Karman* boundary condition and can be expressed as:

$$\psi(\mathbf{r} + N_i \mathbf{a}_i) = \psi(\mathbf{r}); \quad \text{where,} \quad i = 1, 2, 3 \quad (7.8)$$

here the $\mathbf{a}_i$ are the primitive vectors for our Bravais lattice and the N_i are integers giving the number of unit cells in the i-direction and subject to the condition: $N = N_1 N_2 N_3$, which is the total number of unit cells in the crystal, and will be a large number. The boundary condition implies that

$$e^{iN_i \mathbf{k}\cdot\mathbf{a}_i} = 1 \quad (7.9)$$

So when we apply the Bloch theorem we have:

$$\psi_{n\mathbf{k}}(\mathbf{r} + N_i \mathbf{a}_i) = e^{iN_i \mathbf{k}\cdot\mathbf{a}_i} \psi_{n\mathbf{k}}(\mathbf{r}) \quad (7.10)$$

It is now possible to establish that the allowed Bloch wave-vectors can be expressed in the form:

$$\mathbf{k} = \sum_{i=1}^{3} \frac{m_i}{N_i} \mathbf{b}_i \quad (7.11)$$

the m_i are integers and the vectors $\mathbf{b}_i$ are the primitive reciprocal lattice vectors, see Section 3.2. It follows from the above that when the m_i change by one, we generate a new state. We can therefore determine that the volume of k-space occupied by each state will be:

$$\Delta \mathbf{k} = \frac{\mathbf{b}_1}{N_1} \cdot \left(\frac{\mathbf{b}_2}{N_2} \times \frac{\mathbf{b}_3}{N_3} \right) = \frac{1}{N} \mathbf{b}_1 \cdot (\mathbf{b}_2 \times \mathbf{b}_3) \quad (7.12)$$

since the volume of the primitive cell of the reciprocal lattice is $\mathbf{b}_1 \cdot (\mathbf{b}_2 \times \mathbf{b}_3)$, we can assert that *the total number of allowed states is equal to the number of unit cells in the crystal lattice*. While this may appear to be a mere curiosity, it is in fact a very important result and can have important consequences on the nature of a material. Since the volume of the reciprocal lattice primitive unit cell is $(2\pi)^3/v$, where $v = V/N$ is the volume of the unit cell of the direct lattice, we can write (7.12) as:

$$\Delta \mathbf{k} = \frac{(2\pi)^3}{V} \tag{7.13}$$

This is essentially the same result we obtained for the free electron model; see Equation (6.34).

7.4 THE SCHRÖDINGER EQUATION IN A PERIODIC POTENTIAL

We can construct the main components of the problem by substituting the periodic potential along with the Bloch functions of the electrons in the Schrödinger equation. To do this we substitute equations of the form of (7.3) and (7.7) in Equation (7.2):

$$\sum_{\mathbf{k}} \frac{\hbar^2 k^2}{2m} C_{n\mathbf{k}} e^{i\mathbf{k}\cdot\mathbf{r}} + \sum_{\mathbf{G}} V_{\mathbf{G}} e^{i\mathbf{G}\cdot\mathbf{r}} \sum_{\mathbf{k}} C_{n\mathbf{k}} e^{i\mathbf{k}\cdot\mathbf{r}} = E \sum_{\mathbf{k}} C_{n\mathbf{k}} e^{i\mathbf{k}\cdot\mathbf{r}} \tag{7.14}$$

We can rationalize this equation to read:

$$\sum_{\mathbf{k}} \left\{ \left[\left(\frac{\hbar^2 k^2}{2m} \right) - E \right] C_{n\mathbf{k}} + \sum_{\mathbf{G}} V_{\mathbf{G}} C_{n\mathbf{k}-\mathbf{G}} \right\} e^{i\mathbf{k}\cdot\mathbf{r}} = 0 \tag{7.15}$$

This condition is valid for all $\mathbf{r}$. Therefore, the expression inside the { } brackets is independent of $\mathbf{r}$ and must vanish for each $\mathbf{k}$. This can be shown by multiplying by a plane wave and integrating. We can now write:

$$\left[\left(\frac{\hbar^2 k^2}{2m} \right) - E \right] C_{n\mathbf{k}} + \sum_{\mathbf{G}} V_{\mathbf{G}} C_{n\mathbf{k}-\mathbf{G}} = 0 \tag{7.16}$$

A choice of $V_G = 0$ allows us to reproduce the results of the Sommerfeld model, which gives: $E = \hbar^2 k^2 / 2m$. Equation (7.16) represents a set of Schrödinger equations in reciprocal space, which couple the expansion coefficients, C_{nk}, of the Bloch functions, $\psi_{nk}(\mathbf{r})$, whose $\mathbf{k}$ values differ from one another by a reciprocal lattice vector $\mathbf{G}$. In doing this, we have separated the original problem into a set of equations, one for each value of $\mathbf{k}$, of which there will be N.

We have made some general comparisons between the Sommerfeld model and the Bloch theorem, however, we should be careful to note that there is an important difference in terms of the wave-vectors. In the Sommerfeld model, $\mathbf{k} = \mathbf{p}/\hbar$, the wave vector simply defines the momentum, $\mathbf{p}$, of the electron. In the Bloch formulation $\mathbf{k}$ will not explicitly relate the momentum of the electron in the periodic potential. We can demonstrate this by looking at the momentum operator[3], $\mathbf{p} = (\hbar/i)\nabla$, which when acting on the Bloch function, $\psi_{nk}(\mathbf{r})$ returns:

$$\frac{\hbar}{i}\nabla\psi_{nk}(\mathbf{r}) = \frac{\hbar}{i}\nabla\left[e^{i\mathbf{k}\cdot\mathbf{r}}u_{nk}(\mathbf{r})\right] = \hbar\mathbf{k}\psi_{nk}(\mathbf{r}) + e^{i\mathbf{k}\cdot\mathbf{r}}\frac{\hbar}{i}\nabla u_{nk}(\mathbf{r}) \quad (7.17)$$

As such, $\psi_{nk}(\mathbf{r})$, is not a momentum eigenstate and will only conform to being so if $u_{nk}(\mathbf{r})$ is not a periodic function but a constant. Thus returning us to the Sommerfeld condition. We have already met the quantity, $\hbar\mathbf{k}$, which we defined as the crystal momentum; see Section 5.2.2. We will see the significance of the wave-vector in more detail when we consider the motion of electrons in a crystal lattice. This will be met in the following chapter when we discuss electron dynamics. For the moment it is sufficient to think of the $\mathbf{k}$'s as a state the electron can take, which depends on the translational symmetry of the lattice in which it moves.

As we have already discussed, any particular state $\mathbf{k}$, can be represented in the first Brillouin zone. This is because the translational symmetry of the reciprocal lattice permits us to write:

$$\mathbf{k'} = \mathbf{k} + \mathbf{G} \quad (7.18)$$

where $\mathbf{G}$ is a reciprocal lattice vector and $\mathbf{k}$ lie outside the first Brillouin zone, i.e., invariance under translation symmetry in reciprocal - space.

So, given that $e^{i\mathbf{G}\cdot\mathbf{r}} = 1$ for any reciprocal lattice vector, if the Bloch function holds for $\mathbf{k}'$, it must also be true for $\mathbf{k}$.

7.5 BRILLOUIN ZONES AND THE FERMI SURFACE

We have already defined the Brillouin zone when we discussed phonons in Chapter 5. However, it is worth extending a little our discussion, which is relevant for the consideration of energy states in periodic structures. The construction of Brillouin zones is made in the same way as the Wigner - Seitz cell, see Section 2.2.2. We can simply states that the Wigner - Seitz cell of the reciprocal lattice corresponds to the first Brillouin zone. However, there can be more than one Brillouin zone, so we should extend the principle here. The simplest way to do this is by example. Consider a two-dimensional lattice and sectioning up the reciprocal lattice as we considered earlier using the reciprocal lattice vectors. This is illustrated in Figure 7.2 for a 2D square array.

We note that each Brillouin zone occupies the same area of reciprocal space (or volume in the three-dimensional case). As we noted earlier, the Brillouin zone boundaries are particularly important

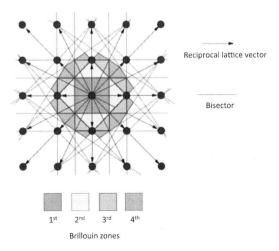

FIGURE 7.2: Construction for Brillouin zones for a two-dimensional square array.

because they mark the Bragg planes for the diffraction of waves by the periodic structure (lattice). The Brillouin zones for three dimensional structures have more complex structures, but are evaluated in the same ways as illustrated for two dimensions. In Figure 7.3, we illustrate the first, second, and third Brillouin zones for the fcc and bcc structures. See also Figure 5.2.

The ground state for a system of N electrons at zero temperature will be obtained by the occupation of all k-states of allowed one-electron levels up to the Fermi energy. In the case of the free electrons, they will have energies defined by: $E(\mathbf{k}) = \hbar^2 k^2/2m$. We will recall from the previous chapter, that the reciprocal space representation for all $\mathbf{k}$-states being occupied is a sphere, which we called the Fermi sphere, the surface of which is referred to as the *Fermi surface*. For a system of Bloch electrons, the ground state will be obtained in a similar fashion, except that we now need to label the states with n as well as $\mathbf{k}$, and the energies, $E_n(\mathbf{k})$, do not have a simple explicit form as for the free electron case. The vector $\mathbf{k}$ must be confined to the first Brillouin zone so the electrons occupy only all the available states and are counted once. In a real solid, a distinct

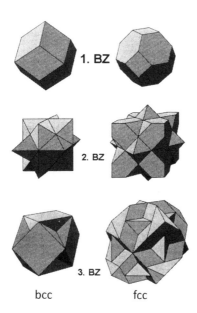

FIGURE 7.3: Brillouin zones for three-dimensional fcc and bcc structures.

behavior will be found depending on whether bands are filled or not. In the case where the uppermost energy band is filled and separated by an energy gap to the next empty band, the Fermi level will sit in the band gap region. The size of the band gap will determine whether the material is an insulator or a semiconductor. For gaps of the order of an eV, thermal energies are generally sufficient at room temperature to promote some charge carriers and the material is semiconducting. For larger band gaps which are well beyond $k_B T$, the material is an insulator. The number of states in each band will be equal to the number of primitive unit cells in the crystal. Since two electrons (one of each spin type) can be accommodated in each level, the band gap situation can only arise for materials where there is an even number of electrons per primitive unit cell. On the other hand, for partially filled bands, the Fermi energy will lie within one or more of the bands. Only in the case where this occurs can a Fermi surface be said to exist. This latter situation is what happens in metals or conductors, see Section 7.11.

The way we represent the Fermi surface depends on the zone scheme we choose. Often it is preferable to use the limit of each branch of a Fermi surface within a particular Brillouin zone. The zone schemes can be repeated or reduced and we will see this in action when we show the results of the various band models in the following sections. We will illustrate some of the Fermi surfaces for certain metals as we proceed, see also Section 8.4.

7.6 THE KRONIG - PENNEY MODEL

The Kronig - Penney model (1930) possibly represents one of the greatest simplifications that can be made regarding the form of a non-zero periodic potential. The one-dimensional periodic potential has two values, one of which is usually take as zero for simplification, while the other has another fixed value. An example of this is shown in Figure 7.4. We can take the upper limit of the potential as V_0 and the lower as being zero. As such we can consider the solutions of the Schrödinger equation in the two regions of potential $V(\mathbf{r}) = 0$ and $V(\mathbf{r}) = V_0$. The width of the potential being b and the periodicity, or lattice constant, is a.

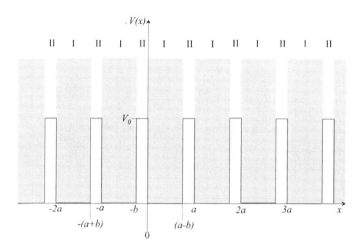

FIGURE 7.4: One dimensional periodic potential used in the Kronig - Penney model.

The one dimensional Schrödinger equation will have the form: $V(\mathbf{r}) = 0$:

$$\frac{d^2\psi}{dx^2} - \alpha^2\psi = 0 \tag{7.19}$$

and for $V(\mathbf{r}) = V_0$:

$$\frac{d^2\psi}{dx^2} + \beta^2\psi = 0 \tag{7.20}$$

where we have:

$$\alpha^2 = \frac{2mE}{\hbar^2}; \quad \beta^2 = \frac{2m(V_0 - E)}{\hbar^2} \tag{7.21}$$

The solutions to the Schrödinger equation must conform to the Bloch theorem and this requires continuity of the wave function and its first derivative throughout the lattice. The general form of the solutions can be expressed as:

In regions indicated as I (see Figure 7.4), $V(x) = 0$:

$$\psi_I(x) = Ae^{i\alpha x} + Be^{-i\alpha x} \tag{7.22}$$

and in regions indicated as II (Figure 7.4), $V(x) = V_0$:

$$\psi_{II}(x) = Ce^{\beta x} + De^{-\beta x} \tag{7.23}$$

where A, B, C and D are constants. We can express the boundary conditions as follows:

At $x = 0$:

$$\psi_I(0) = \psi_{II}(0) \tag{7.24}$$

$$\frac{d\psi_I(0)}{dx} = \frac{d\psi_{II}(0)}{dx} \tag{7.25}$$

At $x = -b$:

$$\psi_I(-b) = \psi_{II}(-b) \tag{7.26}$$

$$\frac{d\psi_I(-b)}{dx} = \frac{d\psi_I(-b)}{dx} \tag{7.27}$$

The Bloch functions can now be expressed as:

$$\psi_I(a-b) = e^{ika}\psi_I(-b) \tag{7.28}$$

$$\frac{d\psi_I(a-b)}{dx} = e^{ika}\frac{d\psi_{II}(-b)}{dx} \tag{7.29}$$

and we obtain the boundary conditions at $x = -b$ as:

$$e^{-ika}\psi_I(a-b) = \psi_{II}(-b) \tag{7.30}$$

$$e^{-ika}\frac{d\psi_I(a-b)}{dx} = \frac{d\psi_{II}(-b)}{dx} \tag{7.31}$$

The application of the boundary conditions and the Bloch functions allow us to evaluate the wave function coefficients as:

$$A + B = C + D \tag{7.32}$$

$$i\beta(A - B) = \alpha(C - D) \tag{7.33}$$

$$e^{-ika}[Ae^{i\beta(a-b)} + Be^{-i\beta(a-b)}] = Ce^{-\alpha b} + De^{\alpha b} \tag{7.34}$$

$$i\beta e^{-ika}[Ae^{i\beta(a-b)} - Be^{-i\beta(a-b)}] = \alpha[Ce^{-\alpha b} - De^{\alpha b}] \tag{7.35}$$

The simultaneous solution of these four equations allows us to obtain, from the determinant, the following solution:

$$\cos(ka) = \frac{(\alpha^2 - \beta^2)}{2\alpha\beta}\sinh(\alpha b)\sin[\beta(a-b)] + \cosh(\alpha b)\cos[\beta(a-b)] \qquad (7.36)$$

Simplification of this equation can be obtained using the limiting case of $b \to 0$; $V_0 \to \infty$, such that $V_0 b$ remains finite and constant. Since we have $\alpha^2 \gg \beta^2$, we find $(\alpha^2 - \beta^2)/2\alpha\beta \to \alpha/2\beta$. In this situation we have $\sinh(\alpha b) \to \alpha b$, $\cosh(\alpha b) \to 1$ and $[(\alpha^2 - \beta^2)/2\alpha\beta]\sinh(\alpha\beta) \to \alpha^2 b/2 = (mV_0/\hbar^2 a)ba$. We can now write:

$$\cos(ka) = \frac{p}{\beta a}\sin(\beta a) + \cos(\beta a) \qquad (7.37)$$

where $p = mV_0 ba/\hbar^2$. From this relationship we can numerically evaluate the allowed values of the solutions of the Schrödinger equation, which correspond to values of the right hand side of the above equation between ± 1, (i.e., when $\cos(ka)$ has a solution), where we are assuming that k is real. The conversion to the energy can easily be made and this reveals the form of the dispersion relation, as shown in Figure 7.5. We note that there naturally occur energy

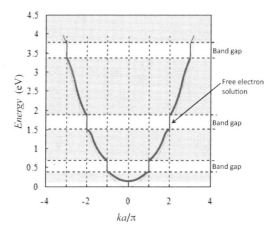

FIGURE 7.5: One dimensional solution of the Kronig - Penney model, showing the existence of energy bands and band gap regions where no solution of the Schrödinger equations exists.

bands, (i.e., regions with real solution of the energy) and band gaps (in which there do not exist real solutions to Equation (7.37)).

7.7 FREE ELECTRONS IN A PERIODIC POTENTIAL

The free electron or *empty lattice model* is a special case of the periodic potential in that we consider the strength of the potential to effectively be vanishing. This means that the electrons are free-like. It may be pertinent to ask why bother if the potential is zero, since it should just give us the free electron like behavior. Indeed, there is some truth to this, however, it can be quite instructive to go through this model, since it provides some significant insights into the behavior of electrons in periodic structures. Essentially the electron does have a free electron like energy, but with the imposition of the structure of the lattice. This can be seen through the free-electron like energies, using the electron states $\mathbf{k} = \mathbf{q} - \mathbf{G}$, which we express as:

$$E^0_{\mathbf{q}-\mathbf{G}} = \frac{\hbar^2 (\mathbf{q}-\mathbf{G})^2}{2m} \tag{7.38}$$

where $\mathbf{q}$ is a vector inside the first Brillouin zone. We can substitute this condition in Equation (7.16), where to denote a different reciprocal lattice vector we write $\mathbf{k} = \mathbf{q} - \mathbf{G}'$, where $\mathbf{G}'$ is another vector of the reciprocal lattice. We can now write:

$$\left\{ \left[\frac{\hbar^2 (\mathbf{q}-\mathbf{G}')^2}{2m} \right] - E \right\} C_{n\mathbf{q}-\mathbf{G}'} + \sum_{\mathbf{G}} V_{\mathbf{G}} C_{n\mathbf{q}-\mathbf{G}'-\mathbf{G}} = 0 \tag{7.39}$$

which is the same as:

$$\left\{ \left[\frac{\hbar^2 (\mathbf{q}-\mathbf{G}')^2}{2m} \right] - E \right\} C_{n\mathbf{q}-\mathbf{G}'} + \sum_{\mathbf{G}''} V_{\mathbf{G}''-\mathbf{G}'} C_{n\mathbf{q}-\mathbf{G}''} = 0 \tag{7.40}$$

where we have written: $\mathbf{G}'' = \mathbf{G} + \mathbf{G}'$. We can simply rewrite this, without any loss of generality, as:

$$(E^0_{\mathbf{q}-\mathbf{G}'} - E) C_{n\mathbf{q}-\mathbf{G}'} + \sum_{\mathbf{G}''} V_{\mathbf{G}''-\mathbf{G}'} C_{n\mathbf{q}-\mathbf{G}''} = 0 \tag{7.41}$$

We can separate the potential into the $\mathbf{G}'' = \mathbf{G}'$ and $\mathbf{G}'' \neq \mathbf{G}'$ terms, such that the above can be expressed as:

$$(E^0_{\mathbf{q}-\mathbf{G}'} - E) C_{n\mathbf{q}-\mathbf{G}'} + V_{\mathbf{G}'-\mathbf{G}'} C_{n\mathbf{q}-\mathbf{G}'} + \sum_{\mathbf{G}'' \neq \mathbf{G}'} V_{\mathbf{G}''-\mathbf{G}'} C_{n\mathbf{q}-\mathbf{G}''} = 0 \quad (7.42)$$

As we earlier noted we can take $V_{\mathbf{G}'-\mathbf{G}'} = V_0 \equiv 0$, such that Equation (7.42) becomes:

$$(E^0_{\mathbf{q}-\mathbf{G}'} - E) C_{n\mathbf{q}-\mathbf{G}'} + \sum_{\mathbf{G}'' \neq \mathbf{G}'} V_{\mathbf{G}''-\mathbf{G}'} C_{n\mathbf{q}-\mathbf{G}''} = 0 \quad (7.43)$$

We can now let the potential go to zero intensity, such that we have a master equation relating the electronic states given as:

$$(E^0_{\mathbf{q}-\mathbf{G}'} - E) C_{n\mathbf{q}-\mathbf{G}'} = 0 \quad (7.44)$$

The non-trivial solution of which is expressed as:

$$E(\mathbf{q} - \mathbf{G}') = E^0_{\mathbf{q}-\mathbf{G}'} = \frac{\hbar^2 (\mathbf{q} - \mathbf{G}')^2}{2m} \quad (7.45)$$

We can represent this result for a one-dimensional lattice in which the energy varies as the square of the wave-vector, as for the free electron. Since we are dealing with a periodic structure, we also see that we can include the shift of the parabolic energy dependence by any number of reciprocal lattice vectors, $\mathbf{G}$. The intersections of the various branches meet at the mid points, which corresponds to the Brillouin zone boundaries. This is illustrated in Figure 7.6.

There is of course the solution to Equation (7.44) above in which we have $C_{n\mathbf{q}-\mathbf{G}'} = 0$, for $\mathbf{G} \neq \mathbf{G}'$. This implies that the summation term in Equation (7.41) causes only small corrections to the electron energy, which should be proportional to the second order in the potential, and since the potential is vanishingly small, will be very minor corrections. We note that in the one dimensional representation of the band structure, the bands intersect at the Brillouin zone boundaries ($\pm n\pi/a$). It is worthy of note that the Brillouin zone boundaries correspond to Bragg conditions, which is rather easy to see for the 1D case: $k_n = \pm n\pi/a$, where we can write

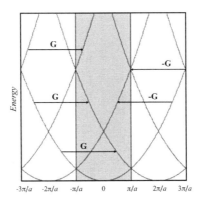

FIGURE 7.6: One dimensional solution of the free electron model, showing the band structure which arises from the solution of the Schrödinger equations for zero potential. The repeated zone scheme can be represented in a single Brillouin zone using the translation reciprocal lattice vectors ($\pm\mathbf{G}$). Note, the first Brillouin zone corresponds to the shaded region.

$|\mathbf{G}| = n\pi/a$, from which we can obtain $2a = n\lambda_n$, which must have $\theta = 90°$. The bands clearly have a parabolic nature and all energy states are represented in the first Brillouin zone, which can be achieved by a translation of the energy states via the reciprocal lattice vector, $\mathbf{G}$, as illustrated by the shaded region in Figure 7.6. No energy gaps exist, and as will be shown in the next section, this is a direct result of having a zero magnitude potential.

We can extend the one-dimensional results to that of a three-dimensional lattice. To do this we must consider the reciprocal lattice vectors for a specific structure. We will do this for the fcc lattice since it is an important structure found in solid state physics and is the basis of many semiconductor materials, which while being diamond in structure have a similar form of reciprocal lattice and hence electronic and band structure. We can remind ourselves that the reciprocal lattice for the fcc real space lattice is a bcc structure, see Chapter 3. The relevant Brillouin zone is illustrated in Figure 7.3 and we apply Equation 7.45 using the the three-dimensional reciprocal lattice vector of the form:

$$\mathbf{G}' = h\mathbf{b}_1 + k\mathbf{b}_2 + l\mathbf{b}_3 \tag{7.46}$$

where for the fcc structure we have:

$$\mathbf{b}_1 = \frac{2\pi}{a}(\hat{\mathbf{x}} - \hat{\mathbf{y}} + \hat{\mathbf{z}}); \quad \mathbf{b}_2 = \frac{2\pi}{a}(\hat{\mathbf{x}} + \hat{\mathbf{y}} - \hat{\mathbf{z}}); \quad \mathbf{b}_3 = \frac{2\pi}{a}(-\hat{\mathbf{x}} + \hat{\mathbf{y}} + \hat{\mathbf{z}}); \quad (7.47)$$

from which we obtain:

$$\mathbf{G}' = \frac{2\pi}{a}[\hat{\mathbf{x}}(h+k-l) + \hat{\mathbf{y}}(-h+k+l) + \hat{\mathbf{z}}(h-k+l)] \quad (7.48)$$

and using the general vector in reciprocal space of the form:

$$\mathbf{q} = \frac{2\pi}{a}[\hat{\mathbf{x}}\xi + \hat{\mathbf{y}}\eta + \hat{\mathbf{z}}\zeta] \quad (7.49)$$

It is possible to generate the various energy bands, where a choice of initial and final vector is used to define the band in question. Given the form of the energy, these bands are parabolic. The lowest energy bands are illustrated in Figure 7.7. We note that the dimensionless quantities, ξ, η and ζ can take any value between zero and unity. This is made under the provision that the $\mathbf{q}$-vector remains inside the first Brillouin zone.

The form of the band structure, at first glance, looks very complex, with a tangle of interweaving lines. We note that each line constitutes an energy band which takes us from one specific point in the Brillouin zone to another. For example, the line: $\Gamma - \Lambda - L$ takes us from the origin of the Brillouin zone (and reciprocal space) to the central point on the hexagonal face at the boundary of the first Brillouin zone, see inset of Figure 7.7. The line corresponds to the energy on the Λ line joining the end states at these points of high symmetry and is parabolic, as indicated by Equation (7.45). Other lines can be explained in a similar manner. The numbers in the figure indicate the degeneracy of the line, i.e., how many equivalent energy bands between the two end states of high symmetry in the first Brillouin zone. Since, as we have previously stated, all energy states can be represented in any Brillouin zone, we only need consider the first Brillouin zone. We note that it is the $\mathbf{q}$ vector which move us along this line and the extremal points will correspond to reciprocal lattice vectors, which we can represent with a $\mathbf{G}$ vector and $\mathbf{q} = 0$.

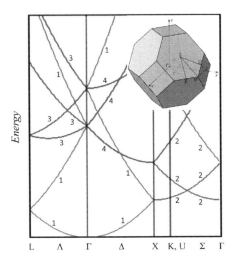

FIGURE 7.7: Empty-lattice or free electron model in three dimensions, showing the band structure as a function of the principal directions in the first Brillouin zone. The directions on the x-axis refer to the relevant directions in k-space as indicated in the inset. Numbers refer to band degeneracy.

We note that at the points of high symmetry (at the origin and on the boundary of the first Brillouin zone, all indicated with a Greek letter, see inset of Figure 7.7), various energy bands converge. There is a conservation of degeneracy at these points, from the bands leading into and out of these points. We further note that at any point of high symmetry, there exist a number of discrete states which corresponds to the quantization of energies with the periodicity in that particular direction of the crystalline structure. It is worth noting that for each of the bands we can express a specific wave-function, which engenders the crystal symmetries as expressed from the Bloch functions.

In general, band structure diagrams are represented as energies versus the directions in the crystalline structure, where the points of high symmetry are indicated with a Greek capital letter. As we will note later, at the points of high symmetry at the boundary of the Brillouin zone, an energy gap will open up. Some of the band crossings will therefore disappear, while others will be maintained. This depends on the symmetry of the wave functions. Crossing between bands of different symmetry can lead to hybridization. Away from the points of high symmetry in the vicinity of the zone boundaries, there is little deviation from the free electron bands. The free electron model, while being very simple in its mathematical construction, illustrates very well the complexities that arise from the crystalline structure.

7.8 THE NEARLY FREE ELECTRON MODEL

The nearly free electron model differs from the free electron model by considering a small value for the magnitude of the periodic potential in the Schrödinger equation. We can use Equation (7.43) which expresses the Schrödinger equation in the relevant form, where we can simplify as:

$$(E - E^0_{\mathbf{q-G'}})C_{\mathbf{G'}} = \sum_{\mathbf{G''} \neq \mathbf{G'}} V_{\mathbf{G''-G'}} C_{\mathbf{G''}} \quad (7.50)$$

We now make the zeroth order approximation and consider the case where $\mathbf{G'} = 0$ and $\mathbf{G}$, which means we are taking the approximations near to the limits of the $\mathbf{G}$ vectors where the changes should be largest. We thus obtain the solutions of the form:

$$(E - E^0_{\mathbf{q}})C_0 = V_0 C_0 + V_{\mathbf{G}} C_{\mathbf{G}} \quad (7.51)$$

$$(E - E^0_{\mathbf{q-G}})C_{\mathbf{G}} = V_{-\mathbf{G}} C_0 + V_0 C_{\mathbf{G}} \quad (7.52)$$

We evaluate the eigenstate E from the 2×2 matrix defined by the simultaneous solution to the above equations, where as we have already stated we can take $V_0 = 0$ as a background potential. This yields:

$$E(\mathbf{q})^{\pm} = \frac{(E^0_{\mathbf{q}} + E^0_{\mathbf{q-G}})}{2} \pm \frac{1}{2}[(E^0_{\mathbf{q}} - E^0_{\mathbf{q-G}})^2 + 4|V_{\mathbf{G}}|^2]^{1/2} \quad (7.53)$$

(We note that $V^*_{\mathbf{G}} = V_{-\mathbf{G}}$.) In a one dimensional crystal we can represent the band structures in the extended, repeat and reduced schemes, as illustrated in Figure 7.8.

The principal difference between the nearly free and the free electron models is the existence of the deviations from the parabolic form of the bands at the Brillouin zone boundaries. This occurs because the nearly free electron model has a small non-zero

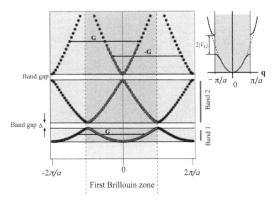

FIGURE 7.8: Repeat zone representation of the nearly free electron energy states as a function of the wave vector. The existence of the potential produces the appearance of a small energy gap (proportional to the magnitude of the potential) at the zone boundaries.

periodic potential, and these correspond to the Bragg condition. The dispersion (diffraction) of the electron waves with energies, $E = \hbar^2 k_{BZ}^2/2m$, where $\mathbf{k}_{BZ} = \pm n\pi/a$, occurs because the electrons have the appropriate energies for diffraction (Bragg) conditions to be satisfied arising from the periodicity of the lattice in the direction of electron motion. The two roots of Equation (7.53) can be more explicitly seen when we consider the point where $\mathbf{q}$ lies on a Bragg plane, where we have $E_{\mathbf{q}}^0 = E_{\mathbf{q}-\mathbf{G}}^0$. This gives:

$$E(q = \pm n\pi/a)^{\pm} = E_{\mathbf{q}}^0 \pm |V_{\mathbf{G}}| \qquad (7.54)$$

This can be interpreted as meaning that at all points on the Bragg plane the two energy solutions correspond to those raised by $|V_{\mathbf{G}}|$ and those lowered by $|V_{\mathbf{G}}|$, thus producing a band gap at that point of $\Delta = 2|V_{\mathbf{G}}|$, see inset of Figure 7.8. It is further possible to demonstrate that the gradient of the energy at the Bragg planes is zero, see Exercise 7.3.

In Figure 7.9, we show the form of the energy bands for the nearly free electron model for an fcc crystal. As with the one-dimensional case, we note that the bands separate at positions around the Brillouin zone boundary and the size of separation is directly proportional to the magnitude of the periodic potential. We see from this that the nearly free electron model produces energy bands that

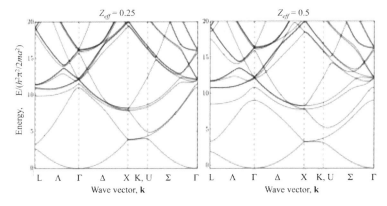

FIGURE 7.9: Repeat zone representation of the nearly free electron energy states as a function of the wave vector. The existence of the potential produces the appearance of a small energy gap (proportional to the magnitude of the potential) at the zone boundaries.

are perturbations of the free electron states. A comparison between Figures 7.7 and 7.9 shows that the general form of the energy bands are very similar, where the degeneracy of the free electron model is lifted by the lattice potential.

7.9 THE TIGHT - BINDING MODEL

The tight binding approximation represents the other extreme to the nearly free electron model. This latter, which was discussed in the previous section, essentially treats the electrons as being in an almost free state and are thus delocalized from the ions that form the crystal lattice. The electronic wave functions thus have a plane wave-like character. This means that the kinetic energy of the electrons is much greater than the potential energy. The tight binding approach, on the other hand, treats electrons as if they are almost entirely bound to these ions, with the electronic wave functions being derived from the atomic like orbitals that occur in isolated atoms. This can be stated as the potential energy dominating the kinetic energy of the electrons.

We will consider a crystal of N atoms located at the Bravais lattice sites. The system is approximated as having atomic like orbitals, whose wave functions are represented as $\psi_{at}(\mathbf{r})$ and have discrete energy states, E_n^{at}, such that we can write the atomic Schrödinger equation in the form:

$$\mathcal{H}_{at}\psi_n^{at}(\mathbf{r}) = E_n^{at}\psi_n^{at}(\mathbf{r}) \tag{7.55}$$

where the atomic Hamiltonian takes the form:

$$\mathcal{H}_{at} = -\frac{\hbar^2}{2m}\nabla^2 + V_{at}(\mathbf{r}) \tag{7.56}$$

The atomic like nature of the wave functions means that they become very small for distances greater than the lattice constant of the crystal. The Hamiltonian for the tightly bound electron is now considered to be a perturbation of the atomic case and will differ from $\mathcal{H}_{at}$ only at distances from the atomic site which exceed the range of the atomic wave function. Given the above, we can state that the wave function $\psi_n(\mathbf{r} - \mathbf{R})$ for all $\mathbf{R}$ of the Bravais lattice, will approximate the stationary like states of the atomic case since the Hamiltonian is mostly atomic like in nature with the imposition of the lattice periodicity via the periodic potential. We now write the corrections to the atomic case in the form of the Hamiltonian for the crystal as:

$$\mathcal{H} = \mathcal{H}_{at} + \Delta V(\mathbf{r}) \tag{7.57}$$

where $\Delta V(\mathbf{r})$ contains the necessary corrections to the atomic potential and can be expressed as: $\Delta V(\mathbf{r}) = V(\mathbf{r}) - V_{at}(\mathbf{r})$ in which we have $V(\mathbf{r}) = V(\mathbf{r} + \mathbf{R})$ to maintain the correct periodicity conditions of the periodic potential, see Figure 7.10. The Schrödinger equation is expressed as:

$$\mathcal{H}\psi_{n\mathbf{k}}(\mathbf{r}) = E_{n\mathbf{k}}\psi_{n\mathbf{k}}(\mathbf{r}) \tag{7.58}$$

To use the above equation we need to define the wave function, which is expressed in the form of a linear combination of atomic orbitals (LCAO), also called a *Wannier function*. This can be written in the form:

$$\psi_{n\mathbf{k}}(\mathbf{r}) = \sum_{\mathbf{R}} e^{i\mathbf{k}\cdot\mathbf{R}}\psi_n^{at}(\mathbf{r} - \mathbf{R}) \tag{7.59}$$

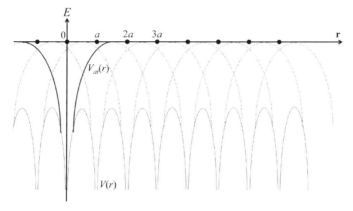

FIGURE 7.10: Schematic representation of the periodic potential (1D).

This satisfies the Bloch condition, as we can show:

$$\begin{aligned}
\psi_{n\mathbf{k}}(\mathbf{r}+\mathbf{R}') &= \sum_{\mathbf{R}} e^{i\mathbf{k}\cdot\mathbf{R}} \psi_n^{at}(\mathbf{r}+\mathbf{R}'-\mathbf{R}) \\
&= e^{i\mathbf{k}\cdot\mathbf{R}'} \sum_{\mathbf{R}} e^{i\mathbf{k}\cdot(\mathbf{R}-\mathbf{R}')} \psi_n^{at}[\mathbf{r}-(\mathbf{R}-\mathbf{R}')] \\
&= e^{i\mathbf{k}\cdot\mathbf{R}'} \sum_{\mathbf{R}} e^{i\mathbf{k}\cdot\mathbf{R}''} \psi_n^{at}(\mathbf{r}-\mathbf{R}'') \\
&= e^{i\mathbf{k}\cdot\mathbf{R}'} \psi_{n\mathbf{k}}(\mathbf{r}) \quad (7.60)
\end{aligned}$$

The energy bands, $E_{n\mathbf{k}}$, are generated via the Schrödinger equation are expressed in Equation (7.58). If we multiply this by an atomic wave function, $\psi_m^{*at}(\mathbf{r})$ and integrate over $\mathbf{r}$, we can write[4]:

$$(E_{n\mathbf{k}} - E_m^{at})\int \psi_m^{*at}(\mathbf{r})\psi_{n\mathbf{k}}(\mathbf{r})\,d\mathbf{r} = \int \psi_m^{*at}(\mathbf{r})\Delta V(\mathbf{r})\psi_{n\mathbf{k}}(\mathbf{r})\,d\mathbf{r} \quad (7.61)$$

We now substitute in the Wannier functions of Equation (7.59) to obtain:

$$\begin{aligned}
(E_{n\mathbf{k}} - E_m^{at})\sum_{\mathbf{R}} e^{i\mathbf{k}\cdot\mathbf{R}} &\int \psi_m^{*at}(\mathbf{r})\psi_n^{at}(\mathbf{r}-\mathbf{R})\,d\mathbf{r} \\
&= \sum_{\mathbf{R}} e^{i\mathbf{k}\cdot\mathbf{R}} \int \psi_m^{*at}(\mathbf{r})\Delta V(\mathbf{r})\psi_n^{at}(\mathbf{r}-\mathbf{R})\,d\mathbf{r} \quad (7.62)
\end{aligned}$$

Given that $\int \psi_m^{*at}(\mathbf{r})\psi_n^{at}(\mathbf{r})\,d\mathbf{r} = \delta_{nm}$, we find:

$$(E_{n\mathbf{k}} - E_m^{at}) = \sum_{\mathbf{R}} e^{i\mathbf{k}\cdot\mathbf{R}} \int \psi_m^{*at}(\mathbf{r})\Delta V(\mathbf{r})\psi_n^{at}(\mathbf{r}-\mathbf{R})\,d\mathbf{r}$$

$$= \int \psi_m^{*at}(\mathbf{r})\Delta V(\mathbf{r})\psi_n^{at}(\mathbf{r})\,d\mathbf{r}$$

$$+ \sum_{\mathbf{R}\neq 0} e^{i\mathbf{k}\cdot\mathbf{R}} \int \psi_m^{*at}(\mathbf{r})\Delta V(\mathbf{r})\psi_n^{at}(\mathbf{r}-\mathbf{R})\,d\mathbf{r} \quad (7.63)$$

It is now a simple matter to express our result in the form:

$$E_{n\mathbf{k}} = E_m^{at} - \alpha - \sum_{\mathbf{R}\neq 0} \beta_{\mathbf{R}} e^{i\mathbf{k}\cdot\mathbf{R}} \quad (7.64)$$

where we have made the following substitutions:

$$\alpha = -\int \psi_m^{*at}(\mathbf{r})\Delta V(\mathbf{r})\psi_n^{at}(\mathbf{r})d\mathbf{r}$$
$$\beta_{\mathbf{R}} = -\int \psi_m^{*at}(\mathbf{r})\Delta V(\mathbf{r})\psi_n^{at}(\mathbf{r}-\mathbf{R})d\mathbf{r} \quad (7.65)$$

The α constant corresponds to the Coulomb energy and will determine the energy shift of the atomic levels, while the β is related to the exchange energy and determines the widths of the energy bands. We can use the fcc structure as an example to illustrate how we can generate the band structure. We refer to the 12 nearest neighbors for this structure as shown in Figure 7.11.[5] These can be expressed by the following position vectors: $\mathbf{R} = \pm\frac{a}{2}(\hat{\mathbf{x}}\pm\hat{\mathbf{y}}); \pm\frac{a}{2}(\hat{\mathbf{x}}\pm\hat{\mathbf{z}}); \pm\frac{a}{2}(\hat{\mathbf{y}}\pm\hat{\mathbf{z}})$. We now write $\mathbf{k} = \hat{\mathbf{x}}k_x + \hat{\mathbf{y}}k_y + \hat{\mathbf{z}}k_z$, such that we have:

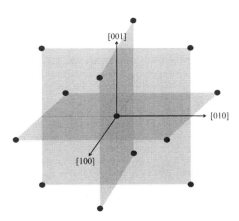

FIGURE 7.11: Illustration of the 12 nearest neighbors in the fcc structure.

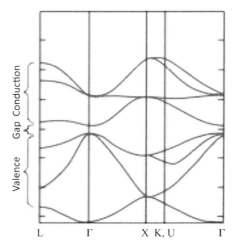

FIGURE 7.12: Energy bands for the fcc structure using LCAO in the tight binding model.

$$\mathbf{k} \cdot \mathbf{R} = \frac{a}{2}(\pm k_i \pm k_j) \tag{7.66}$$

where $i, j = x, y; y, z; z, x$. Substituting into Equation (7.64) we obtain:

$$E_{n\mathbf{k}} = E_m^{at} - \alpha - 4\beta \left[\cos\left(\frac{a}{2}k_x\right) \cos\left(\frac{a}{2}k_y\right) + \cos\left(\frac{a}{2}k_y\right) \cos\left(\frac{a}{2}k_z\right) \right.$$
$$\left. + \cos\left(\frac{a}{2}k_z\right) \cos\left(\frac{a}{2}k_x\right) \right] \tag{7.67}$$

The outcome of this calculation gives the variation of the lowest energy band. The energy bands for the fcc structure are illustrated in Figure 7.12. We see that the general form of this variation has many features in common with the other models we have considered, compare Figures 7.7 and 7.9 for the bands structures using the empty lattice and nearly free electron models.

7.10 OTHER MODELS: POTENTIALS AND WAVE-FUNCTIONS

There are many other models that are more sophisticated than the approximations outlined above. In many situations the tight-binding method is sufficient for the representation of the band structure of the ion core levels. The nearly free electron model, as we have seen makes a small modification of the free electron description, but does not realistically describe the band structure of a real solid. The complex problem of the band structure is an active area of research used to explain the electronic properties of solids. A fuller description of such methods goes beyond the scope this book. We will here only give a brief outline of some of the main principles of some of the most common methods.

In reality the question of the band structure of a solid requires us to take into account the structure of the solid, the periodic potential produced by the ions in the solid with positions described by the Bravais lattice and the wave functions of all the electrons in the material. In addition to this we also need to account for the interactions of the electrons among themselves due to the Coulomb forces. It would seem then that this is an almost impossibly difficult problem to resolve and recourse to simplifying assumptions is necessary to make the problem more manageable. Self-consistent methods, such as those introduced by Hartree and Fock are often employed to find an approximate potential, which requires an iterative method to obtain a working potential, often beginning with a guess at the potential. This is then inserted into the Schrödinger equation to evaluate the wave functions and then the potential is computed. This can then be repeated a number of times. When the computed potential is sufficiently close to the input potential, one says that *self-consistency* has been achieved. The principal components of the Schrödinger equation are the Hamiltonian and the wave-function, which must take the required mathematical form to allow us to evaluate the form of the energy states in the particular solid.

The models that we outlined above demonstrate the existence of energy bands with forbidden regions (energies). These analytical models, despite giving a reasonable qualitative agreement, are not sufficiently precise when compared to experimental results. The nearly free electron models, for example, displays an exaggerated planar wave aspect to its wave functions, while the tight binding model exaggerates the atomic-like character of the electrons. Other models use a perturbative expansion near the critical points in the Brillouin zone. One such technique is called the **k** · **p** method. This model considers a small variation of the wave-vectors around the extrema of the bands to develop the energy dispersion relation, $E_{n\mathbf{k}}$, in these regions. This can be expressed in a modified Bloch function of the form:

$$\psi_{n\mathbf{k}}(\mathbf{r}) = e^{i\mathbf{k}_0 \cdot \mathbf{r}} e^{i\Delta\mathbf{k} \cdot \mathbf{r}} u_{n\mathbf{k}_0 + \Delta\mathbf{k}}(\mathbf{r}) \tag{7.68}$$

where $\Delta\mathbf{k} = \mathbf{k} - \mathbf{k}_0$, and $\mathbf{k}_0$ is an extremal wave-vector. The energy dispersion relation is usually expressed as a modification of the unperturbed form and allows the introduction of the so-called *effective mass*, which we will discuss in more detail in the next chapter.

Another important class of model is the *pseudopotential* method, which accounts for the potential of the ion cores as well as the Coulomb repulsion between core electrons. In this model the upper valence electrons are evaluated using the nearly free electron model, while the inner core electron states are calculated from the atomic core states. Thus a distinction between the two types of electron is made. The pseudopotential method requires a certain knowledge of the band structure to allow an estimation of the initial potential, which is then refined by an iterative evaluation of the band structure which is then compared with experimental data until an acceptable agreement is achieved.

To obtain an even more accurate calculation of the band structure, in addition to the inclusion of the Coulomb energy of the electrons, the exchange and correlation energies should also be taken into account in a self-consistent manner. Such terms arise from *many-body* interactions of the electrons. It is possible to show that these depend only on the local density. *Density functional theory* uses such an approach and requires the use of sophisticated computational tools to obtain accurate band structure calculations.

7.11 METALS, SEMICONDUCTORS, AND INSULATORS

The electronic properties of solids are used as a form of classification of materials. These properties are a direct result of the availability of charge carriers in their lattice and electronic structure of their bands. The number of charge carriers available will depend on the manner in which electrons are used in the formation of bonds between the atoms in the solid (see Chapter 1). Delocalized electrons are available to conduct electricity while localized electrons are occupied in bonds, or are bound in atomic core states. Another way of envisaging this scenario is in the band structure of the solid and the specific energy distribution of the electrons. In the previous chapter, we saw that the electrons, which are fermions and subject to the Pauli exclusion principle, can have energies up to the Fermi level (E_F) at zero Kelvin. If the Fermi level falls within the region of the band gap region of the band structure, then all energy states in the band below the Fermi level, called the *valence band*, will be filled, while all states above the Fermi level, the *conduction band*, will be empty. In this case there will be no electrical conduction in the solid since there are no available states in the valence band for electrons to move into and no electrons in the conduction band which can contribute to electrical conduction. Such a material, at absolute zero of temperature, will be a perfect insulator. In reality and at finite temperatures the situation will be slightly different. The finite temperatures will allow a certain number of electrons to move to energies above the Fermi level, leaving some empty states available below the Fermi energy and thus produce a small distribution of the energy occupancy around the Fermi level, (see Figure 6.2). If the forbidden energy region is large compared to the distribution of the energy states around the Fermi level, then there will be very few electrons which can pass from the valence to conduction bands. Such a material is called an *insulator*. If the band gap is relatively small, the material is referred to as a *semiconductor*. In this representation, the only difference between an insulator and a semiconductor is the size of the band gap. If the Fermi level is within an energy band and/or the conduction and valence bands overlap, then there will be energy states and electrons available for the conduction of charge, and the material is a *conductor* or metal. In Figure 7.13, we illustrate these main points.

In Figure 7.14, we illustrate the band structure for silicon. In the figure, the band gap is indicated. The region below the band gap is the valence band, while above it is the conduction band. The labelling in the figure refers to the positions of high symmetry in the Brillouin zone as indicated previously. The band gap itself corresponds to the energy between the top (maximum energy) of the

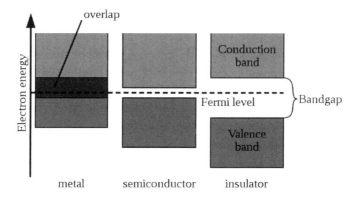

FIGURE 7.13: Illustration the classification of materials through the structure of their energy bands.

valence band and the bottom (minimum energy) of the conduction band. It will be seen that these two points, in the case of Si, do not correspond to the same points in reciprocal space. Such a feature is called an *indirect band gap*. In the semiconductor GaAs, these the top of the valence band and the bottom of the conduction band occur at the same position in the Brillouin zone, and this is called a *direct band gap* material or semiconductor.

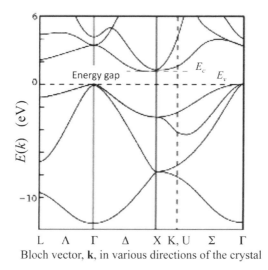

FIGURE 7.14: Band structure for silicon.

The promotion of electrons from the valence to the conduction band is an important process in semiconductor materials and we will return to this in Chapter 9. In this case the energy gap is of the order of about 1 eV, while that for insulators is several eV. This means at normal working temperatures, there will be some electrons promoted to the conduction band in semiconductors, while in insulators there will be a negligible change in the electron distributions. The potential for shifting electrons from the valence to conduction bands is one of the principal reasons why these materials occupy such an important role in the electronics industry. In fact, it is not only from thermal excitation that electrons can me promoted, but also from the incidence of radiation, such as light. This gives rise to optoelectronic devices, which will also be discussed in Chapter 9. When an electron is promoted in this way, not only do we gain a free charge carrier (an electron) in the conduction band, but the empty state remaining in the valence band can also act as a charge carrier, since an electron in the valence band can now move into this available energy state. Such a state is referred to as a *hole*. Holes play an important role in semiconductor physics, where electrons and holes are charge carriers which are both equally responsible for the electrical conduction in semiconducting devices. We note that the hole acts like a particle of positive charge moving in the valence band.

7.12 SUMMARY

The motion of electrons in solids is intimately related to the physical distribution of the atoms in the solid. For crystalline materials we can evaluate the available energy states by considering the symmetry of the atomic arrangement of atoms. Of particular importance, as we have seen, is the Bragg condition, which we find corresponds to the boundaries of the so-called Brillouin zones. At such points, in reciprocal space, the Bragg scattering means that the electrons with these energies are strongly scattered and thus no stationary energy states can be established. This gives rise to forbidden energy gaps in the energy spectrum. To be more precise, we need to introduce the concept of the periodic potential, which arises from the atomic potentials of the regularly spaced atoms in the crystal structure.

The energy states of the electrons are derived from the consideration of the kinetic and potential energies of the electrons in the solid. This is usually obtained via the Schrödinger equation. The principal components of this approach are the forms of the periodic potential and the wave functions which describe the electrons in the solid. In this chapter, we have considered a number of models that allowed us to evaluate the allowed energy states for the electrons. The *empty lattice model* considers the electrons to move in a potential free crystalline structure, where all energy states are available, but the band structure is imposed by the lattice of atoms. Once we introduce a weak potential, as in the *nearly free electron model*, there appear small energy gaps in the energy spectrum which occur at the Brillouin zone boundaries. In fact, the appearance of such band gaps is a direct result of the periodic potential, as was confirmed in the *Kronig - Penney model*. While the nearly free electron model over estimates the plane wave character of the electrons by allowing an excess in the kinetic energy, the *tight binding model* exaggerates the localized or orbital character of the electron wave functions. Despite this both models produce a similar band structure with forbidden energy regions. A more realistic band structure requires the use of more sophisticated approximations. Often such models require a knowledge of some of the properties of the materials as a starting point. Frequently recourse to numerical methods is necessary to undertake such complex calculations.

REFERENCES AND FURTHER READING

Basic Texts

- J. S. Blakemore, *Solid State Physics*, Cambridge University Press, Cambridge (1985)

- H. P. Myers, *Introductory Solid State Physics*, Taylor and Francis, London (1998)

- D. W. Snoke, *Solid State Physics: Essential Concepts*. Addison - Wesley, (2009)

Advanced Texts

- W. A. Harrison, *Solid State Theory*, Dover Publications, (1979)
- W. A. Harrison, *Elementary Electronic Structure*, World Scientific, (2004)
- M. A. Wahab, *Solid State Physics: Structure and Properties of Materials*, Alpha Science International Ltd., Harrow (2007)
- N. W. Ashcroft and N. D. Mermin, *Solid State Physics*, Saunders College, Philadelphia (1976)
- J. Singleton, *Band Theory and Electronic Properties of Solids*, Oxford University Press, Oxford (2001)
- H. Ibach and H. Lüth, *Solid State Physics: An Introduction to Principles of Materials Science*, Springer, Berlin (2009)

EXERCISES

Q1. What is the physical argument behind the *adiabatic approximation*?

Q2. Show explicitly the steps involved in obtaining Equation (7.15).

Q3. Demonstrate that the gradient of the energy bands at the Bragg plane are zero. Use the nearly free electron model.

Q4. What are the consequences of the above result on the form of Fermi surfaces for metals? Why does this not apply to the case of semiconductors?

Q5. Develop the free electron bands for a 2D square lattice.

Q6. Construct the first Brillouin zone for the bcc structure, check with Figure 7.3.

Q7. Derive the formula for the simplified Kronig - Penney model, given by Equation (7.37).

Q8. Use the result of Question 6 to plot the energy bands for the 1D Kronig - Penney model.

Q9. Consider how the results for the tight-binding model would alter from the fcc structure to that for Si and GaAs.

Q10. Evaluate the first energy band for the bcc structure using the tight-binding model.

NOTES

[1] The reciprocal lattice is discussed in detail in Chapter 3.

[2] The theorem was first demonstrated by Flouquet and also goes under the name of the *Flouquet theorem*.

[3] We notice that the momentum operator forms part of the Hamiltonian operator, $\mathcal{H}$, see Equation (7.2), where the kinetic energy portion is: $(\mathbf{p})^2/2m$.

[4] We have used the fact that: $\int \psi_m^{*at}(\mathbf{r}) \mathcal{H}_{at} \psi_{n\mathbf{k}}(\mathbf{r}) d\mathbf{r} = \int (\mathcal{H}_{at} \psi_m^{at}(\mathbf{r}))* \psi_{n\mathbf{k}}(\mathbf{r}) d\mathbf{r} = E_m^{at} \int \psi_m^{*at}(\mathbf{r}) \psi_{n\mathbf{k}}(\mathbf{r}) d\mathbf{r}$

[5] Note that we only need to use the nearest neighbors, since the wave-functions become negligible beyond these distances.

CHAPTER 8

ELECTRON DYNAMICS AND TRANSPORT PHENOMENA

"Science is built up of facts, as a house is built of stones; but an accumulation of facts is no more a science than a heap of stones is a house."

—Henri Poincaré

"Physics is mathematical not because we know so much about the physical world, but because we know so little; it is only its mathematical properties that we can discover."

—Bertrand Russell

8.1 INTRODUCTION

The motion of charge carriers in solids is governed by the restrictions imposed on them by the structure of the solid and the interactions between electrons and the ions in the solids as well as between the charge carriers themselves. The movement of charge carriers under the action of an applied electric or magnetic field will therefore depend on the form of the electronic band structure of the material in question. This applies to electrons in the conduction band and holes in the valence band. The concept of the hole was introduced in the previous chapter and arises from the fact that an empty state in the valence band is available for electrons to move into. So, for example, under the action of an applied electric field, an electron will move in the direction opposed to the direction of the electric field, **E**, due to our sign convention for the electronic charge being negative: $\mathbf{F} = q\mathbf{E} = -e\mathbf{E}$. Applying Newton's second law to this situation leads to

$$\mathbf{F} = m\frac{d\mathbf{v}}{dt} = -e\mathbf{E} \qquad (8.1)$$

Therefore we see that the direction of the velocity is opposite to that of the electric field due to the negative sign:

$$d\mathbf{v} = -\frac{e}{m}\mathbf{E}dt \qquad (8.2)$$

Since an electron which fills the empty state leaves behind it an empty state, the situation is effectively equivalent to the motion of the empty state, which we call the *hole*. The hole then acts like a quasi-particle of positive electric charge, since the absence of an electron corresponds to a charge, $+e$. Due to the Pauli exclusion principle applicable to fermions, the filled electron states will be those of lowest energy, meaning that the empty states will be, on average, at the top of the valence bands. In fact, the picture we have for holes has a correspondence with that for electrons. The (free) electrons in the conduction band occupy the lowest energy states available and are hence found at the bottom of the band. We can think of the situation as an analogy, in which electrons act like stone in water and sink to the bottom (of conduction band), while hole are like bubbles which float to the top (of the valence band).

It is interesting to note that the physical significance of the band gap between the valence and conduction bands in a solid corresponds to the effective binding energy of an electron to an ionic core, or its bonding state between atoms. The localized electronic states correspond to electrons in the valence band, which can be freed from the localized ionic states by the electron acquiring energy, say from a phonon or photon, which will overcome the binding energy and free the electron to move in the solid. We then say that the electron has "gained" energy and enters the conduction band. Of course the electron is not entirely free since it is still subject to the periodic potential of the solid. To become free from the solid it requires additional energy, which is called the *work function* of the material.

In this chapter, we will consider the motion of charge carriers in solids subject to the restrictions imposed by the solid under the influence of electric and magnetic fields. The application of these fields produce a response by the charge carrier which can be described by a force. The charge carriers will then follow specific trajectories due to the direction of these forces and the imposition of the band structure of the solid. This can give rise to some complex behavior, which forms the main subject of this chapter.

8.2　ELECTRON DYNAMICS IN CRYSTALS

As mentioned above, the motion of an electron in a solid will depend on the strength and direction of an applied external field as well as the restriction to motion which results from the internal structure of the solid. The regular disposition of ion cores in a solid crystal allows us to find the general rules for scattering due to interference or diffraction effects; this is the Bragg scattering we met in Chapter 3. We take this into account in the wave-like description we apply to the charge carrier and more specifically in the form of the wave or Bloch function we apply to it. We can see that the effect of a crystal on the motion of an electron is more complex than the free space situation by applying the quantum mechanical momentum operator $(\mathbf{p} = -i\hbar\nabla)$ to a general Bloch function:

$$\begin{aligned}\mathbf{p}\psi_k(\mathbf{r}) &= -i\hbar\nabla\psi_k(\mathbf{r}) = -i\hbar\nabla[e^{i\mathbf{k}\cdot\mathbf{r}}u_k(\mathbf{r})] \\ &= \hbar\mathbf{k}\psi_k(\mathbf{r}) + e^{i\mathbf{k}\cdot\mathbf{r}}\nabla u_k(\mathbf{r})\end{aligned} \qquad (8.3)$$

We see that the quantity $\hbar\mathbf{k}$ is not an eigenvalue of the operator and hence the crystal structure is imposing some further restriction on the motion of the charge carrier. To describe the motion of an electron in a solid we need to define its wave vector. However, in doing so, we must recall that a perfectly defined wave-vector implies a perfectly defined momentum and hence complete uncertainty in its position.

We are confronted with the same problem in free space, where such a defined momentum leads to a description of a plane wave which extends to infinity, i.e., there is no knowledge of its position. If we consider the localizaton of the electron, to say a position within space Δx, then we have a corresponding uncertainty in the momentum and hence wave-vector of $\Delta k \sim (\Delta x)^{-1}$. The localizaton of the electron can be expressed using the construction of the wave packet as a superposition of time-dependent Bloch functions of the form:

$$\Phi_{n\mathbf{k}_0}(\mathbf{r},t) = \int a_{n\mathbf{k}} \psi_{n\mathbf{k}}(\mathbf{r},t) d^3k \qquad (8.4)$$

where the time-dependent Bloch functions are written:

$$\psi_{n\mathbf{k}}(\mathbf{r},t) = e^{i\mathbf{k}\cdot\mathbf{r}} u_{n\mathbf{k}}(\mathbf{r}) e^{-i(E_{n\mathbf{k}}/\hbar)t} \qquad (8.5)$$

here, $E_{n\mathbf{k}}$ is the energy eigenvalue of the Bloch state. Due to the uncertainty in the wave vector, where we have: $\mathbf{k} = \mathbf{k}_0 + \Delta \mathbf{k}$, we can write the energy and spatial component of the Bloch functions as Taylor expansions about $\mathbf{k}_0$:

$$E_{n\mathbf{k}} = E_{n\mathbf{k}_0} + \Delta \mathbf{k} \cdot \nabla_{\mathbf{k}_0} E_{n\mathbf{k}_0} + \ldots \qquad (8.6)$$

and

$$u_{n\mathbf{k}}(\mathbf{r}) = u_{n\mathbf{k}_0}(\mathbf{r}) + \Delta \mathbf{k} \cdot \nabla_{\mathbf{k}_0} u_{n\mathbf{k}_0}(\mathbf{r}) + \ldots \qquad (8.7)$$

Substitution of these into Equation (8.4), where we limit the result to leading terms, yields:

$$\Phi_{n\mathbf{k}_0}(\mathbf{r},t) = e^{i\mathbf{k}_0\cdot\mathbf{r}} u_{n\mathbf{k}_0}(\mathbf{r}) e^{-i(E_{n\mathbf{k}_0}/\hbar)t} \int a_{n\Delta\mathbf{k}} e^{-i\Delta\mathbf{k}\cdot[\mathbf{r}-(\nabla_{\mathbf{k}_0}E_{n\mathbf{k}_0}/\hbar)t]} d^3(\Delta\mathbf{k}) \qquad (8.8)$$

This represents a Bloch function for an electron in state $\mathbf{k}_0$ which is modulated with an envelope function given by the integral over $\Delta \mathbf{k}$. The envelope function will have the same value for all $\mathbf{r}$ and t under the condition:

$$\mathbf{r} - \left(\frac{\nabla_{\mathbf{k}_0} E_{n\mathbf{k}_0}}{\hbar}\right) t = \text{constant} \qquad (8.9)$$

We can therefore regard the description of the electron as a wave packet travelling with a group velocity:

$$\mathbf{v}_g = \frac{1}{\hbar} \nabla_{\mathbf{k}_0} E_{n\mathbf{k}_0} = \nabla_{\mathbf{k}_0} \omega_{n\mathbf{k}_0} \tag{8.10}$$

This shows that the motion of the electron is dependent on the gradient of the energy with respect to the wave vector and that this in intimately related to the band structure. This is given by $E_{n\mathbf{k}_0}$, which is the energy environment in which the electron finds itself within the solid. For a free electron the energy relation is simply; $E_{\mathbf{k}} = \hbar^2 k^2 / 2m$, from which we obtain $\mathbf{v}_g = \hbar \mathbf{k}/m = \mathbf{p}/m$. We note that the phase velocity of the perfectly free electron is given by $\mathbf{v}_p = \omega/\mathbf{k} = \hbar \mathbf{k}/2m$ and is thus half the group velocity. The situation in a crystal, as we shall see, depends on where the electron finds itself in the energy band. So while for a free electron the direction of the group and phase velocities is always the same, this is not necessarily the case for a electron in an energy band of a solid, which can have constant energy surfaces which are non-spherical in k-space. This is illustrated in Figure 8.1.

We can see that the position, i.e., energy, of the electron in k-space will determine its group velocity. This will also be true for the case of holes in the valence band of the material. In Figure 8.2,

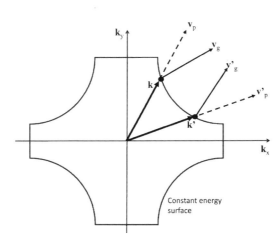

FIGURE 8.1: The difference between the group and phase velocities can be seen when a constant energy surface is warped by the band structure of the solid.

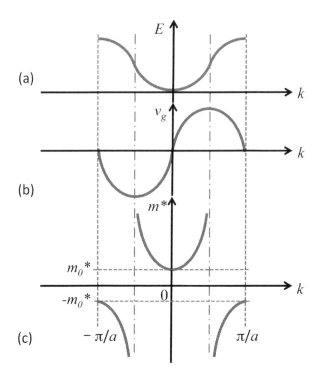

FIGURE 8.2: Schematic variation of (a) an energy band, (b) the group velocity, and (c) the effective mass of an electron in an energy band.

we show a schematic view of the variation of the group velocity of an electron in an energy band. We note that the position in k-space corresponds to a particular energy state of the charge carrier. If an electron is scattered inelastically, it will reappear at a different location in k-space. However, when the electron is subject to an electric field, it will progress through k-space, and real space, at a rate determined by the rate of change of energy. The effect of a magnetic field will be different due to the form of the force produced by a magnetic field, and the direction of motion (and its corresponding position in k-space) will change without any change in the energy. It thus turns out that the effect of a magnetic field will be to move the electron over a surface of constant energy in k-space. We will discuss this and other topics related to the effect of an applied magnetic field in Section 8.8.

8.3 THE EFFECTIVE MASS

If we consider the force on a charge carrier due to say an electric field, the charge carrier will change its momentum according to Equation (8.1). In fact, we can express this change by referring to the change of the wave vector for the charge carrier, where for an electron we can write:

$$\mathbf{F} = -e\mathbf{E} = \hbar \frac{d\mathbf{k}}{dt} \tag{8.11}$$

It is then a simple matter to show that we can write:

$$\mathbf{F} \cdot \mathbf{v}_g = \hbar \frac{d\mathbf{k}}{dt} \cdot \left[\frac{1}{\hbar} \nabla_\mathbf{k} E_{n\mathbf{k}} \right] = \frac{dE_{n\mathbf{k}}}{dt} \tag{8.12}$$

The acceleration of the electron due to the external electric field can be expressed as:

$$\mathbf{a} = \frac{d\mathbf{v}_g}{dt} = \frac{1}{\hbar} \frac{d}{dt} [\nabla_\mathbf{k} E_{n\mathbf{k}}] = \frac{1}{\hbar} \nabla_\mathbf{k} \left(\frac{dE_{n\mathbf{k}}}{dt} \right)$$

$$= \frac{1}{\hbar} \frac{d}{d\mathbf{k}} \hbar \frac{d\mathbf{k}}{dt} \left(\frac{1}{\hbar} \nabla_\mathbf{k} E_{n\mathbf{k}} \right) = \frac{1}{\hbar} \frac{d}{d\mathbf{k}} \hbar \frac{d\mathbf{k}}{dt} \left(\frac{1}{\hbar} \frac{dE_{n\mathbf{k}}}{d\mathbf{k}} \right) \tag{8.13}$$

Using Equation (8.11) we can now write this as:

$$\mathbf{a} = \frac{1}{\hbar^2} \frac{d}{d\mathbf{k}} \mathbf{F}(\nabla_\mathbf{k} E_{n\mathbf{k}}) = \frac{1}{\hbar^2} \mathbf{F} \left(\frac{d^2 E_{n\mathbf{k}}}{d\mathbf{k}_i d\mathbf{k}_j} \right) \tag{8.14}$$

Using the classical force relation, $\mathbf{F} = m\mathbf{a}$, we can now express the above equation in the following form of the so called *inverse effective mass tensor*:

$$\frac{1}{m^*_{ij}} = \frac{1}{\hbar^2} \left(\frac{d^2 E_{n\mathbf{k}}}{d\mathbf{k}_i d\mathbf{k}_j} \right) \tag{8.15}$$

The ij subscripts are important in the formation of the tensor and refer to the directions (x, y, z) in k-space. Since the effective mass tensor, m^*_{ij}, and its inverse, $(m^*_{ij})^{-1}$, are symmetric, they can be transformed into principal axes, giving only leading diagonal elements in the tensor. In the simplest case in which the three effective masses (i.e., in the three principal directions) are equivalent, we have:

$$\frac{1}{m^*} = \frac{1}{\hbar^2}\left(\frac{d^2 E_{n\mathbf{k}}}{d\mathbf{k}^2}\right) \quad (8.16)$$

Such a case would be for parabolic bands which can be approximated in the relation:

$$E_{n\mathbf{k}} = E_n \pm \frac{\hbar^2}{2m^*}(k_x^2 + k_y^2 + k_z^2) \quad (8.17)$$

In the case of the semiconductors, such as Si, the minimum of energy occurs along the (100) axes, at six equivalent points in reciprocal space in the Brillouin zone. The surfaces of constant energy will be six prolate ellipsoids centered at these locations, see Figure 8.3. The electron energy is a minimum when, $\mathbf{k} = \mathbf{k}_0$, and in the appropriate locations equivalent to $\mathbf{k}_0$. The three principal components of m_{ij}^* are usually positive in the vicinity of $\mathbf{k}_0$ and its counterparts in equivalent regions of k-space. Since the effective mass can

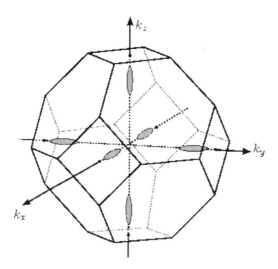

FIGURE 8.3: Constant energy surfaces for the conduction band in silicon.

be considered to be fairly energy independent in the region of $\mathbf{k}_0$, or E_{k_0}, such that we can write:

$$E_{1\mathbf{k}} = E_{1\mathbf{k}_0} \pm \frac{\hbar^2}{2}\left[\frac{(k_x - k_{0x})^2}{m_l^*} + \frac{(k_y - k_{0y})^2 + (k_z - k_{0z})^2}{m_t^*}\right] \quad (8.18)$$

In regions of the Brillouin zone for which the electron energy is a maximum, all three components of the effective mass are generally negative and we can write:

$$E_{2\mathbf{k}} = E_{2\mathbf{k}_0} \pm \frac{\hbar^2}{2}\left[\frac{(k_x - k_{0x})^2}{m_l'^*} + \frac{(k_y - k_{0y})^2 + (k_z - k_{0z})^2}{m_t'^*}\right] \quad (8.19)$$

An electron in such an energy state with negative mass is commonly treated as a particle of positive charge, i.e., the hole states we discussed earlier. We note that the prolate ellipticity of the energy bands can be characterized by the ratio of the long and short axes and will be equal to $\sqrt{m_l^*/m_t^*}$.

The use of the effective mass is a convenience tool, in which we can use the effective mass to treat more easily the movement of the electron (or hole) in the energy bands of the solid. We can use this concept with the description of the free electron gas in a metal. The form of Equation (8.15) shows that the effective mass depends not only on the energy of the charge carrier but also on the direction of its motion in the solid. Another way of looking at the effective mass of a charge carrier is to note that it depends on the curvature of the energy bands. The variation of the effective mass for the conduction band illustrated in Figure 8.2(c). We note that the variation is very distinctive, where at the center and edges of the band, the effective mass tends to a constant value, m_0^*, while at the inflections points of the bands the effective mass tends to $\pm\infty$, the points at which the group velocity is a maximum or minimum and the acceleration goes to zero. The plus or minus sign depends on the direction from which we take the gradient of the energy and we see that this is related to whether the curvature of the bands is positive or negative.

8.4 THE FERMI SURFACE

The filling of energy bands, at absolute zero of temperature, with available electrons continues up to the Fermi energy, E_F, as discussed in Chapter 5. In terms of the corresponding wave vectors, electrons with en energy $E_k = E_F$ will have a wave vector k_F. In other words the Fermi level marks the boundary between filled and unfilled states at zero temperature. For free electrons the value of the wave vector will be equal in all directions in reciprocal space and the surface of constant energy defined by electrons with wave vector, k_F, will form a sphere, see Figure 8.4(a). Such a surface is called the Fermi surface. Inside a three dimensional (real) solid the Fermi sphere can intersect with several bands. This will cause a deformation of the Fermi sphere and the resulting surface can take many forms depending of the number of electrons available and the shape of the band structure. This deformation is caused by the Bragg reflections that occur at the Brillouin zone boundaries. In Figure 8.4(b), we illustrate the deformation of the Fermi sphere to Fermi surface at the crossing of the Bragg plane at the edge of the Brillouin zone. The fact that the Brillouin zones are delimited by Bragg planes therefore has important consequences for the behavior of the electrons (and holes) as they move through the solid under the influence of electric and magnetic fields.

In the cases where the Fermi sphere crosses the Brillouin zone boundary, there will be different shapes of Fermi surface in the

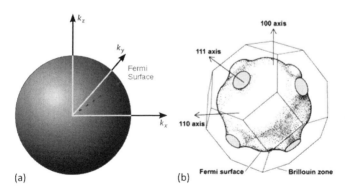

FIGURE 8.4: (a) Free electron Fermi sphere. (b) Modification of the Fermi sphere in Cu producing the Fermi surface in the first Brillouin zone.

different Brillouin zones. The degree of deformation of the Fermi sphere will depend on the strength of the periodic potential. In the case of the free electron model, i.e., for zero potential, there will be no deformation of the sphere at the Bragg planes and the Fermi surface will be the region of the Fermi sphere within the Brillouin zone of interest. A schematic illustration of the crossing of a Bragg plane by a Fermi sphere is illustrated in Figure 8.5. The region of deformation at the Bragg plane can be characterized by the parameter Δk, which from the nearly free electron model can be estimated as:

$$\Delta k \simeq \frac{2}{\hbar}\sqrt{m^* |V_G|} \qquad (8.20)$$

where $|V_G|$ indicates the magnitude of the periodic potential in the nearly free electron model, see Section 7.8. Clearly for the case of vanishing potential, $|V_G| \to 0$, then $\Delta k \to 0$ and there is no deformation in the Fermi surface. This doesn't mean that the Fermi surface in the first Brillouin is a sphere, since it will still be cut at the zone boundary, see gray line in Figure 8.5.

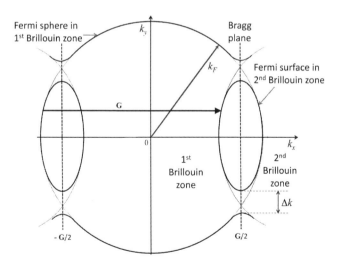

FIGURE 8.5: Fermi sphere deformation at a Bragg plane or zone boundary. For simplicity we only show one such boundary. The deformation of the Fermi surface can only occur for $k_F > G/2$. The Fermi surface for the second Brillouin zone is also show and is formed by its repetition into the first Brillouin zone (i.e., that portion of the surface to the left of the Bragg plane). Note that the first and second Brillouin zones are separated by the Bragg plane. The magnitude of the deformation is indicated by Δk.

We can show the form of the Fermi surface in the extended zone scheme, which is useful when considering the motion of the electrons in the solid. Such a representation is shown in Figure 8.6 for the case of the metal Cu, here given in the first Brillouin zone only, see Figure 8.4(b) for comparison. We note that the Fermi surface is continuous through the Brillouin zone boundary forming characteristic necks. We can also observe that there enclosed areas of the surface, which correspond to inverse or hole closed orbits. In the case of Cu, these orbits are referred to as *dog bone* structures, (bold line in Figure 8.6). Orbits can be the normal closed orbits and even open orbits, which do not return to the initial position. We will discuss the various types of orbital motion when we consider the effects of magnetic fields on the motion of electrons.

The form of the Fermi surface in higher order Brillouin zones can be generated in the same way as for the first zone, where only the portion of the Fermi sphere in that specific Brillouin zone is repeated (translated via reciprocal lattice vectors) to form the relevant surface. This process is illustrated in Figure 8.7, where we see the form of the Fermi surfaces for the first four Brillouin zones for a 2D square lattice. We note that in this case there is no Fermi surface in the first Brillouin zone since in this case we have $k_F > |\mathbf{G}|/2$. In effect, the periodic potential will alter the shape of the Fermi surfaces by rounding off the sharp edges of the extended zone represented free electron surfaces. We can further note that if $k_F < |\mathbf{G}|/2$,

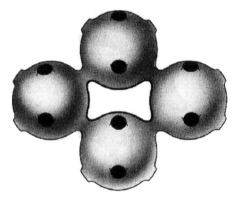

FIGURE 8.6: First Brillouin zone Fermi surface of Cu in the extended zone scheme. The open orbit *dog bone* structure can be seen in the center of the figure.

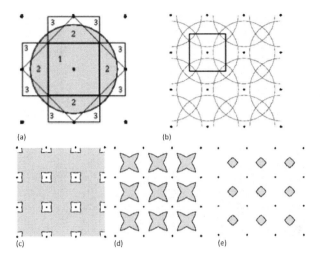

FIGURE 8.7: Formation of the Fermi surface in various Brillouin zones shown as section in two dimensions. (a) The (free electron) Fermi sphere (blue) extends beyond the first Brillouin zone and cuts into the second, third and fourth zones, where it has a surface. (b) Extended zone scheme in which we repeat the Fermi sphere at each reciprocal lattice point. (c) The Fermi surface in the second Brillouin zone. (d) The Fermi surface in the third Brillouin zone and (e) The Fermi surface in the fourth Brillouin zone.

then the Fermi surface will be spherical and only exist in the first Brillouin zone.

It is worth noting here that for the case of intrinsic semiconductors and insulators, the Fermi level at zero temperature sits in the middle of the band gap. As such there will be no Fermi surface for the solid since no electrons will be present at the Fermi energy and no electrons exist at $k = k_F$. While in these cases a Fermi surface does not exist, we can still represent constant energy surfaces in reciprocal space, as illustrated in Figure 8.3. In fact, these constant energy surfaces must be enclosed within the first Brillouin zone.

8.5 POSITIVE CHARGE CARRIERS: HOLES

We have previously noted that the absence of an electron in the valence band of a semiconductor can be thought of as a positive charge carrier. This is referred to as a *hole*. As we shall see later in this chapter, the Hall effect permits us to determine the nature of charge carriers and can distinguish between electrons (negative charge carriers) and holes. In most metallic solids, we expect the charge carriers to be negative, since the current is produced by the motion of electrons only. However, the current in semiconductors can be carried by both electrons and holes, and the Hall coefficient can in fact be positive for *doped semiconductors* (see Chapter 9). To fully appreciate the change of sign for the charge carrier we need to consider the behavior of electrons in energy bands which are almost full. (We note that both empty and full bands cannot contribute to electrical conductivity.) In Figure 8.8, we illustrate the formation of a hole state in the valence band with the excitation of an electron from here to the conduction band. The application of an electric field will result in the motion of both the hole in the valence band and the electron in the conduction band. In order to consider the motion of the two charge carriers we need to determine the corresponding k-states. The valence band, which we will consider to be initially full and containing N electrons, with equal

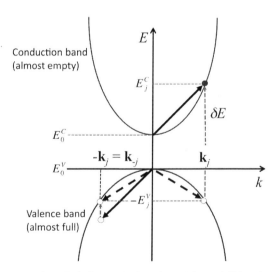

FIGURE 8.8: Creation of a hole state in the valence (almost full) band. An electron of state $+\mathbf{k}_j$ in the valence band is promoted (via the absorption of an appropriate quantity of energy) to the conduction band. Therefore the conduction band gains this value of k, while the valence band loses the same amount of k. The result is that the valence band hole-state has a k-value of $-\mathbf{k}_j$. We note that both energy and momentum (k) are conserved in this excitation.

numbers of positive and negative k-states; $\pm k$. Therefore, for our filled band we have:

$$\sum_{i=\pm 1}^{\pm N/2} \mathbf{k}_i = 0 \tag{8.21}$$

Therefore the filled band has a total k of zero. Clearly, the conduction band will also have zero k, since there are no charge carriers in its empty state. Once we excite an electron from the valence to the conduction band, the above situation will alter. Since we have removed an electron with say a positive value of k, let us say k_j, there will be more electrons with negative k and hence the valence band will now have a total k of:

$$\sum_{i=\pm 1}^{\pm N/2} \mathbf{k}_i = -\mathbf{k}_j = \mathbf{k}_{-j} \tag{8.22}$$

Equilibrium is maintained since the k of the conduction band is now $+\mathbf{k}_j$, i.e momentum is conserved. The specifics of the transition depends on the shape of the bands in question and the available energy states.

We now consider the energy associated with the hole state, $\mathbf{k}_h$, which as we saw above is related to the electron state, $\mathbf{k}_h = -\mathbf{k}_e$. We note that as $|\mathbf{k}_h|$ increases, the hole state moves to lower energies in the valence band, see Figure 8.8, and the energy $E_e(\mathbf{k}_e)$ of that state decreases. However, the total energy of the electron system that occupies the valence band will increase by the same amount, since while the hole state falls from its initial state of higher to lower energy, an occupied state makes the reverse transition. This leads to a definition of the hole energy, $E_h(\mathbf{k}_h)$, to be the negative of $E_e(\mathbf{k}_e)$:

$$E_h(\mathbf{k}_h) = -E_e(\mathbf{k}_e) \tag{8.23}$$

Now given that $\mathbf{k}_h = -\mathbf{k}_e$, we can write:

$$E_h(\mathbf{k}_h) = -E_e(-\mathbf{k}_h) = -E_e(\mathbf{k}_h) \tag{8.24}$$

since for any state $-\mathbf{k}_h$, there exists a state of the same energy with wave vector $\mathbf{k}_h$. From this equation we expect E_e to be a decreasing function of its argument. For the simple case of spherical parabolic bands we can express the energy variations in the conduction and valence bands as:

$$E_e(\mathbf{k}_e) = E_0^C + \frac{\hbar^2 k_e^2}{2m_C^*} = E_C(\mathbf{k}_e) \tag{8.25}$$

$$E_e(\mathbf{k}_e) = E_0^V + \frac{\hbar^2 k_e^2}{2m_e^*} \tag{8.26}$$

$$E_h(\mathbf{k}_h) = -E_0^V - \frac{\hbar^2 k_h^2}{2m_e^*} = -E_0^V + \frac{\hbar^2 k_h^2}{2m_h^*} = E_V(\mathbf{k}_h) \tag{8.27}$$

where E_0^C marks the bottom of the conduction band and E_0^V the top of the valence band. It is clear from the above that $m_h^* = -m_e^*$. (N.B. in Figure 8.8, we have illustrated the situation with $E_0^V = 0$.)

We can now write the group velocity for the hole from our definition given in Section 8.2, which we express as:

$$\mathbf{v}_{gh} = \frac{1}{\hbar} \nabla_{\mathbf{k}_h} E_h(\mathbf{k}_h) \tag{8.28}$$

The corresponding expression for electrons is

$$\begin{aligned}\mathbf{v}_{ge} &= \frac{1}{\hbar} \nabla_{\mathbf{k}_e} E_e(\mathbf{k}_e) \\ &= -\frac{1}{\hbar} \nabla_{\mathbf{k}_h} E_e(-\mathbf{k}_h) = -\frac{1}{\hbar} \nabla_{\mathbf{k}_h} E_e(\mathbf{k}_h) \\ &= \frac{1}{\hbar} \nabla_{\mathbf{k}_h} E_h(\mathbf{k}_h) = \mathbf{v}_{gh}\end{aligned} \tag{8.29}$$

where we have used $E_h(\mathbf{k}_h) = -E_e(\mathbf{k}_h)$.

The application of an electric field, ε, will cause a displacement of the charge carrier in the bands, providing there are available states for them to move into. We can evaluate the corresponding current associated with this movement of charge, where we have $\mathbf{J}_{e(h)} = n_{C(V)} e_{C(V)} \mathbf{v}_{ge(h)}$ for electrons (e) and holes (h), respectively.

This can be expressed in terms of the effective mass of the charge carriers in the form:

$$\mathbf{J}_{e(h)} = n_{C(V)} e_{C(V)} \frac{\hbar \sum_{i=\pm 1}^{\pm N/2} \mathbf{k}_i}{m^*_{e(h)}} = n_{C(V)} e_{C(V)} \frac{\hbar \mathbf{k}_{j(-j)}}{m^*_{e(h)}} \tag{8.30}$$

We can also indicate that the equation of motion due to the electric field, for electrons in the conduction band, can be expressed as:

$$\hbar \frac{d\mathbf{k}_e}{dt} = e_C \mathcal{E} = -e\mathcal{E} \tag{8.31}$$

while for holes in the valence band we have:

$$\hbar \frac{d\mathbf{k}_h}{dt} = e_V \mathcal{E} = e\mathcal{E} \tag{8.32}$$

This also shows that the charge for the hole is positive; $e_V = +e$. It is worth noting that the Fermi surface of a metal contains both electrons and holes, which arises due to the fact that this is the interface between the two, due to the definition of the Fermi level, i.e., both electron and hole states at E_F.

8.6 DRIFT AND DIFFUSION OF CHARGE CARRIERS

The motion of electrons in a solid under the action of an electric field is referred to as *drift*. This subject was already introduced in Chapter 6, see Section 6.4. As we saw, this motion of charge results in an electrical current.

Another important source of charge motion is caused when there is a spatial gradient of charge carriers in the solid. Such concentration gradients will persist until the concentration is uniform in the solid, where the tendency is for the charge carriers to move from regions of high concentration to low concentration. This process is referred to as *diffusion*. It is a relatively simple matter to illustrate that the overall flux of charge carriers moves from regions of high to low concentrations. To do this we can consider the one dimensional case where the charge concentration, $n(x)$ varies with position, as shown in Figure 8.9. We consider the sample to be at uniform temperature such that the thermal energy of the charge carriers does not alter with position and only $n(x)$ will vary.[1] We now consider the number of charge carriers (electrons say) that pass a plane at position $x = 0$, per unit area per unit time. Due to finite temperature, the electron's motion is random, with thermal velocity, v_{th}, and the mean free path, $\lambda = v_{th}\tau$, so electrons on either side of the plane at $x = 0$ within this distance can pass across the plane with a probability of $1/2$, within the relaxation time τ. The average rate of flow of electrons per unit area, F_+, crossing the plane at $x = 0$ from the left can be expressed as:

$$F_+ = \frac{1}{2}\frac{n(-\lambda)\lambda}{\tau} = \frac{1}{2}n(-\lambda)v_{th} \qquad (8.33)$$

In the same way we can express the average rate of flow of electrons per unit area, F_-, crossing the plane at $x = 0$ from the right as:

$$F_- = \frac{1}{2}n(\lambda)v_{th} \qquad (8.34)$$

Therefore the overall flux can be written as:

$$F_x = F_+ - F_- = \frac{1}{2}[n(-\lambda) - n(\lambda)]v_{th} \qquad (8.35)$$

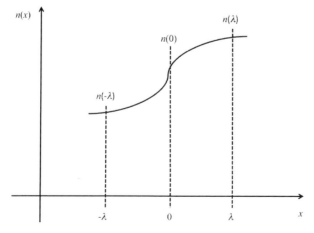

FIGURE 8.9: Variation of concentration of charge carriers as a function of position in the solid.

We now make Taylor expansions (up to the first two terms only) of the two concentrations around the zero plane, which allows us to write:

$$F_x = \frac{1}{2} v_{th} \left\{ \left[n(0) - \lambda \frac{dn}{dx} \right] - \left[n(0) + \lambda \frac{dn}{dx} \right] \right\}$$
$$= -v_{th} \lambda \frac{dn}{dx} = -D_n \frac{dn}{dx} \qquad (8.36)$$

where we have used $D_n \equiv v_{th} \lambda$, which we define as the *diffusivity*. We note that Equation (8.36) has the form of *Fick's first law* in one dimension. The three dimensional expression has the form:

$$F = -D_n \nabla n \qquad (8.37)$$

Given that electrons have a charge of $-e$, the flux of these charge carriers will give rise to a current, such that we can write the current density as:

$$J_{nx} = -eF_x = eD_n \frac{dn}{dx} \qquad (8.38)$$

$$\mathbf{J}_n = -e\mathbf{F} = eD_n \nabla n \qquad (8.39)$$

This is called the *diffusion current*. Above we have shown both the one and three dimensional forms of the current density. Since the diffusion current depends on the random thermal motion of the charge carriers, we can use equations of the form (6.7) and (6.17) for the thermal velocity of the electrons and the corresponding mobility to obtain:

$$\mathbf{J}_n = -e\mathbf{F} = eD_n \nabla n = e \left(\frac{k_B T}{e} \mu_n \right) \nabla n \qquad (8.40)$$

This means that we can relate the diffusivity and mobility as:

$$D_n = \left(\frac{k_B T}{e} \right) \mu_n \qquad (8.41)$$

This is know as *Einstein's relation* and relates two important constants that characterize the transport of charge carriers by diffusion and drift mechanisms. In the presence of an electric field and a concentration gradient, the current will be composed of two parts (diffusion and drift), giving a total current density which we can express as:

$$\mathbf{J}_n^{TOT} = ne\mu_n \mathbf{E} + eD_n \nabla n \qquad (8.42)$$

here n denotes the electron concentration. Using the Einstein relation this equation takes the form:

$$\mathbf{J}_n^{TOT} = \mu_n(ne\mathbf{E} + k_B T \nabla n) \qquad (8.43)$$

In the above discussion, we have only considered the motion of electrons. The displacement of holes will have an associated current density which can be written as:

$$\mathbf{J}_p^{TOT} = pe\mu_p \mathbf{E} - eD_p \nabla p \qquad (8.44)$$

Using the Einstein relation this equation takes the form:

$$\mathbf{J}_p^{TOT} = \mu_p(pe\mathbf{E} - k_B T \nabla p) \qquad (8.45)$$

here p refers to the (positive) hole charge concentration. Such considerations are important in the motion of charge carriers in semiconducting materials in which both types of charge carrier can be present simultaneously. We will discuss these materials in more detail in the following chapter. In this case, the total current density must account for the movement of all charges which contribute to the current, such that we must use:

$$\mathbf{J}^{TOT} = \mathbf{J}_n^{TOT} + \mathbf{J}_p^{TOT} \qquad (8.46)$$

This equation is very important when we consider the transport properties in semiconductor devices. We will develop this further in Chapter 9 when we consider other processes which contribute to the variation of the total charge concentration in semiconducting materials.

8.7 ELECTRON SCATTERING IN BANDS

When an electron undergoes an inelastic scattering process, it will abruptly vanish from one point in reciprocal space and reappear at another position. The change in its wave vector will depend on that specific scattering process and the change of its momentum, which also includes its direction. It is in fact these collision processes in the solid which give rise to the electrical resistance of materials. The absence of these effects would be characterized by the flow of electrical charges without resistance and lead to the phenomenon of *superconductivity*. In such a situation, a current would continue to flow indefinitely, in accord with the equations of motion considered above. Superconductivity has been observed in a number of materials at low temperatures. This occurs under specific conditions in certain materials and will be discussed in Chapter 11. In normal solids, however, electrical resistance is observed.

In Chapter 6, we introduced the Drude model for the conduction of electrons in metals. We noted that despite the relative success of the model there are a number of shortcomings, which essentially lead to the abandonment of this model for a more sophisticated approach. Of particular importance is the scattering of electrons in solids, which as we stated above produces the electrical resistance of materials. In the Drude model, it is assumed that the scattering is produced from the positive ion cores of the crystal lattice, implying a mean free path of a few Å, which falls well short of the value found in metals at room temperature. Actually, the periodic potential that is produced by the lattice does not scatter the electrons inelastically. This is apparent since the Bloch waves traversing the solid are stationary states of the Schrödinger equation and $|\psi|^2$ is time independent. In terms of the one-electron approximation that was used for the most part of Chapter 7, perturbations of the Bloch stationary states can occur via the electron scattering due to the deviation of the perfect periodic lattice. Such deviations are generally characterized as being time-independent in the form of fixed defects (see Chapter 4) or time-dependent variations of the lattice structure in the form of lattice vibrations or phonons (see Chapter 5). The one-electron approximation neglects the electron - electron interaction which is englobed within the non-interacting Fermi gas concept.

In fact, such effects can also cause a perturbation of the Bloch states, though in general these are less important than the effects of deviations of the perfect periodic potential.

We note that elastic scattering, as produced for example via diffractive effects in solids only results in a change of wave vector direction and not magnitude. In quantum mechanics, the scattering process is characterized by the scattering probability, $W_{k,k'}$, which defines the initial and final Bloch vectors, k and k', respectively, along with the Hamiltonian for the perturbation, $\mathcal{H}'$, and can be expressed in a general form as:

$$W_{k,k'} = \left| \int \psi_{k'}^*(\mathbf{r}) \mathcal{H}' \psi_k(\mathbf{r}) d\mathbf{r} \right|^2 \tag{8.47}$$

If the Hamiltonian is constant in time, then we expect an elastic scattering process of the Bloch waves with the conservation of energy. If, however, the Hamiltonian is a time varying potential, such as for a lattice excitation (phonon), the scattering will be inelastic. In this case the energy conservation condition can be expressed as:

$$E_{k'} - E_k = \hbar \omega(\mathbf{q}) \tag{8.48}$$

Such scattering from a phonon, with wave vector $\mathbf{q}$, gives rise to a scattering matrix element of the form:

$$\int u_{k'}^*(\mathbf{r}) u_k(\mathbf{r}) e^{i(\mathbf{k} - \mathbf{k}' + \mathbf{q})} d\mathbf{r} \tag{8.49}$$

Since the functions $u_{k'}^*(\mathbf{r})$ and $u_k(\mathbf{r})$ have the periodicity of the lattice and can be expanded in a Fourier series in terms of the reciprocal lattice vectors, the matrix elements will only be non-zero for:

$$\mathbf{k}' - \mathbf{k} = \mathbf{q} + \mathbf{G} \tag{8.50}$$

This has a form which resembles that of the von Laue relation for diffraction. Actually, it only differs in the fact that some momentum is transferred from the electron to some form of excitation and by an amount corresponding to a reciprocal lattice vector, since it is only in these discrete quantities that conservation of momentum (and energy) is preserved. It should be clear that while the von Laue relation corresponds to an elastic scattering process, Equation (8.50) refers to an inelastic event. The use of the conservation laws is a convenient way for expressing the scattering of Bloch-state electrons.

The conservation of energy and momentum can also serve as a guiding principle in the study of electron-electron scattering. We can write the collision between two electrons in the form: $E_1 + E_2 = E_3 + E_4$, where each term, $E_i = E(\mathbf{k}_i)$ denotes a one-particle energy in the non-interacting Fermi gas. The corresponding conservation of momentum can be expressed as:

$$\mathbf{k}_1 + \mathbf{k}_2 = \mathbf{k}_3 + \mathbf{k}_4 + \mathbf{G} \tag{8.51}$$

While one might expect a high scattering probability due to the electron density in a metal since the Coulomb repulsion can be expected to be reasonably strong, in fact this will be prohibited to a large extent due to the Pauli exclusion principle. This means that, to a good approximation, the Pauli exclusion principle allows us to treat the electrons in a solid as non-interacting. It is therefore that most treatments of electrical conduction in solids only considers scattering processes from defects and phonons.

8.8 MAGNETIC FIELD EFFECTS

We will now consider the effect of a magnetic field on charge carriers in a solid. We can express this by using the force equation, $\mathbf{F} = \hbar d\mathbf{k}/dt$, and the Lorentz force, such that we can write:

$$\hbar \frac{d\mathbf{k}}{dt} = e_{C(V)}[\mathbf{v}_{ge(h)} \times \mathbf{B}] \tag{8.52}$$

Since the vector product has a direction perpendicular to the two vectors, we note that $d\mathbf{k}/dt$ will be perpendicular to the group velocity, $\mathbf{v}_{ge(h)}$, and hence to $\nabla_\mathbf{k} E_{n\mathbf{k}}$. We can denote this condition of orthogonality in the form:

$$\frac{d\mathbf{k}}{dt} \cdot \nabla_\mathbf{k} E_{n\mathbf{k}} = 0 \tag{8.53}$$

This means that the charge carriers will move in orbits of constant energy, which can be seen more explicitly if we expand the above condition:

$$\frac{d\mathbf{k}}{dt} \cdot \nabla_\mathbf{k} E_{n\mathbf{k}} = \frac{dk_x}{dt}\frac{dE_{n\mathbf{k}}}{dk_x} + \frac{dk_y}{dt}\frac{dE_{n\mathbf{k}}}{dk_y} + \frac{dk_z}{dt}\frac{dE_{n\mathbf{k}}}{dk_z} = \frac{dE_{n\mathbf{k}}}{dt} = 0 \quad (8.54)$$

This has important implications for the motion of electrons and holes in the presence of a magnetic field. This fact is frequently exploited to study Fermi surfaces in metals and other transport properties of solids. Some of these topics will be discussed in the following sections.

8.8.1 The Hall Effect

The Hall effect was discovered in 1879 by Edwin Hall (1855 - 1938), and can be described as the appearance of a transverse electrical potential difference across a conductor (or semiconductor), when an applied magnetic field is present in a direction perpendicular to an electrical current flowing in the material. The Hall effect is a direct consequence of the Lorentz force on the charge carriers in the solid and can be exploited to characterize the electronic properties of materials. A schematic illustration of the measurement of the Hall effect is shown in Figure 8.10.

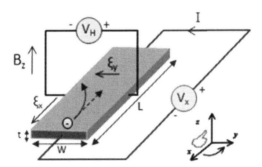

FIGURE 8.10: Schematic illustration of the Hall effect measurement, where a sample has a current I (current density, J) flowing in a direction perpendicular to the orientation of an applied magnetic field. The Hall effect is the appearance of the potential difference at the lateral edges of the sample, demoted by V_H.

The equation of motion for a charge carrier, with drift velocity, v, in the presence of a magnetic field, $\mathbf{B}$, can be expressed as:

$$m^*\left(\frac{d\mathbf{v}}{dt}+\frac{\mathbf{v}}{\tau}\right)=e_c(\mathbf{E}+\mathbf{v}\times\mathbf{B}) \tag{8.55}$$

From the figure above, we set the magnetic field along the z-direction with the electric field and hence drift velocity in the x-direction. Since the Lorentz force produces a lateral force on the charge carriers, these will be deflected from their original path in the x-direction and the charge accumulation, i.e., the Hall effect, will give rise to a component of the electric field in this direction. As such we can write: $\mathbf{B}=(0,0,B)$ and $\mathbf{E}=(E_x,E_y,0)$. Equation (8.55) will now take the form:

$$m^*\left(\frac{dv_x}{dt}+\frac{v_x}{\tau}\right)=e_c(E_x+v_yB) \tag{8.56}$$

$$m^*\left(\frac{dv_y}{dt}+\frac{v_y}{\tau}\right)=e_c(E_y-v_xB) \tag{8.57}$$

which is expressed in component form. The deflection of charge carriers will continue until the accumulation at the sample edge is such that an equilibrium is established between the electric and magnetic forces on the charge carriers and the system will be in steady-state. This means that the derivatives: $dv_x/dt=dv_y/dt=0$. It is a simple matter to establish:

$$v_y=\frac{e_c\tau}{m^*}\frac{(E_y-e_cB\tau E_x/m^*)}{[1+(e_cB\tau E_x/m^*)^2]}\simeq\frac{e_c\tau}{m^*}\left(E_y-\frac{e_cB\tau E_x}{m^*}\right) \tag{8.58}$$

where the approximation regards the case for low magnetic fields. Such an equation is valid for each type of charge carrier (i.e., for electrons and holes). Given that the net transversal current must be zero, we can write:

$$\sum_i n_iv_{yi}e_{ci}=0 \tag{8.59}$$

where i refers to either electrons or holes. Using Equations (8.58) and (8.59) we can write the following relation:

$$\frac{n_c e_c^2 \tau_c}{m_c^*}\left(E_y - \frac{e_c B \tau_c}{m_c^*} E_x\right) + \frac{p e_h^2 \tau_h}{m_h^*}\left(E_y - \frac{e_h B \tau_h}{m_h^*} E_x\right) = 0 \quad (8.60)$$

where the c subscript refers to electrons in the conduction band and h to holes in the valence band. Using the expression for charge carrier mobility; $\mu_i = |e_{ci} \tau_c / m_c^*|$, the above equation takes the form:

$$E_y(n_c \mu_c + p \mu_h) = E_x(p \mu_h^2 - n_c \mu_c^2) B \quad (8.61)$$

Writing $e_c = -e$ and $e_h = e$, we can write:

$$E_x = \frac{j_x}{\sigma} = \frac{j_x}{|e|(n_c \mu_c + p \mu_h)} \quad (8.62)$$

Finally we equate this with (8.61), which gives:

$$R_H = \frac{E_y}{j_x B} = \frac{(p \mu_h^2 - n_c \mu_c^2)}{|e|(n_c \mu_c + p \mu_h)^2} \quad (8.63)$$

The quantity R_H is called the Hall coefficient and is used to characterize the Hall effect in a sample. It will be readily seen that for metal, the Hall coefficient will be negative, with $R_H = -1/ne$. One of the principle applications of the Hall effect is to measure the the type and concentration of charge carriers in semiconductors. The other very important application of this effect is in the measurement of magnetic fields, since a known sample can be easily calibrated, providing a relatively accurate and simple experimental tool. In fact, the Hall probe is almost universally found in research laboratories where electromagnets are employed. The more recent discoveries of the *quantum* or *quantized Hall effect* (QHE) and the *fractional quantum Hall effect* (FQHE), will be discussed at the end of this section.

8.8.2 Cyclotron Resonance

As a first approach to the consideration of the effect of a magnetic field on an electron or charge carrier, we can look at the Lorentz force due to the magnetic field applied to a free charge carrier: $\mathbf{F} = |e_c|(\mathbf{v} \times \mathbf{B})$. From classical physics, we know that the magnetic field has the effect of deviating the charged particle from its rectilinear trajectory, when no field is applied. For a uniform magnetic field this then results in a circular or orbital motion of the charge carrier, as in the case of the classical Thomson experiment (1897) to measure the charge-to-mass ratio (e/m). The force will be a maximum when the velocity has a direction perpendicular to the applied magnetic field; $F_{max} = |e_c|vB$. Equating with Newton's second law for the centripetal force $(F = mv^2/r)$, we can obtain the radius or the circular orbit as:

$$r = \frac{mv}{|e_c|B} \tag{8.64}$$

where m is the mass of the charged particle. The period for one orbit can be obtained from the velocity as: $T = 2\pi r/v$. from which we obtain the angular velocity of the motion:

$$\omega_c = \frac{2\pi}{T} = \frac{v}{r} = \frac{|e_c|B}{m} \tag{8.65}$$

For charge carriers in solids this is called the *cyclotron frequency*. Often the mass will be replaced by the effective mass, $m*$. In fact, the cyclotron resonance is used to measure the effective mass of electrons and holes in semiconductors. In terms of experimental techniques, the measurement of cyclotron resonance is performed in a microwave spectrometer in which the sample is subject to an applied static magnetic field as well as a the electromagnetic field from a microwave source. As the frequency of the radiation is fixed, the magnetic field is varied over a certain range. As the condition for resonance is approached (defined by both the frequency and field conditions of the experiment), the carriers will absorb energy from the microwaves. The experiment consists of the measurement of this absorption as a function of the applied magnetic field. We can commence analysis by applying Equation (8.55) to our situation, where we have: $\mathbf{B} = (0,0,B)$, $\mathbf{E} = (E_x^0, 0, 0)e^{-i\omega t}$ and $\mathbf{v} = (v_x^0, v_y^0, 0)e^{-i\omega t}$. The sample is situated such that it is in a region where the electric field component of the microwave radiation is a maximum while that for the magnetic field component (of the microwave radiation) is a minimum, and negligible. We now write the equation of motion in component form as:

$$m^*\left(-i\omega + \frac{1}{\tau}\right)v_x = |e_c|(E_x + v_y B) \quad (8.66)$$

$$m^*\left(-i\omega + \frac{1}{\tau}\right)v_y = -|e_c|v_x B \quad (8.67)$$

Eliminating v_y from the above and using: $j_x = n_c|e_c|v_x$, where n_c is the carrier concentration, we obtain:

$$\sigma(\omega) = \frac{j_x}{E_x} = \frac{n_c|e_c|v_x}{E_x} = \frac{\sigma_0(1+i\omega\tau)}{1+(\omega_c^2-\omega^2)\tau^2 + 2i\omega\tau} \quad (8.68)$$

This is the complex conductivity of the material, where σ_0 is the static value. In the cyclotron resonance experiment, the absorbed power is proportional to the real part of the complex conductivity;

$$\text{Re}[\sigma(\omega)] = \frac{\sigma_0[1+(\omega_c^2+\omega^2)\tau^2]}{[1+(\omega_c^2-\omega^2)\tau^2]^2 + 4\omega^2\tau^2} \quad (8.69)$$

Experimentally, the absorption is measured as a function of the applied magnetic field strength, where peaks in the absorption indicate the presence of the charge carriers with specific resonance conditions depending on their effective mass, as given by the cyclotron frequency: $\omega_c = |e_c|B/m^*$. Therefore electrons in different energy bands as well as heavy and light holes will appear as distinct absorption peaks in the absorption spectrum. The width of the resonance lines being determined by the relaxation time, τ, which is related to the collision rate of the charge carriers with defects in the solid. To be able to observe the cyclotron resonance of a charge carrier, it is necessary that the mean free path be sufficiently long to be able to move in its orbit between collisions. This condition can be expressed as: $\omega_c\tau > 1$. This condition is most easily satisfied by performing measurements at low temperature, though high frequencies and high fields will also assist in reaching this condition. Early measurements of cyclotron resonance in semiconductors were performed by

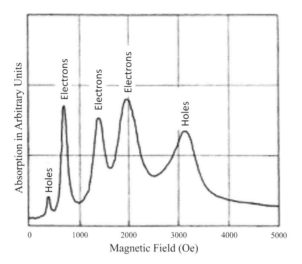

FIGURE 8.11: Cyclotron resonance spectrum for germanium, for a frequency of 24 GHz at 4K. Spectrum was taken with the magnetic field applied in the (110) plane at 60 from the [100] axis. (Reprinted figure with permission from: Dresselhaus et al. Phys. Rev.**98**, 368 (1955). Copyright 1955 by the American Physical Society.)

Dresselhaus and co-workers[2]. In Figure 8.11, we show the absorption spectrum for Ge, where the various resonance peaks are indicated as electrons or holes.

By taking measurements of the cyclotron resonance as a function of the orientation of the magnetic field with respect to the crystalline axes, it is possible to observe the effect of the electronic band structure of solid, since the effective mass of the charge carriers is intimately related to inverse second derivative of the energy (see Section 8.3). Allowing the magnetic field to vary, say in the $x - z$ plane: $\mathbf{B} = B(\sin\theta, 0, \cos\theta)$, we can now write:

$$i\omega v_x - \omega_t v_y \cos\theta = 0 \qquad (8.70)$$

$$\omega_t v_x \cos\theta + i\omega v_y - \omega_t v_z \sin\theta = 0 \qquad (8.71)$$

$$\omega_l v_y \sin\theta + i\omega v_z = 0 \qquad (8.72)$$

where $\omega_t = eB/m_t^*$ and $\omega_l = eB/m_l^*$, are the transversal and longitudinal cyclotron frequencies, due to the transversal and longitudinal effective masses (see Equation 8.18). Solving this set of simultaneous equations, we obtain:

$$\omega^2 = \omega_c^2 = \omega_t^2 \cos^2\theta + \omega_t \omega_l \sin^2\theta \tag{8.73}$$

Here ω_c is the cyclotron frequency; $\omega_c = eB/m_c^*$, where the cyclotron effective mass, m_c^*, takes the form:

$$\left(\frac{1}{m_c^*}\right)^2 = \frac{\cos^2\theta}{m_t^{*2}} + \frac{\sin^2\theta}{m_t^* m_l^*} \tag{8.74}$$

The comparison of theory and measurement, see Figure 8.12, allows the determination of the effective masses for the charge carriers in the sample. The values for Ge and Si are given in Table 8.1.

8.8.3 Magnetoresistance

The phenomenon of the change of electrical resistance in the presence of a magnetic field is termed *magnetoresistance*, and is a general

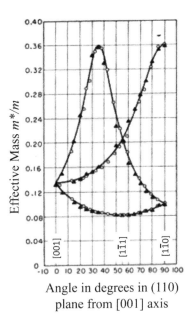

Angle in degrees in (110) plane from [001] axis

FIGURE 8.12: Effective mass of electrons in Ge at 4K for the magnetic field in the (110) plane. Theoretical curves were obtained using Equation (8.74). Note that the different symbols are for different experimental runs. (Reprinted figure with permission from: Dresselhaus et al. Phys. Rev. **98**, 368 (1955). Copyright 1955 by the American Physical Society.)

TABLE 8.1: Effective masses for electrons and holes in Ge and Si as measured by cyclotron resonance. All units are given in units of the free electron mass. Here subscripts refer to the following: *l*- longitudinal, *t*- transversal, *lh*- light hole and *hh*- heavy hole. (Data taken from Dresselhaus et al. Phys. Rev. **98**, 368 (1955).)

	m_l^*	m_t^*	m_{lh}^*	m_{hh}^*
Ge	1.58	0.082	0.043	0.34
Si	0.97	0.19	0.16	0.52

observation in a majority of metals. We again start with Equation (8.55) and taking the magnetic field to be along the z-direction we can write, for the steady-state condition in the sample; i.e., where $dv_{x,y}/dt = 0$:

$$v_x = \frac{|e_c|\tau}{m^*} E_x + \frac{|e_c| B\tau}{m^*} v_y \tag{8.75}$$

$$v_y = \frac{|e_c|\tau}{m^*} E_y - \frac{|e_c| B\tau}{m^*} v_x \tag{8.76}$$

$$v_z = \frac{|e_c|\tau}{m^*} E_z \tag{8.77}$$

which we can express, using the cyclotron frequency, ($\omega_c = |e_c| B/m^*$, and the mobility, $\mu = |e_c|\tau/m^*$, as:

$$v_x = \mu E_x + \omega_c \tau v_y \tag{8.78}$$

$$v_y = \mu E_y - \omega_c \tau v_x \tag{8.79}$$

$$v_z = \mu E_z \tag{8.80}$$

We can express these results in condensed form using the conductivity tensor, $\tilde{\sigma}$, where $\mathbf{J} = |e_c| n_c \mathbf{v} = \tilde{\sigma}\mathbf{E}$, and we obtain:

$$\tilde{\sigma} = \begin{pmatrix} \sigma_{xx} & \sigma_{xy} & 0 \\ -\sigma_{xy} & \sigma_{xx} & 0 \\ 0 & 0 & \sigma_0 \end{pmatrix} \tag{8.81}$$

where

$$\sigma_{xx} = \frac{\sigma_0}{1+\omega_c^2\tau^2} \quad (8.82)$$

$$\sigma_{xy} = \frac{\sigma_0\omega_c\tau}{1+\omega_c^2\tau^2} \quad (8.83)$$

here $\sigma_0 = n_c e_c^2 \tau / m^*$ and $\omega_c = |e_c| B / m^*$, are the zero-field static conductivity and the cyclotron frequency, respectively. The magnetoresistance phenomenon is common in metals and generally arises from a combination of electrons in anisotropic bands with different values of effective mass and different relaxation times. As such, these electrons will move in different orbits under the action of a magnetic field. This means that the total current from the different electron contributions will reduce with an increase of magnetic field and introduce the increase of electrical resistance that is observed experimentally. Data are usually presented in terms of resistivity rather than conductivity, and we can use the inverse tensor of the conductivity to obtain the resistivity tensor, which will have components of the form:

$$\rho_{xx} = \frac{\sigma_{xx}}{\sigma_{xx}^2 + \sigma_{xy}^2} \quad (8.84)$$

$$\rho_{xy} = -\frac{\sigma_{xy}}{\sigma_{xx}^2 + \sigma_{xy}^2} \quad (8.85)$$

$$\rho_{zz} = \frac{1}{\sigma_{zz}} = \frac{1}{\sigma_0} \quad (8.86)$$

The magnetoresistance is expressed as:

$$\frac{\Delta\rho}{\rho} = \frac{[\rho(B) - \rho(0)]}{\rho(0)} \quad (8.87)$$

One particularly important consideration in the magnetoresistance is the type of orbit the charge carriers will follow under the effect of an applied magnetic field. Since the charge carriers in question will be at the Fermi surface, the type of orbit can be open or closed, depending on the direction in which the magnetic field is applied with respect to the crystalline lattice. From the form of the conductivity tensor elements, as the magnetic field $B \to \infty$, $\omega_c \tau \to \infty$, and since τ is constant, this means that the electron will make many orbits on the Fermi surface before scattering. Thus the average velocity of the electron, in the plane perpendicular to the B field, will tend to zero, and therefore the leading conductivity elements will vary as B^{-2} at high magnetic fields. This results in a B^2 - dependence of the transverse magnetoresistance and saturates at very high fields. The open type orbit, which the electron does not return to the same position on the Fermi surface, the electron velocity will not average to zero, and the magnetoresistance continues to vary as B^2, even as $B \to \infty$. Magnetoresistive effects will be stronger in ferromagnetic metals due to the imbalance in spin-up and spin-down electron concentrations. Furthermore, in magnetically structured materials a giant magnetoresistive effect can be observed. Such topics will be discussed in Chapter 10.

8.8.4 Magnetic Sub-Bands and Oscillatory Phenomena in Solids

In the free electron model, we obtain the energy relation: $E(\mathbf{k}) = \hbar^2 k^2 / 2m$. This can be easily derived from the Schrödinger equation, (see Section 6.6). However, if a magnetic field is applied in the z-direction, the electron motion in the $x - y$ plane will be affected, but not that in the z-direction. The resultant motion will be a combination of the linear motion in the z-direction plus the cyclotron motion in the $x - y$ plane; i.e., the electron will describe a helical path.

The Schrödinger equation will be modified to take the form:

$$\left\{ \frac{\hbar^2}{2m} \nabla^2 + \frac{m\omega_c^2}{8}(x^2 + y^2) - i\hbar\omega_c \left[x \frac{\partial}{\partial y} + y \frac{\partial}{\partial x} \right] \right\} \psi = E\psi \quad (8.88)$$

The magnetic field is incorporated in the cyclotron frequency, $\omega_c = |e_c| B/m$. The solution of this equation, which was first performed by Lev Landau (1908 - 1968), takes the form:

$$E = \frac{\hbar^2 k_z^2}{2m} + \hbar\omega_c(\nu + 1/2) \quad (8.89)$$

where ν is an integer. Comparing the free electron solution with the Landau energies we can write:

$$k_x^2 + k_y^2 = \frac{\omega_c m}{\hbar}(2\nu+1) = \frac{eB}{\hbar}(2\nu+1) \qquad (8.90)$$

This equation has the form of the equation of a circle, where the radius is $eB(2\nu+1)/\hbar$ in k-space. Since there is no restriction in the z-direction, these will describe cylinders or tubes in k-space, as illustrated in Figure 8.13. The various tubes arise from the discrete values of ν in Equation (8.90), these are known as *magnetic sub-bands* or *Landau levels*.

The electron energies are quantized in to Landau levels in the plane perpendicular to the direction of the magnetic field, while along B, the energies are unrestricted. The energy and density of states for electrons will now take on a discrete form, as shown in Figure 8.14(a) and (b). The oscillatory dependence of the magnetic susceptibility and the conductivity (magnetoresistance) arises when the electronic density of states moves through the chemical potential. These effects are respectively known as the *de Haas - van Alphen* or *dHvA effect* and the *Shubnikov - de Haas effect*. Since the energy between levels can be quite small, typically it is necessary to perform such measurements at low sample temperatures. At finite temperatures, the Fermi - Dirac distribution function becomes smeared in the region of the chemical

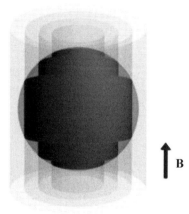

FIGURE 8.13: Landau tubes superposed on to the Fermi sphere for free electrons.

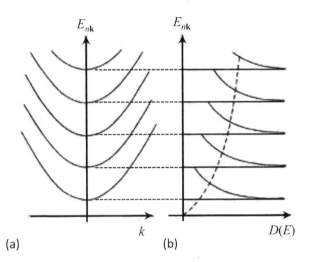

FIGURE 8.14: Energy sub-bands and density of states for Landau levels.

potential, so low temperatures are required to observe the oscillatory phenomena; i.e., $\hbar\omega_c > \sim k_B T$. The conductivity becomes vanishingly small whenever the highest Landau level is full. This can be expressed for a two-dimensional system as:

$$N_s = \nu \frac{2eB}{h} \tag{8.91}$$

where ν is an integer and N_s expresses the electron density per unit area in a two-dimensional electron gas (2DEG). A conductivity maximum will occur for:

$$N_s = \left(\nu + \frac{1}{2}\right)\frac{2eB}{h} \tag{8.92}$$

From which we can evaluate the the periodicity of the conductivity oscillations as a function of the magnetic field:

$$\Delta(1/B) = \frac{2e}{hN_s} \tag{8.93}$$

Oscillatory phenomena in magnetic fields such as these are frequently used to measure the Fermi surface of materials since the motion of the charge carriers in a magnetic field occurs on the Fermi surface itself, as discussed above.

8.8.5 The Quantum and Fractional Quantum Hall Effects

The experimental set-up for the measurement of the quantum Hall effect (QHE) is usually the same as that for the conventional Hall effect and as such the resistivity tensor components will be as given in Equations (8.84) and (8.85), with $\sigma_{xx} = \sigma_{yy} \sim I/V_x$ and $\sigma_{yx} = -\sigma_{xy} \sim -I/V_H$, where I is the current through the sample. This is all much as we had for the conventional Hall effect. However, under conditions of high magnetic fields, $\omega_c \tau \gg 1$, from Equations (8.82) and (8.83), we find $|\sigma_{xy}| \gg |\sigma_{xx}|$, which means that we can write:

$$\rho_{xx} \simeq \frac{\sigma_{xx}}{\sigma_{xy}^2} \tag{8.94}$$

$$\rho_{xy} \simeq \frac{1}{\sigma_{xy}} \tag{8.95}$$

In the situation for high magnetic fields when the highest Landau level is filled, there will be no conductivity, since filled band do not contribute. Therefore, the conductivity component σ_{xx} goes to zero, and from Equation (8.94) above, so does the corresponding resistivity component ρ_{xx}. This is an unusual situation, where both conductivity and resistivity both effectively vanish. If we now consider the Hall voltage, at values of the applied magnetic field where ρ_{xx} goes to zero, the Hall voltage becomes independent of the field, i.e., the Hall resistivity reaches a plateau. As the field further increases, the resistivity can increase as carriers become available and the Hall voltage jumps up to the next value. We can no consider the QHE by considering the variation of the conductivity as a function of the applied magnetic field. We saw earlier that the conductivity ρ_{xx} goes to zero when the Landau levels are full, at which point the electron density of a 2DEG is given by Equation (8.91). The corresponding Hall resistivity can be expressed as:

$$\rho_{xx} = R_H B = \frac{B}{N_s e} = \frac{1}{\nu} \frac{h}{2e^2} \tag{8.96}$$

For our considered case of a 2DEG, ρ_{xx} will be measured in Ω. In Figure 8.15, we illustrate the variation of the Hall resistivity

Electron Dynamics and Transport Phenomena • **267**

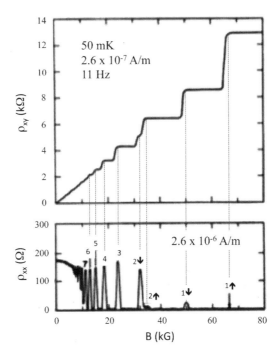

FIGURE 8.15: The quantum Hall effect measured at 50 mK in a two-dimensional electron gas formed by a GaAs/AlGaAs heterostructure. (a) Hall resistivity, ρ_{xy}, as a function of the applied magnetic field. (b) Variation of the sample resistivity, ρ_{xx}, as a function of the applied field. This corresponds to the case of oscillations in the magnetoresistance or the Shubnikow - de Haas effect. (Reprinted figure with permission from: M. A. Paalanen, D. C. Tsui, and A. C. Gossard. Phys. Rev. B **25**, 5566 (1982). Copyright 1982 by the American Physical Society.)

and the magnetoresistance oscillations or the Shubnikow - de Haas effect in a GaAs/AlGaAs heterostructure.

The quantum Hall effect was discovered in 1980 by von Klitzing, who was awarded the 1985 Nobel Prize for this discovery. Indeed, the importance of this discovery resides in the fact that the Hall resistivity is really essentially determined by the fundamental constants e and h, Equation (8.96). The QHE can be used to determine the fine structure constant;

$$\alpha = \frac{e^2 \mu_0 c}{2h} \tag{8.97}$$

One of the complexities inherent in the QHE is the existence of the resistance plateaus, which can extend over significant regions of the magnetic field. This is usually explained in terms of localized states between the Landau levels. These states, which are derived from lattice imperfections, accept electrons and trap them such that they cannot contribute to the current. This results in a plateau in the Hall resistance, as observed experimentally. In Figure 8.15, we note that there is spin-splitting in the Landau levels. This can be taken into account using the following Landau level energies:

$$E = \frac{\hbar^2 k_z^2}{2m} + \hbar \omega_c (\nu + 1/2) \pm \frac{1}{2} g * \mu_B B \qquad (8.98)$$

here the $\pm$ will take the spin of the electrons into account and $g*$ is the effective g-factor. For low fields, the spin splitting becomes negligible, which explains why the spin degeneracy lifting is not observed for the higher states, see Figure 8.15.

The fractional quantum Hall effect (FQHE) was discovered very soon after the QHE, also referred to as the integral quantum Hall effect (IQHE). The Nobel Prize was also awarded for this discovery, with H. L. Störmer, D. C. Tsui and R. B. Laughlin being the recipients of the 1998 prize. The FQHE is only observed in good quality epitaxial 2DEG systems, where additional plateaus can be observed for the condition:

$$\rho_{xy} = \frac{h}{\nu e^2} \qquad (8.99)$$

where $\nu = p/q$, are non-integer rational fractions, such as 1/3, 2/3, 3/5 etc. The FQHE is understood in terms of the highly correlated motion of many electrons in a 2DEG subject to a magnetic field and imply that the Landau levels have some internal structure. Ultimately, the magnetic flux imparts a 2π phase twist on the wavefunction describing the many-electron states. These are termed *vortices* and represent a charge deficit. The vortices are delocalized in the sample (2D) plane and are related to flux quanta. Given that the vortices correspond to a charge deficit, they attract electrons, allowing considerable energy gain to be achieved by placing vortices onto electrons. For the $\nu = 1/3$ state, there are three times as many vortices as electrons and each vortex represents a local charge deficit of $e/3$. The plateaus in the Hall resistivity occur for the effective filling factor, $\nu^* = N_s h / eB^*$, which corresponds to:

$$\nu^* = \frac{N_s h}{e(B - 2jN_s h / e)} \qquad (8.100)$$

where $B* = B - 2j\Phi_0 N_s$ is the effective field accounting for the flux quantum of number j.

8.9 SUMMARY

The motion of charge carriers in solids is derived principally from the application of electric and magnetic fields, though thermal motion is also a contributory factory as is a concentration gradient. In the former cases, the equations of motion can be established from the force produced by an electric or magnetic field. The band structure of a material can put severe limits on how the electrons and holes can move once a field has been applied, and in any case, the direction of the force (electric or magnetic) will determine the resulting motion of the charge carriers in the bands. To assist, or simplify, the treatment, the effective mass concept can be used. This will allow the effect of the band structure to be condensed in to the effective mass and permits the equations of motion to be greatly simplified.

The highest electron energy states at zero temperature occur at the Fermi energy. For a free electron the electron energy will be independent of direction and we can hence define a constant energy surface in reciprocal space, which will be spherical. For electrons or energy equal to the Fermi energy, this constant energy surface is called the Fermi sphere. In a solid, where this sphere crosses the first Brillouin zone, there will be a deformation of the surface due to Bragg scattering. Thus a Fermi surface in a solid will not generally be spherical. It is worth noting that since there are no charge carriers at the Fermi energy in a semiconductor or insulator, these materials will not strictly have a Fermi surface. The specific form of the Fermi surface in a metal will depend on its crystalline structure and the strength of the periodic potential.

At non-zero temperatures, electrons can be excited from valence to conduction bands. The missing electrons in the valence band act as positive charge carriers called holes. These have their own specific mobilities and effective masses, which differ from the electrons in the conduction band. As we stated above, the effective mass of a charge carrier will be intimately dependent on the band structure. This will have a profound effect on how the charge carriers can move through the solid, and in a crystal, will generally be anisotropic. Since the effect of a magnetic field produces transversal motion of charge carriers, their motion will occur on constant energy surfaces. This effect can be exploited to measure the motion of electrons and hence experimentally elucidate the form of the Fermi surface in metals. The oscillatory motion of electrons can be observed in measurements of the cyclotron resonance, de Haas - van Alphen and Shubnikov - de Haas effects. These reveal important information about the nature of charge carrier motion in solids. Furthermore, we can understand the Hall effect and integer quantum Hall effect in terms of the electric and magnetic fields in the solid.

REFERENCES AND FURTHER READING

Basic Texts

- J. S. Blakemore, *Solid State Physics*, Cambridge University Press, Cambridge (1985)

- H. P. Myers, *Introductory Solid State Physics*, Taylor and Francis, London (1998)

- J. Singleton, *Band Theory and Electronic Properties of Solids*, Cambridge University Press, Cambridge (2003)

Advanced Texts

- M. A. Wahab, *Solid State Physics: Structure and Properties of Materials*, Alpha Science International Ltd., Harrow (2007)

- N. W. Ashcroft and N. D. Mermin, *Solid State Physics*, Saunders College, Philadelphia (1976)

- J. M. Ziman, *Electrons and Phonons*, Oxford University Press, Oxford (1960)

- H. Ibach and H. Lüth, *Solid State Physics: An Introduction to Principles of Materials Science*, Springer, Berlin (2009)

EXERCISES

Q1. Using the momentum operator $\mathbf{p} = -i\hbar\nabla$, find the expectation value of the velocity of a Bloch electron.

Q2. Consider a parabolic energy band of the form: $E = E_0 + \alpha_1 k_1^2 + \alpha_2 k_2^2 + \alpha_3 k_3^2$. Show that the effective mass tensor can be expressed as:

$$m_{ij}^* = \frac{\hbar^2}{2\alpha_i} \frac{1}{\delta_{ij}} \qquad (8.101)$$

Q3. Construct the effective mass tensor for the energy bands given in Equation (8.18).

Q4. Consider a two-dimensional solid with the following dispersion relation:

$$E(k) = E_0 + A\cos(k_1 a_1) + B\cos(k_2 a_2) \qquad (8.102)$$

Determine the effective mass tensor.

Q5. Consider the energy relation given in exercise Q4, which has just one electron in the energy band. A a certain moment an electric field, ε, is applied along the direction of α_1. How do the electron's position (initially at $x = 0$), velocity and effective mass in this direction vary as a function of time?

Q6. Sketch the form of the Fermi surface for the solid in question Q4 for the case where there is an almost empty band. How would this change if the band was almost full?

Q7. Consider the motion of an electron under the influence of a magnetic field. Show that the projection of the electron's real-space orbit in the plane perpendicular to the magnetic field, is the reciprocal-space orbit rotated by $\pi/2$ radians about the direction of the field and is scaled by a factor of $\hbar e/B$.

Q8. Describe how you would expect the magnetoresistance of copper to vary in the (100) and (110) planes. Hint: Consider the form of the Fermi surface for copper, as illustrated in Figures 8.4 and 8.6.

Q9. Consider the cyclotron resonance experiment, where the electron in the solid is described by the effective mass, m^*, and relaxation time, τ. Derive the equations of motion for the electron drift velocity in a static magnetic field, $\mathbf{B} = (0, 0, B)$ and a high-frequency electric-field, $\mathbf{E} = (\xi_0 e^{i\omega t}, \xi_0 e^{i(\omega t + \pi/2)}, 0)$.

Q10. Derive the Hall coefficient, R_H, for an intrinsic semiconductor, where both electrons and holes contribute equally to the current in the sample.

Q11. Evaluate the conductivity and resistivity tensors for an intrinsic semiconductor, taking into account the longitudinal and transversal effective masses. What is the effect of applying a high magnetic field, $\omega_c \tau \gg 1$, to the sample?

NOTES

[1] A temperature gradient can also be expected to produce a net flow of charge since electrons in regions of higher temperature will have a greater thermal velocity. Thermalization processes will transfer energy to the crystal lattice in the form of phonons and act to eliminate any temperature gradient. This is analogous to the diffusion process, which *acts* to remove concentration gradients.

[2] G. Dresselhaus A. F. Kip and C. Kittel, Phys. Rev. **92**, 827 (1953) and Phys. Rev. **98**, 368 (1955)

CHAPTER 9

SEMICONDUCTORS

"I tore myself away from the safe comfort of certainties through my love for truth - and truth rewarded me."

—Simone de Beauvoir

"Nothing in life is to be feared, it is only to be understood. Now is the time to understand more, so that we may fear less."

—Marie Curie

9.1 INTRODUCTION

In previous chapters, we have referred to materials in which there exist two types of charge carrier: the electron and the hole. These refer to the charge carriers in the conduction and valence bands, respectively. In this chapter, we will consider materials in which there are significant contributions to the electrical properties from these charge carriers. These materials are called semiconductors, and are distinguished from metals and insulators in their specific electronic properties. The principal way in which we distinguish these types of solids is via their electrical resistivity. In Table 9.1, we show some electrical resistivities of solids in these categories of materials. Typically metals have resistivities of the order of $10^{-8}(\Omega\text{-m})$, while insulators have $\rho > 10^{10}(\Omega\text{-m})$. Semiconductors can have a broad range of values for the electrical resistivity, depending on impurities, defects and ambient conditions.

TABLE 9.1: Electrical resistivities of selected materials at room temperature

Material	Resistivity at room temperature $\rho(\Omega\text{-m})$	Category
Copper	1.68×10^{-8}	Conductor (metal)
Silver	1.59×10^{-8}	
Gold	2.44×10^{-8}	
Germanium	4.6×10^{-1}	Semiconductor
Silicon	6.4×10^{2}	
Glass	$10^{10} - 10^{13}$	Insulator
Air	10^{16}	
Teflon	$10^{22} - 10^{24}$	

As can be seen, the semiconductor class of solids has a resistivity in between those of metals and insulators. Actually we shall see that there are some fundamental differences between metals and semiconductors, which is principally related to the electronic structure of the solids and the availability of the two types of charge carrier. In metals, electrons occupy energy levels within the conduction band, which is partially filled, while the valence band is completely full. This can be otherwise thought of as the Fermi level being located somewhere inside the highest occupied energy band. In semiconductors, the Fermi level lies between the valence and conduction bands, thus at zero temperature, all states in the valence band are occupied while in the conduction band they are completely empty. An insulator has the same situation. The principal difference being the size of the energy gap between the bands. In the case of semiconductors the energy difference between the top of the valence band and the bottom of the conduction bands is roughly about 1 eV, while for insulators it is typically above 4 eV. This difference, though academic, is sufficient to reduce the effective number of charge carriers at normal operating conditions of temperature, such that the electrical properties can be distinguished loosely in this way. We will discuss these considerations further in this chapter. The metal on the other hand can be thought of as having zero band gap. Actually there is usually an overlap in terms of energy between the conduction and valence bands in metals. In Figure 9.1, we show the band structure for Si. Illustrated are the conduction and valence bands as

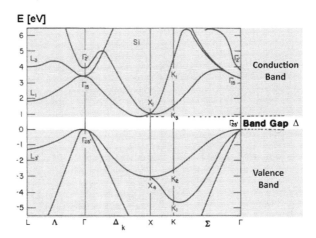

FIGURE 9.1: Band structure for Si. Electron states do not exist in the band gap region for a semiconductor and insulator. The Fermi energy will lie in this region.

well as the band gap, Δ. The Fermi level for pure semiconductors and insulators lies exactly half way between the valence and conduction bands at absolute zero of temperature.

Semiconducting materials hold a very special place in the electronics industry. This is because the principal electronic components used in devices for computing purposes are based on semiconductors. Such devices as the diode and the transistor are of particular importance. We shall see how these devices are constructed and how they function.

So why are materials, which do not electrically conduct as well as metals, so important in electronic devices? We can answer this question by stating that it is the capacity for manipulating their electronic properties in a very precise manner, which allows them to be made to perform very specific tasks under well defined conditions. We shall see how their thermal behavior and in the presence of photons differs from other materials. We will also discuss the possibility to introduce impurities in a controlled manner in semiconductors to control their conductivities. Furthermore, the manipulation of the composition of semiconductors in the fabrication process allows the delicate control of the materials electronic structure and permits the design of materials with specific electronic and optoelectronic properties. The combination of different types of semiconductor further allows the development of a multitude of devices for optical and electronic applications.

9.2 SEMICONDUCTING MATERIALS

While there are naturally occurring semiconductors, such as silicon and germanium, compound materials can be formed with semiconducting properties, such as GaAs and CdSe. In Chapter 1, we introduced the periodic table of elements, where we see the specific grouping of elements according to their electronic shell configurations. The semiconducting elements Si and Ge both occur in the group IV period, each having four electrons in their outer (p) shell, which contribute to the valence band at low temperatures. These elements crystallize in the diamond structure. In fact, diamond is formed by carbon and also has 4 p-electrons. The only reason that diamond is not a semiconductor is that it has a larger energy band gap between the valence and conduction bands. On either side of this column in the periodic table, are the group III and group V elements, which in compound form make up the III - V semiconductors, see Figure 1.5. The most common being the binary compound GaAs, though other III - V compounds can be formed also in ternaries and quarternies, with elements such as In, P and Sb. These materials have a particularly important role in optoelectronic devices based on quantum well structures, as will be discussed later. The III - V semiconductors typically form in a zinc-blende type structure, which is closely related to that of diamond, see Chapter 2. The nature of the bonding between atoms was discussed in Chapter 1, see Section 1.4 for more details.

Beyond the III - V materials lie the II - VI type compounds, such as CdS, CdTe and ZnSe. As with the III - V compounds, the exchange of electrons between the components gives rise to partial covalent and ionic bonds, with four valence electrons. The semiconductors of the II - VI type can crystallize in the wurtzite like structure, such as CdS, or in the zinc-blend structure, as is the case of CdTe. Both structures are based on the tetragonal type units that also occur in both the diamond crystal type lattice.

TABLE 9.2: Properties of some common semiconductors at low temperature.

Semiconductor	Group	Band gap, Δ (eV)	Lattice type	Lattice constant (Å)
Si	IV	1.17	Diamond	5.431
Ge	IV	0.74	Diamond	5.658
GaAs	III - V	1.52	Zinc-Blende	5.653
GaP	III - V	2.35	Zinc-Blende	5.45
InP	III - V	1.42	Zinc-Blende	5.869
ZnSe	II - VI	2.82	Zinc-Blende	5.661
CdTe	II - VI	1.61	Zinc-Blende	6.423
CdS	II - VI	2.58	Wurtzite	$a = 4.139$ $c/a = 1.62$
CdSe	II - VI	1.84	Wurtzite	$a = 4.309$ $c/a = 1.63$

As we have stated above, one of the principal properties of the semiconductors is the energy gap (band gap) between the top of the valence band and the bottom of the conduction band. In Table 9.2, we list some of the band gap energies of the principal semiconductors at low temperature. We note that the properties at room temperature will be different from those listed in the table. For example, the band gap in Si reduces to 1.12 eV at room temperature, while that for Ge is 0.67 eV at 300 °C.

The properties of pure (or *intrinsic*) semiconductors can be significantly modified by the introduction of controlled quantities of impurities. Such an introduction of foreign atoms into the semiconductor is called *doping* and the resulting material is said to be an *extrinsic semiconductor*. These will be discussed in more detail later in this chapter.

In addition to the distinctions between types of semiconductor given above, there is another important classification, which relates to the fundamental band structure. For band structures in which the minimum of the conduction band and the maximum in the valence band occur at the same value of wave vector, we designate as a *direct band-gap semiconductor*. When this is not the case we refer to the semiconductor as an *indirect band-gap* material. GaAs is a well-known example of a direct band-gap semiconductor, while Si, Ge and GaP are examples of indirect band-gap semiconductors. This property is of great importance in the transitions of electrons between the valence and conduction bands, since a change in the value of the wave-vector implies a change of momentum and direction of the electron. Direct transitions are therefore more efficient, making direct band-gap semiconductors more attractive for optical device applications. In Figure 9.2, we show the band structures for (a) GaAs and (b) GaP, which are respectively, direct and indirect semiconductors. In the latter case, we note that the minimum energy transition between the valence and conduction bands, corresponding to $\Delta^X = 2.432$ eV, requires a change in the wave-vector, Δk. The direct transition is given as $\Delta^\Gamma = 2.75$ eV, see Figure 9.2(b).

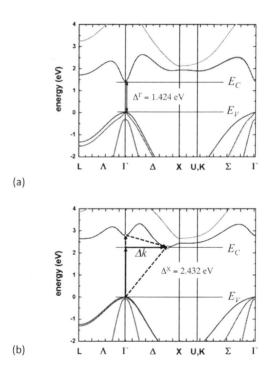

FIGURE 9.2: Band structure for (a) GaAs and (b) GaP. In the case of GaAs we see that the minimum in the conduction band is directly above the maximum in the valence band, making it a direct type semiconductor. In the case of GaP, the minimum in the conduction band is shifted from the maximum of the valence band and the semiconductor is of the indirect type.

9.3 EQUILIBRIUM STATISTICS: ELECTRONS AND HOLES

In order to evaluate the electrical conductivity of semiconductors, it is necessary to determine the concentration of the various charge carriers in the valence and conduction bands. In addition to the doping of semiconductors, the number of charge carriers in the solid depends on the excitation of electrons from the valence to the conduction band, via thermal excitation or the absorption of photons. In describing these processes, we need to consider the thermodynamic equilibrium in terms of the materials' temperature. Additionally, excess charge carriers can by electrically injected into a semiconductor by an external source. In the following we will consider the evaluation of the concentration and distribution of charge carriers in the valence and conduction bands of semiconductors, where it is helpful to distinguish between intrinsic and extrinsic type semiconductors.

9.3.1 Intrinsic Semiconductors

We will consider a pure or intrinsic semiconductor as one in which there are no impurity atoms with a band gap of Δ between the valence and conduction bands. The distribution of charge carriers among the available energy states will be determined by the sample temperature and can be evaluated with the use of the Fermi - Dirac distribution function. It will be noted that since the Fermi energy will be at the center of the band gap for intrinsic semiconductors, the number of electrons and holes in the semiconductor will be equal. This is logical since *free* electrons will only be produced by the excitation of electrons from the valence to the conduction band. It is worthwhile noting here the true meaning of the concept of the band gap. Since there is a region of energy states that cannot be occupied, this means that to excite an electron from the valence band to the conduction band requires a minimum energy of Δ, which is the binding energy of the valence electron to an atom in the lattice from its normal bonding state. This energy can be furnished in the form of thermal energy (via the sample temperature), to produce an agitation (phonon) of the atom, or via the absorption of a photon of sufficient energy. In whatever case, the energy can only be transferred to the electron to free it from the influence of the atom if this energy is larger than the binding energy. Once freed from the atom it can, at least for a certain amount of time, travel through the crystal lattice. Should it be attracted back to an atom, it is said to relax and transit from the conduction to the valence band, and in the process will transfer energy either back to the lattice (in the form of a phonon) or with the emission of electromagnetic radiation (a photon). These transitions between bands are obviously not just restricted to intrinsic semiconductors and are a general physical process, one which plays a very important role in the applications of semiconducting materials and devices. We note that the motion of these free (conduction band) electrons are subject to all the scattering laws discussed in previous chapters, such as the Bragg law. Before evaluating the concentrations of the charge carriers in the semiconductor, we note that the conductivity of the material will be determined by both types of charge carrier, which we can express as:

$$\sigma(T) = |e_c|[n(T)\mu_n + p(T)\mu_p] \qquad (9.1)$$

where μ_n and μ_p are the mobilities of the electrons and holes, which have concentrations n and p, respectively. (Compare this equation with Equation (6.16).) We note that in the previous chapter we used the subscripts e and h to denote electrons and holes. In this chapter, we use n and p, since they relate to the type of semiconductor, as will be discussed shortly. Since electrons and holes have opposite signs of charge and the drift velocities of the two types of charge carrier will also be opposite, they both contribute with the same sign to the conductivity, σ. Here we have neglected any energy or k-dependence of the charge carriers on the mobilities, which is generally a good approximation for not too high temperatures. Explicitly indicated in Equation (9.1) is the fact that the conductivity for a semiconductor is highly temperature dependent. This principally arises from the carrier concentration being sensitive to thermal excitations of charge carriers, as mentioned above. This dependence is due to the Fermi-Dirac distribution function, $f_{FD}(E, T)$, which we introduced in Chapter 6, Section 6.5.

We express the carrier concentration for electrons in the conduction band as:

$$n(T) = \int_{E_c}^{\infty} g_c(E) f_{FD}(E,T) dE \qquad (9.2)$$

here $g_c(E)$ represents the density of electrons states in the conduction band. The corresponding expression for holes in the valence band takes the form:

$$p(T) = \int_{-\infty}^{E_v} g_v(E)[1 - f_{FD}(E,T)] dE \qquad (9.3)$$

we note that the factor $[1 - f_{FD}(E, T)]$ corresponds to the hole (missing electron) distribution. The ranges of the integrals have been extended to infinity to aid evaluation, and this can be done since the Fermi - Dirac function decreases sufficiently rapidly. The densities of states, see Section 6.7, in the conduction and valence bands can be expressed (in the parabolic band approximation) as:

$$g_c(E) = \frac{1}{2\pi^2} \left(\frac{2m_c^*}{\hbar^2} \right)^{3/2} \sqrt{E - E_c}; (E > E_c) \qquad (9.4)$$

and

$$g_v(E) = \frac{1}{2\pi^2} \left(\frac{2m_v^*}{\hbar^2} \right)^{3/2} \sqrt{E_v - E}; (E < E_v) \qquad (9.5)$$

This is so far quite similar to what we found in Section 6.7. The form of the carrier concentrations can be simplified for low temperatures, subject to the condition: $k_B T \ll (E_c - E_F)^1$, where we have: $f_{FD}(E, T) \rightarrow f_{MB}(E, T)$. The condition is thus stated as the difference between the Fermi level and the conduction band edge being much larger than the thermal energy. We note that the thermal energy at room temperature is about 1/40 eV, so for a band gap of around 1 eV, the energy difference $(E_c - E_F)$ will be about 0.5 eV, therefore the thermal energy at room temperature represents about 5 % of this difference. When the above approximation is valid, the semiconductor is referred to as a *non-degenerate semiconductor*. When this approximation is not valid, we are dealing with a *degenerate semiconductor* and we must use the full expressions given by Equations (9.2) and (9.3). Now for the parabolic bands:

$$E_{ck} = E_c + \frac{\hbar^2 k^2}{2m_c^*} \qquad (9.6)$$

$$E_{v\mathbf{k}} = E_v + \frac{\hbar^2 k^2}{2m_v^*} \qquad (9.7)$$

we obtain the following approximations for the carrier concentrations:

$$n(T) \simeq e^{(E_F - E_c)/k_B T} \int_0^\infty g_c(E) e^{E/k_B T} \, dE \qquad (9.8)$$

$$p(T) \simeq e^{(E_v - E_F)/k_B T} \int_{-\infty}^0 g_v(E) e^{E/k_B T} \, dE \qquad (9.9)$$

Using standard integrals[2] we can evaluate these expressions to yield:

$$n(T) = 2\left(\frac{k_B T m_c^*}{2\pi\hbar^2}\right)^{3/2} e^{(E_F - E_c)/k_B T} = N_c(T) e^{(E_F - E_c)/k_B T} \qquad (9.10)$$

$$p(T) = 2\left(\frac{k_B T m_v^*}{2\pi\hbar^2}\right)^{3/2} e^{(E_v - E_F)/k_B T} = P_v(T) e^{(E_v - E_F)/k_B T} \qquad (9.11)$$

where $N_c(T)$ and $P_v(T)$ are called the *effective density of states* for the conduction and valence bands, respectively. It is convenient to write these as:

$$N_c(T) = 4.83 \times 10^{21} \left(\frac{m_c^*}{m}\right) T^{3/2} \qquad (9.12)$$

$$P_v(T) = 4.83 \times 10^{21} \left(\frac{m_v^*}{m}\right) T^{3/2} \qquad (9.13)$$

which are given in units of m^{-3}. Given that the effective mass of charge carriers divided by the free electron mass is typically of the order of unity, we can estimate that the density of carriers in an intrinsic semiconductor will be of the order of around 10^{25} m^{-3} or 10^{19} cm^{-3}. It is worth noting that the density of states effective mass for ellipsoidal bands, as is the case for Si and Ge, can be obtained using the longitudinal and transversal values of the effective mass (see Equations 8.18 and 8.19 of the previous chapter, for example), and we can write: $m_{DOS}^* = (m_l^* m_t^{*2})^{1/3}$. Similarly, for multiple valence bands, which we denote with effective masses for *heavy holes*, m_{hh}^* and *light holes*, m_{lh}^*,[3] which depends on the curvature of the energy bands, we can write: $m_v^* = (m_{hh}^{*3/2} + m_{lh}^{*3/2})^{2/3}$.

9.3.2 The Law of Mass Action

We will now multiply the expressions for the concentrations of electrons and holes, which we can express in the following manner:

$$\begin{aligned}n(T)p(T) &= N_c(T)P_v(T)e^{(E_v-E_F)/k_BT}e^{(E_F-E_c)/k_BT} \\ &= N_c(T)P_v(T)e^{(E_v-E_c)/k_BT} \\ &= N_c(T)P_v(T)e^{-\Delta/k_BT} = n_i^2 \end{aligned} \quad (9.14)$$

where $\Delta = E_c - E_v$ is the band gap and n_i is the intrinsic charge carrier concentration. Equation (9.14) is commonly known as the *law of mass action*. This relation shows that the carrier concentration is entirely dependent on the material properties; Δ, m_c^*, m_v^* and the sample temperature. In Table 9.3, we show some examples for the common semiconductors.

The law of mass action thus states that the product of carrier concentrations in a non-degenerate semiconductor is constant at a specific temperature. The form given here requires that the Fermi energy is far from the band edges, which is the required condition for a non-degenerate semiconductor. The thermal excitation of electrons in the intrinsic semiconductor means that we have: $n = p$. We thus can write:

$$n = p = n_i = \sqrt{N_c P_v}e^{-\Delta/2k_BT} = 2\left(\frac{k_BT}{2\pi\hbar^2}\right)(m_c^*m_v^*)^{3/4}e^{-\Delta/2k_BT} \quad (9.15)$$

On this assumption that $n = p$ we also obtain the following relation:

$$N_c(T)e^{(E_{F_i}-E_c)/k_BT} = P_v(T)e^{(E_v-E_{F_i})/k_BT} \quad (9.16)$$

TABLE 9.3: Band gaps and intrinsic carrier concentrations for selected materials at room temperature (300 K).

Semiconductor	Δ (eV)	n_i (m^{-3})
Si	1.12	1.5×10^{16}
Ge	0.67	2.4×10^{19}
GaAs	1.43	5×10^{13}

where we have indicated the *intrinsic Fermi level* as E_{F_i}, which from Equation (9.16) and the above, we can express as:

$$E_{F_i} = E_v + \frac{\Delta}{2} + \frac{3}{4} k_B T \ln\left(\frac{m_v^*}{m_c^*}\right) \tag{9.17}$$

This shows that for the intrinsic case, the Fermi level at 0K is exactly in the center of the band gap. As the temperature increases the Fermi level is displaced, moving up in energy towards the conduction band if $m_c^* < m_v^*$, or down towards the valence band for $m_c^* > m_v^*$. Only if $m_c^* = m_v^*$ will the Fermi energy remain constant as a function of temperature. We can illustrate some of the main physical quantities and dependencies, as shown in Figure 9.3. We note that for an imbalance in the effective masses between the conduction and valence bands, the position of the Fermi level can move up or down. In either case, the number of holes must still be equal to the number of electrons.

9.3.3 Extrinsic Semiconductors: Doping

If the impurities in a semiconductor contribute significantly to the number of charge carriers, the semiconductor is said to be *extrinsic*. The *doping* of a semiconductor consists in the controlled

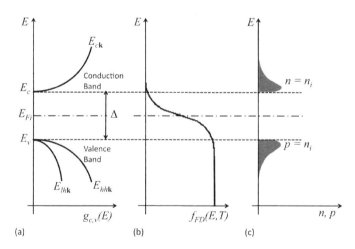

FIGURE 9.3: Intrinsic semiconductor: (a) Density of states for the conduction and valence bands, with heavy and light hole bands indicated. (b) Fermi - Dirac function at finite temperature. (c) Charge concentrations for electrons and holes in the conduction and valence bands, respectively.

introduction of a specific concentration of impurity atoms, which can either act as *donors* or as *acceptors*. In the case of donors, the impurity atoms *give up* electrons when they form bonds with the surrounding semiconductor atoms. Actually these atoms become ionized and in the process an excess electron (not occupied in bonding is freed and thus "enters" the conduction band). Such atoms will have more electrons in the outer p-states, coming from the group to the right in the periodic table, for example, As atoms in Ge act as donors. An extrinsic semiconductor of this nature is said to be an *n-type* semiconductor. Acceptor atoms are similar, but do the opposite. They have less electrons in the outer p-states than the host semiconductor atoms and readily ionize by removing an electron from a host atom, which effectively introduces a hole state in the valence band. An example of this would be in the case of a B atom impurity in a Si lattice. Such semiconductors are labelled *p-type*. These are so labelled to reflect the majority charge carriers in the semiconductor; n for negative (electrons) and p for positive (holes). In terms of energies, the donor and acceptor levels are close to the conduction and valence band edges, respectively, thus facilitating the ionization of impurities at relatively low temperatures. We will now describe the detailed neutrality conditions which allow us to determine the carrier concentrations in the doped semiconductor. In Figure 9.4, we show a schematic illustration of the impurity and Fermi levels.

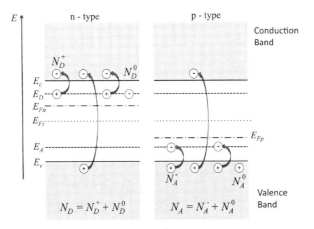

FIGURE 9.4: Extrinsic semiconductor: Energy state definitions and concentrations of donors and acceptors.

Indicated are the notations used to discuss the concentrations of the donors, N_D, and acceptors N_A, also shown are the proportions which are ionized; N_D^+, N_A^-, and neutral (or un-ionized); N_D^0, N_A^0. In addition to the charge carriers introduced via the doping process, thermal excitation between the conduction and valence bands will also occur as in the case of the intrinsic semiconductor. We note that the law of mass action is still valid even though we have: $n - p = \Delta n \neq 0$.

In the following, we discuss the explicit case of the n-type semiconductor. The corresponding relations for p-type materials can be obtained in a similar manner by substituting the acceptor concentration, N_A, instead of that of the donors, N_D. In the case of n-type semiconductors (which we denote using the n subscript), we can write the number of ionized donors as:

$$N_D^+ = \frac{N_D}{1 + 2e^{(E_{F_n} - E_\delta)/k_B T}} \quad (9.18)$$

where $E_\delta = E_c - E_D$, see Figure 9.4, is the energy required to ionize a donor atom. For the case where $(E_{F_n} - E_\delta) \gg k_B T$, virtually all the donors will be ionized and we can write: $n_n \simeq N_D \simeq N_D^+$. Actually we can be more precise and take into account the interband excitation, such that:

$$n_n = p_n + N_D^+ \simeq p_n + N_D \quad (9.19)$$

where the majority carriers are electrons, denoted as n_n, and the minority carriers are holes designated as p_n. Given that the law of mass action can be applied, we can write:

$$n_n = \frac{n_i^2}{n_n} + N_D \quad (9.20)$$

from which we obtain, using the physically meaningful solution:

$$n_n = \frac{1}{2}[N_D + (N_D^2 + 4n_i^2)^{1/2}] \quad (9.21)$$

From this equation we can determine the various concentration regimes: i) $N_D = n_i \Rightarrow n_n \simeq n_i$; ii) $N_D \gg n_i \Rightarrow n_n \simeq N_D$; iii) $N_D \ll n_i \Rightarrow n_n \simeq n_i$.

A more general analysis can be reached expressing Equation (9.18) in the form:

$$N_D^+ \simeq n_n = \frac{N_D}{1 + 2e^{(E_{F_n} - E_c + E_D)/k_B T}} \qquad (9.22)$$

Using Equation (9.10) we can eliminate the factor $(E_{F_n} - E_c)$, which yields:

$$\frac{n_n^2}{(N_D - n_n)} = \frac{N_c}{2} e^{-E_D/k_B T} \qquad (9.23)$$

This quadratic equation can be solved to give:

$$n_n = \frac{N_c}{4} e^{-E_D/k_B T} \left[\sqrt{1 + 4\frac{N_D}{N_c} e^{E_D/k_B T}} - 1 \right]$$

$$= 2N_D \left[1 + \sqrt{1 + 4\frac{N_D}{N_c} e^{E_D/k_B T}} \right]^{-1} \qquad (9.24)$$

From this expression we can identify the following limiting cases:

i) *Freeze-out range*:

This corresponds to the low temperature regime, where we have the condition: $4(N_D/N_c)e^{E_D/k_B T} \gg 1$. In this situation, Equation (9.24) gives:

$$n_n \simeq \sqrt{N_D N_c}\, e^{-E_D/2k_B T} \qquad (9.25)$$

In this region there can still be significant numbers of donor atoms which are not ionized, and with decreasing temperature the charge carriers (electrons) from the conduction band are *frozen-out*. It should be noted that the form of Equation (9.25) bears a strong resemblance to Equation (9.15). This is no coincidence, and the same thermal excitation process is taking place, but in different temperature ranges. The important point here is that the donor binding energies are significantly smaller than the band gap energy, and so the exponential dependence occurs over a different temperature range for the two Boltzmann factors.

ii) *Saturation range*:

For the condition: $4(N_D/N_c)e^{E_D/k_B T} \ll 1$, Equation (9.24) approximates to:

$$n_n \simeq N_D = const. \quad (9.26)$$

This result shows that in this regime, all donor impurity atoms are ionized and are thus saturated. In a first approximation we neglect the contribution from electrons excited from the valence band. For reasonable donor/acceptor concentration levels, this is a valid assumption, since the proportion of interband excited carriers will be negligible.

iii) *Intrinsic range*:

For higher temperatures the number of electrons excited from the valence band will start to become significant and we enter the third regime, which is the intrinsic range, since this is the same process that occurs for intrinsic semiconductors, depending on the intrinsic properties of the semiconductor in question.

We can visualize the different concentration regimes in Figure 9.5. In the figure are shown various curves for different concentrations of impurities of both donors and acceptors. The three concentration regimes discussed above are also indicated.

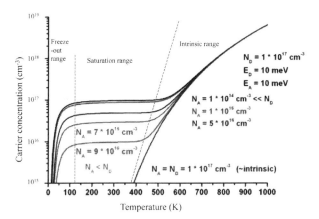

FIGURE 9.5: Carrier concentration temperature dependence illustrating the various concentration regimes. Shown are the variations for different doping concentrations for both donors and acceptors. For $N_D \gg N_A$ the behavior is that of an n-type semiconductor. As the N_A concentration increases, the saturation level reduces proportionally until $N_D = N_A$, where the semiconductor has an intrinsic like behavior. Actually this is called the compensated state, where the donor excess electrons are matched by the acceptor excess holes. (Figure adapted from source: *www.nextnano.de/nextnano3/tutorial/1Dtutorial doped semiconductors.htm*.)

It is instructive to equate expression (9.24) and (9.10), where we can solve for the Fermi energy, to obtain:

$$E_{F_n} = E_c - k_B T \ln\left\{\frac{N_c}{2N_D}\left[1 + \sqrt{1 + 4\frac{N_D}{N_c}e^{E_D/k_B T}}\right]\right\} \quad (9.27)$$

In the low temperature range, where we have: $4(N_D/N_c)e^{E_D/k_B T} \gg 1$, this simplifies to:

$$E_{F_n} \simeq E_c - \frac{E_D}{2} + \frac{k_B T}{2}\ln\left(\frac{N_D}{N_c}\right) \quad (9.28)$$

The variation of the Fermi level with temperature can thus be understood as a function of temperature and impurity concentration, see Figure 9.6.

At zero temperature (0K), the Fermi level starts at a position between the conduction band edge and the impurity (donor) level. With an increase of temperature there will be a small increase in E_F, but with a further increase of temperature the impurity levels become denuded and excitation of electrons between the valence and conduction bands occurs and the Fermi level approaches the intrinsic value, E_{F_i}.

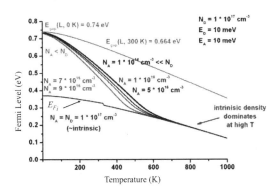

FIGURE 9.6: Temperature dependence of the Fermi level. Shown are the variations for different doping concentrations for both donors and acceptors. For $N_D \gg N_A$ the behavior is that of an n-type semiconductor. As the N_A concentration increases, the Fermi level falls from near to the conduction band edge to the intrinsic fermi level, E_{F_i}. At $N_D = N_A$, which the semiconductor has an intrinsic like behavior, where $E_F = E_{F_i}$. Also shown is the variation of the band gap as a function of temperature. The values used are based on germanium. (Figure adapted from source: *www.nextnano.de/ nextnano3/tutorial/1Dtutorial doped semiconductors.htm*.)

In the case of impurity atoms which are acceptors, the situation is similar, though we need to convert what we have stated for electrons above for the case of holes in a p-type material. The concentration of holes, which will now be the majority charge carrier, denoted as p_p, will have an analogous dependence on the concentration of impurities, N_A, as electrons in the n-type semiconductor on N_D. The minority charge carrier in the p-type semiconductor is the electron, which has a concentration, n_p. In more general terms, we can express the free carrier concentration in the form:

$$n, p = \frac{N_D, N_A}{1 + g e^{\pm(E_{F_{n,p}} - E_{\delta,\alpha})/k_B T}} \quad (9.29)$$

where g is the degeneracy of the impurity level. Equation (9.29) can now be employed to evaluate the concentration of both types of charge carrier. In the case of electrons, we must use N_D, E_{F_n}, E_δ and the positive sign in the exponential, while for the hole concentration p we use N_A, E_{F_p} and E_α, with a negative sign in the exponential.

9.3.4 Compensated Semiconductors

In many cases the semiconductor will have both types of impurity and the type of semiconductor will depend on the dominant number of impurities. So if $N_A > N_D$, the semiconductor will be p-type, while for $N_A < N_D$, it will be n-type. Semiconductors with both donors and acceptors are called *compensated semiconductors*. This is because electrons from the donors will be transferred to the acceptors even at $T = 0$ K. Full compensation will only occur for $N_A = N_D$. The charge neutrality condition for compensated semiconductors can be expressed as:

$$n + N_A^- = p + N_D^+ \quad (9.30)$$

For semiconductors with both types of impurity, we can express a more general form of Equation (9.23):

$$\frac{n_n(n_n + N_A)}{(N_D - N_A - n_n)} = \frac{N_c}{2} e^{-E_D/k_B T} \quad (9.31)$$

The solution takes the form:

$$n_n = \frac{2(N_D - N_A)}{1 + \frac{N_A}{N_c^{eff}} + \left[\left(1 - \frac{N_A}{N_c^{eff}}\right)^2 + 4\frac{N_D}{N_c^{eff}}\right]^{1/2}} \qquad (9.32)$$

where $N_c^{eff} = \frac{N_c}{2} e^{-E_D/k_BT}$. In the low temperature limit this expression simplifies to:

$$n_n \simeq \frac{(N_D - N_A)}{N_A} N_c^{eff} = \frac{(N_D - N_A)}{2N_A} N_c e^{-E_D/k_BT} \qquad (9.33)$$

This shows that the electron concentration will decrease with an increase of N_A. For the other extreme of temperature, where both N_A and N_D are less than N_c^{eff} we have:

$$n_n = N_D - N_A \qquad (9.34)$$

which corresponds to the case where all impurities are ionized. Alternatively, we can use the law of mass action along with the neutrality condition to establish the following relation:

$$n_n + N_A = \frac{n_i^2}{n_n} + N_D \qquad (9.35)$$

The solution to this quadratic equation then takes the form:

$$n_n = \frac{1}{2}\left\{(N_D - N_A) + [(N_D - N_A)^2 + 4n_i^2]^{1/2}\right\} \qquad (9.36)$$

and

$$p_n = \frac{n_i^2}{n_n} \qquad (9.37)$$

The corresponding relations for the p-type partially compensated semiconductor, where the majority carriers are holes and the minority carriers are electrons, are written:

$$p_p = \frac{1}{2}\left\{(N_A - N_D) + [(N_A - N_D)^2 + 4n_i^2]^{1/2}\right\} \qquad (9.38)$$

and

$$n_p = \frac{n_i^2}{p_p} \qquad (9.39)$$

In general, the magnitude of the concentration of impurities is greater than the intrinsic concentration; $|N_D - N_A| < n_i$, such that we have:

$$n_n \simeq N_D - N_A; \quad \text{for} \quad N_D > N_A \qquad (9.40)$$

and

$$p_p \simeq N_A - N_D; \quad \text{for} \quad N_A > N_D \qquad (9.41)$$

9.4 NON-EQUILIBRIUM DISTRIBUTIONS

Frequently, when semiconductors are employed in device applications, they function in regimes which are not in thermal equilibrium. As such, the charge carrier distributions will be modified beyond the conditions we discussed in the previous section. We will discuss some simple cases of non-equilibrium conditions in the following sections.

9.4.1 Carrier Injection: Injection Levels

At thermal equilibrium the condition $np = n_i^2$ is valid. However, conditions of *non-equilibrium* can exist when an imbalance is created by the introduction of excess charge carriers, such that $np > n_i^2$. Such a condition can be produced by the process of *carrier injection*. There are several ways in which to inject charge carriers into a semiconductor, such as the incidence of electromagnetic radiation of sufficient energy; $\hbar\omega > \Delta$, or via the application of using a forward bias across the junction between p- and n-type materials. In the former case, the absorption of photons via the transfer of energy to an electron in the valence band. This liberates the electron to allow it to move in the conduction band, thus forming an electron - hole pair and abruptly increasing the number of charge carriers above the equilibrium values, which are then free to contribute to electrical conduction processes. Such charge carriers are referred to as *excess carriers*.

The amount of excess charge carriers relative to the concentration of majority carriers is determined by the level of injection. This is best illustrated by way of example. Let us consider a sample of Si which is doped with $N_D = 10^{21}$ m^{-3} donors and at thermal equilibrium. At room temperature we can expect a majority of these donors to be ionized, such that at thermal equilibrium we have $n_{n0} \simeq N_D = 10^{21}$ m^{-3}, where the "0" in the subscript is used to indicate thermal equilibrium. This situation is illustrated in Figure 9.7 (a), where we show the positions of the intrinsic concentration, n_i and the minority carrier density, p_{n0}. This latter is calculated using the law of mass action; $p_{n0} = n_i^2/n_{n0}$.

If we introduce excess charge carriers via the excitation of electrons from the valence to the conduction band, we will have equal number of each type of excess charge carrier. For this we can write in our n-Si example, $\Delta n_n = \Delta p_p$. In the case of low-level injection we

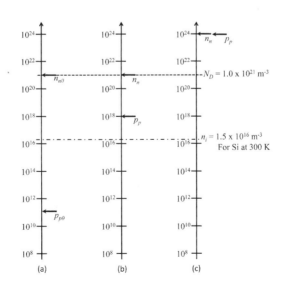

FIGURE 9.7: Carrier concentrations for (a) Thermal equilibrium in an n-Si sample with $N_D = 10^{21}$ m^{-3}. (b) Low-level carrier injection: $\Delta n_n = \Delta p_n = 10^{18}$ m$^{-3} \ll N_D$. (c) High-level carrier injection: $\Delta n_n = \Delta p_n = 10^{24}$ m$^{-3} \gg N_D$.

have the condition; $\Delta n_n = \Delta p_p \ll N_D$. In Figure 9.7 (b), we illustrate the case for $\Delta n_n = 10^{18}$ m^{-3}. Since we have $\Delta n_n \gg p_{n0}$, the minority carrier level is dominated by excess charge carriers, such that: $p_n = \Delta p_n$. However, low-level injection is characterized by $\Delta n_n \ll N_D$ and therefore $n_n \simeq n_{n0}$. Thus we see for low-level injection, only the minority carrier concentration is significantly affected.

High-level carrier injection corresponds to the case where $\Delta n \gg N_D$. This means that the carrier levels, both majority and minority, are dominated by injection and we have: $\Delta n_n = \Delta p_p \simeq n_n \simeq p_n$. For high-level injection, therefore, both majority and minority carrier concentrations are affected, see Figure 9.7 (c).

9.4.2 Generation and Recombination Processes

When we perturb a system out of its thermal equilibrium state ($np \neq n_i^2$), relaxation processes will come into play to re-establish thermal equilibrium; i.e., $np \to n_i^2$. Typically this process of relaxation is via the recombination of electrons and holes (from conduction to valence band), and thus reduces the number of excess charge carrier until equilibrium is established. Depending on the nature of the transition, energy will be transferred in the process, giving rise to the emission of photons (of energy: $E = \hbar\omega = \Delta$), if the process is radiative or via the production of phonons. In the former, the process is said to be *radiative recombination*, while in the latter it is termed *non-radiative recombination*.

Recombination processes can be classified as *direct* or *indirect*. In a direct process, the electron simply loses energy to other forms, while in an indirect process, energy is transferred in combination with a change of momentum; i.e., $\Delta k \neq 0$. The nature of transitions is very much related to whether the semiconductor is a direct or indirect band-gap material. Indirect processes generally occur via recombination or trapping centers, which have energies situated somewhere in the forbidden band gap region.

We will consider the direct process in more detail, as it is instructive for understanding some of the basic relaxation mechanisms in semiconductors. Let us consider a semiconductor in thermal equilibrium, where the thermal vibrations (phonons) of lattice atoms allow some of the electrons normally occupied in bonding to be freed, thus generating an electron - hole pairs. Such a process of an electron passing from the valence to the conduction band is referred to as *carrier generation*, which we will denote as G_{th}. The reverse process corresponds to the transition of the electron from the conduction to the valence band, in the process annihilating an electron and a hole. This is called recombination, R_{th}. Therefore, at thermal equilibrium the rates of generation and recombination are matched and we will have: $G_{th} = R_{th}$, such that the condition: $np = n_i^2$ is maintained. This situation is schematically represented in Figure 9.8 (a).

When excess charge carriers are introduced into a direct semiconductor, there is a high probability for direct recombination, since it is a more efficient process. In this case the recombination rate will be proportional to the number of electrons available in the conduction band and the number of available holes in the valence band. As such we can write:

$$R = \beta np = \beta n_i^2 \tag{9.42}$$

where β is a constant of proportionality. In the case of thermal equilibrium, the rates of recombination and generation will be equal, and we obtain:

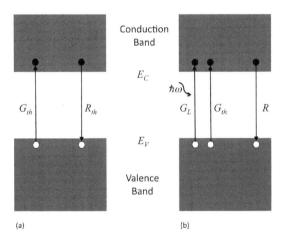

FIGURE 9.8: Direct generation and recombination of electron - hole pairs: (a) Thermal equilibrium and (b) Under illumination with light; $\hbar\omega \geq \Delta$.

$$G_{th} = R_{th} = \beta n_{n0} p_{n0} = \beta n_i^2 \tag{9.43}$$

If we now illuminate the semiconductor with radiation sufficient to produce electron - hole pairs, with a generation rate of G_L, the carrier concentrations will rise above the equilibrium values, see Figure 9.8 (b). This gives the rates of recombination and generation as:

$$R = \beta n_n p_n = \beta (n_{n0} + \Delta n)(p_{n0} + \Delta p) \tag{9.44}$$

and

$$G = G_L + G_{th} \tag{9.45}$$

We expect $\Delta n = \Delta p$, to maintain charge neutrality. We can express the rate of change of the minority charge carriers as:

$$\frac{dp_n}{dt} = G - R = G_L + G_{th} - R \tag{9.46}$$

Under constant illumination the system will eventually arrive at a steady-state condition, at which we will have: $dp_n/dt = 0$, such that the above equation gives:

$$G_L = R - G_{th} \equiv U \tag{9.47}$$

where U is the global recombination rate. Using the above relations for G_{th} and R, we obtain:

$$U = \beta [\Delta n p_{n0} + \Delta p n_{n0} + \Delta n \Delta p]$$
$$\simeq \beta [p_{n0} + n_{n0} + \Delta p] \Delta p \tag{9.48}$$

where we have assumed that $\Delta n = \Delta p$. For the case of low-level injection; $p_{n0} \ll n_{n0}$ and Equation (9.48) gives:

$$U \simeq \beta n_{n0} \Delta p$$
$$= \beta n_{n0} (p_n - p_{n0}) \tag{9.49}$$

Given that U has the units of $m^{-3}s^{-1}$, the above expression can be interpreted if we express it in the following form:

$$U = \frac{(p_n - p_{n0})}{\tau_p} \quad (9.50)$$

where $\tau_p = 1/\beta n_{n0}$. Now we see that this quantity is related to the relaxation of the excited excess charge carriers and is called the *lifetime* of the minority excess charge carriers. We can demonstrate this with a simple example. Let us consider an n-type semiconductor under illumination with light, such that there is an excess of majority carriers (electrons), Δp, and minority (holes) carriers, Δn. Once a certain level of excess carriers has been generated, a stationary state will be established, such that:

$$G_L = U = \frac{(p_n - p_{n0})}{\tau_p} \quad (9.51)$$

which we can express as:

$$p_n = p_{n0} + \tau_p G_L \quad (9.52)$$

We now consider at time, $t = 0$, the light is switched off, which we can express as an initial or boundary condition: $p_n(t=0) = p_n(0) = p_{n0} + \tau_p G_L$. A second boundary condition can be established for long times: $p_n(t \to \infty) = p_{n0}$. This means that for after a certain time we expect the thermal equilibrium to be re-established and the charge carriers will return to the normal concentration. We can now write Equation (9.46) as:

$$\frac{dp_n}{dt} = G - R = -U = -\frac{(p_n - p_{n0})}{\tau_p} \quad (9.53)$$

The solution of this differential equation, considering the boundary conditions, takes the form:

$$p_n(t) = p_{n0} + \tau_p G_L e^{-t/\tau_p} \quad (9.54)$$

This situation is illustrated in Figure 9.9, where the minority charge carriers (holes) decay in time due to the recombination with the majority carriers (electrons).

Thus far we have only discussed the direct type of transition, i.e., where the electrons are excited and relax back between the valence

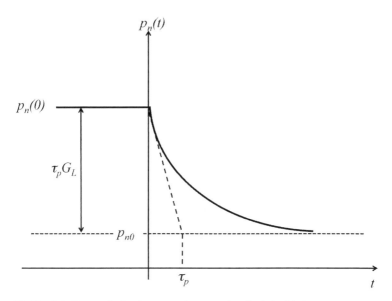

FIGURE 9.9: Decay of minority excess charge carriers (holes) with decay constant, τ_p.

and conduction bands. However, we only expect such direct processes to occur in direct semiconductors. The more common types of semiconductor, such as Si and Ge, are indirect types, and the excitation and relaxation process will take place via so-called *trapping centers*. These are localized states which are somewhere in the band-gap of the semiconductor. The fact that transition probabilities depend on the energy difference between initial and final states means that these trapping centers enhance the recombination and excitation processes. The trapping centers are typically defects and impurities in the crystal lattice and can be controlled in the fabrication process of the semiconductor. The treatment of indirect processes can be made by considering the trapping centers for electrons and holes using the usual occupation probability statistics, and then evaluate the centers as capture or emission sites of charge carriers, depending on whether the are occupied or not.

9.4.3 The Continuity Equations

As we have seen in previous section, there are several physical processes that can occur in a semiconductor, which will affect the transport of charge carriers and hence the electrical properties of semiconductors. In this section we will join these together in to a single guiding equation, which, depending on the specific conditions, can be applied in a number of useful situations to determine the motion of charge carriers in the system. This central relation is called the *continuity equation*.

To derive the form of the continuity equation we will take the one-dimensional case and then extend to three dimensions. We start by considering an element of the semiconductor, say an n-type, with volume: $A\mathrm{d}x$, where A is the cross-sectional area and $\mathrm{d}x$ is the width of the element at position x. The number of electrons in this volume due to the movement into and out of the volume, as well as the generation and recombination inside the volume must be considered simultaneously. The rate of change of concentration of electrons will therefore be the difference of those entering and leaving the element at x and $x + \mathrm{d}x$ as well as the difference in the rates of generation and recombination. In the case of the electrons entering and leaving the element, these are proportional to the current on either side of the element, while the generation and recombination are G_n and R_n, respectively. The net rate of change of electron concentration in our elemental volume can be expressed as:

$$\frac{\partial n}{\partial t} A\mathrm{d}x = \left[\frac{J_n(x)A}{-e} - \frac{J_n(x+\mathrm{d}x)A}{-e} \right] + (G_n - R_n) A\mathrm{d}x \quad (9.55)$$

We now make a Taylor expansion of $J_n(x + \mathrm{d}x)$ at $x + \mathrm{d}x$:

$$J_n(x+\mathrm{d}x) = J_n(x) + \frac{\partial J_n(x)}{\partial x}\mathrm{d}x + \ldots \quad (9.56)$$

We can thus write the basic form of the continuity equation as:

$$\frac{\partial n}{\partial t} = \frac{1}{e}\frac{\partial J_n}{\partial x} + (G_n - R_n) \quad (9.57)$$

Clearly a consideration of the variation of holes in the same element will follow the same rationale, and we can write the corresponding hole continuity equation as:

$$\frac{\partial p}{\partial t} = -\frac{1}{e}\frac{\partial J_p}{\partial x} + (G_p - R_p) \quad (9.58)$$

A more detailed version of the continuity equation can be obtained if we introduce the current density, $J_n(x)$, as expressed in Equation (8.42). Considering the variation of minority charge carriers can now be expressed, for low-level injection, by taking the derivative of the current density with position from the first term in the above. For the case of electrons in a p-type semiconductor we obtain:

$$\frac{\partial n_p}{\partial t} = n_p \mu_n \frac{\partial \varepsilon}{\partial x} + \mu_n \varepsilon \frac{\partial n_p}{\partial x} + D_n \frac{\partial^2 n_p}{\partial x^2} + G_n - \frac{(n_p - n_{p0})}{\tau_n} \quad (9.59)$$

The corresponding expression for minority carriers (holes) in an n-type material has the form:

$$\frac{\partial p_n}{\partial t} = -p_n \mu_p \frac{\partial \varepsilon}{\partial x} - \mu_p \varepsilon \frac{\partial p_n}{\partial x} + D_p \frac{\partial^2 p_n}{\partial x^2} + G_p - \frac{(p_n - p_{n0})}{\tau_p} \quad (9.60)$$

where ε is the electric field. A three-dimensional form of these equations can be written in the form:

$$\frac{\partial n_p}{\partial t} = n_p \mu_n \nabla \cdot \boldsymbol{\varepsilon} + \mu_n \boldsymbol{\varepsilon} \cdot \nabla n_p + D_n \nabla^2 n_p + G_n - \frac{(n_p - n_{p0})}{\tau_n} \quad (9.61)$$

$$\frac{\partial p_n}{\partial t} = -p_n \mu_p \nabla \cdot \boldsymbol{\varepsilon} - \mu_p \boldsymbol{\varepsilon} \cdot \nabla p_n + D_p \nabla^2 p_n + G_p - \frac{(p_n - p_{n0})}{\tau_p} \quad (9.62)$$

In addition to the continuity equations, the Poisson equation should also be satisfied, where we have:

$$\frac{d\varepsilon}{dx} = \frac{\rho_s}{\epsilon_s}, \quad \text{for} \quad 1D \quad \text{and} \quad \nabla \cdot \boldsymbol{\varepsilon} = \frac{\rho_s}{\epsilon_s}, \quad \text{for} \quad 3D \quad (9.63)$$

where ϵ_s is the dielectric constant of the semiconductor and ρ_s the charge density. In principle Equations (9.61) - (9.63) can be used in conjunction with the relevant boundary conditions to solve most practical situations. However, the complexity of these equations at times requires some form of simplification. Some examples will be considered in the exercise section.

9.5 THE P - N JUNCTION

When we join a p-type semiconductor with an n-type semiconductor, the result is a rather complex non-linear asymmetric device, known as the *p-n junction*. Its principal property is that of *rectification*, in which, under ideal circumstances, electrical current can only pass in one direction. Such a device is called a *diode*. In practice, there is usually some *leakage current* in reverse bias. However, it is very small and usually negligible when compared to the forward bias current. At relatively high reverse bias voltage, the p-n junction will suffer *breakdown*, at a specific value called the *breakdown voltage*, V_B. At this point the diode can become permanently damaged due to avalanche effects, where electrons are accelerated by the reverse bias potential and collide with atoms and free further electrons. At this point the current rapidly increases and rectification is lost.

The p-n junction plays a very important role in semiconductor applications and is at the heart of many electronic devices, such as transistors, LEDs, lasers and photovoltaic devices or solar cells. Given the importance of the p-n junction in semiconductor electronics, we will consider the physics of the junction in some detail.

9.5.1 Thermal Equilibrium

In the previous section, we saw that the Fermi level of a doped semiconductor shifts with respect to the center of the gap, for and undoped semiconductor. In semiconductors which are not heavily doped, in general, the n-type semiconductor has its Fermi energy close to the conduction band edge, while the p-type has its Fermi energy just above the valence band edge. Once we join the two materials, in the formation of the p-n junction, there will be a strong concentration gradient of charge carriers and diffusion will occur as a result. Holes from the p-side will diffuse across the interface towards the n-side, while electrons will make the opposite journey. The diffusion of charge carriers in the region of the interface between the two types of semiconductor will mean that the n-side adjacent to the interface will have a net positive charge, since the donors will be left ionized positively; N_D^+. On the p-side the opposite will be true, where a net negative charge will remain due to ionized acceptors; N_A^-. The means that in the region of the interface, there will be a region of *space-charge* distribution, which causes an electric field to permanently exist in the region. This electric field will cause the movement of charge carriers as a drift current. In total, we will have diffusion currents from the concentration gradient and drift currents from the electric field which continue until an equilibrium of charge "re-distribution" is established. The thermal equilibrium of the system will be a stationary state at a specific temperature, T, and zero net current will flow. Therefore we can establish the partial currents for the charge carriers as a sum of the diffusion and drift components, such that:

$$J_n = e\mu_n n\varepsilon + eD_n \nabla n = 0 \quad (9.64)$$

and

$$J_p = e\mu_p p\varepsilon - eD_p \nabla p = 0 \quad (9.65)$$

In one dimension we can write these equations as:

$$J_n = e\mu_n n \left(\frac{1}{e}\frac{dE_{F_i}}{dx} \right) + k_B T \mu_n \frac{dn}{dx} = 0 \quad (9.66)$$

and

$$J_p = e\mu_p p \left(\frac{1}{e}\frac{dE_{F_i}}{dx} \right) - k_B T \mu_p \frac{dp}{dx} = 0 \quad (9.67)$$

where we have used: $e\varepsilon = dE_{F_i}/dx$, in which ε is the electric field. We have also introduced the Einstein relation, Equation (8.41). We can express the concentration of electrons using the following relation:

$$n = n_i e^{(E_F - E_{F_i})/k_B T} \quad (9.68)$$

Using this expression, we can write the gradient, in one dimension, of the electron concentration as:

$$\frac{dn}{dx} = \frac{n}{k_B T} \left(\frac{dE_F}{dx} - \frac{dE_{F_i}}{dx} \right) \quad (9.69)$$

Substituting this result in Equation (9.66) we obtain:

$$J_n = \mu_n n \frac{dE_F}{dx} = 0 \quad \Rightarrow \quad \frac{dE_F}{dx} = 0 \qquad (9.70)$$

The same analysis can be performed using the hole concentration. Equation (9.70) shows that at thermal equilibrium the Fermi level is constant and independent of position. In Figure 9.10, we illustrate the principle features of the p-n junction at thermal equilibrium. The result given in Equation (9.70) has the consequence of causing the bands to bend as a function of position through the junction region, since the Fermi level must remain constant at all positions across the p-n junction. The magnitude of the bending can be expressed as the quantity, eV_{bi}, where V_{bi} is referred to as the *built-in potential*. We can also note from the figure that the shift of the Fermi level from its intrinsic value can be associated with the electrostatic potential: n-side - $E_F - E_{F_i} = e\psi_n$ and p-side - $E_{F_i} - E_F = e\psi_p$. The exchange of charge carriers at the interface between the n and p

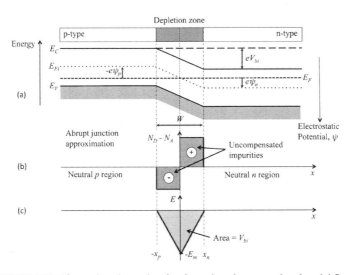

FIGURE 9.10: The p-n junction, using the abrupt junction approximation. (a) Band structure diagram at thermal equilibrium. The Fermi level is constant across the junction. (b) Space-charge distribution, showing the distribution of uncompensated impurities on the p and n sides. The extension of this region is called the *depletion zone*, where there are no un-ionized impurities. (c) The electric field distribution at the junction, which results from the space charge in the depletion zone. The area of this shaded region corresponds to the built-in potential, V_{bi}.

semiconductors produces a complete ionization of the impurities in the region of the junction, and is hence called the *depletion zone*. The fixed space charge will introduce a net electric field in this region, in the direction from the n- to the p-side. The area under the curve of the electric field will correspond to the built-in potential.

9.5.2 The Depletion Zone

The space charge distribution and electrostatic potential are related via the Poisson relatoion:

$$\frac{d^2\psi}{dx^2} = -\frac{d\varepsilon}{dx} = -\frac{\rho_s}{\epsilon_s} = -\frac{e}{\epsilon_s}(N_D - N_A + p - n) \quad (9.71)$$

In the regions away from the junction, the charge neutrality of the semiconductor is maintained and the space-charge is zero. Therefore in these neutral zones we have:

$$\frac{d^2\psi}{dx^2} = 0 \quad \text{and} \quad N_D - N_A + p - n = 0 \quad (9.72)$$

In the neutral n-region, $N_A = 0$ and $n \gg p$. The electrostatic potential can be obtained from the above in conjunction with Equation (9.68) and putting: $n = N_D$:

$$N_D = n_i e^{-e\psi_n/k_B T} \quad (9.73)$$

which gives:

$$\psi_n = -\frac{1}{e}(E_{F_i} - E_F)_{x \geq x_n} = \frac{k_B T}{e} \ln\left(\frac{N_D}{n_i}\right) \quad (9.74)$$

We can obtain the corresponding relation in the p-region as:

$$\psi_p = -\frac{1}{e}(E_{F_i} - E_F)_{x \leq -x_p} = -\frac{k_B T}{e} \ln\left(\frac{N_A}{n_i}\right) \quad (9.75)$$

From this we can obtain the built-in potential:

$$V_{bi} = \psi_n - \psi_p = \frac{k_B T}{e} \ln\left(\frac{N_D N_A}{n_i^2}\right) \quad (9.76)$$

The extension of the depletion region can be evaluated using the Poisson relation in the space-charge region, where we have:

$$\frac{d^2\psi}{dx^2} = \frac{eN_A}{\epsilon_s}; \quad -x_p \leq x \leq 0 \qquad (9.77)$$

$$\frac{d^2\psi}{dx^2} = -\frac{eN_D}{\epsilon_s}; \quad 0 \leq x \leq n_n \qquad (9.78)$$

To maintain charge neutrality we require: $N_A x_p = N_D x_n$. We also note that the width of the depletion zone is: $W = x_p + x_n$. The electric field distribution in the depletion region is evaluated from the first integration of Equations (9.77) and (9.78):

$$\varepsilon(x) = -\frac{d\psi}{dx} = -\frac{eN_A}{\epsilon_s}(x+x_p); \quad -x_p \leq x \leq 0 \qquad (9.79)$$

$$\varepsilon(x) = -\frac{d\psi}{dx} = -\varepsilon_m + \frac{eN_D}{\epsilon_s} = \frac{eN_D}{\epsilon_s}(x-x_n); \quad 0 \leq x \leq n_n \qquad (9.80)$$

where ε_m is the maximum of the electric field at $x = 0$, which we can write as:

$$\varepsilon_m = \frac{eN_D}{\epsilon_s}x_n = \frac{eN_A}{\epsilon_s}x_p \qquad (9.81)$$

It is now a simple matter to evaluate the built-in potential as:

$$V_{bi} = \frac{e}{2\epsilon_s}(N_A x_p^2 + N_D x_n^2) \qquad (9.82)$$

Recognizing this as being the area under the curve in Figure 9.10 (c); $V_{bi} = \varepsilon_m W/2$, it is possible to obtain:

$$W = \left[\frac{2\epsilon_s}{e}\frac{(N_A + N_D)}{N_A N_D}V_{bi}\right]^{1/2} \qquad (9.83)$$

It is further possible to show that the application of an external electric potential to the junction, will alter the width of the depletion region such that:

$$W = \left[\frac{2\epsilon_s}{e}\frac{(N_A + N_D)}{N_A N_D}(V_{bi} - V)\right]^{1/2} \qquad (9.84)$$

Where for a forward bias (+V), the width of the depletion zone diminishes, while for reverse bias (−V) it will increase.

9.5.3 Junction Capacitance

Since the depletion region has no available charge carriers it can act as a capacitance, with electrical capacity:

$$C_{pn} = \frac{dQ}{dV} \qquad (9.85)$$

where dQ represents an increment in the charge of the depletion zone per unit area with increase dV in the applied voltage. This in turn will produce an increase of the electric field associated with the junction region, due to the increased numbers of charge carriers in both the n and p regions in equal amounts. We can express the increase of electric field with charge as: $d\varepsilon = dQ/\epsilon_s$. The increase of the potential can be thus given as: $dV = W d\varepsilon = W dQ/\epsilon_s$. It is now possible to express the junction capacitance as:

$$C_{pn} = \frac{dQ}{W dQ/\epsilon_s} = \frac{\epsilon_s}{W} \qquad (9.86)$$

In units of Fcm^{-2}. This equation has the form of the parallel plate capacitor, where the plate separation is given as the depletion region thickness. The above derivation considers that the polarization of the potential is in the reverse bias direction only. For forward biases there will be a large current through the junction.

9.5.4 Current - Voltage Characteristics

Once a potential is applied to the p-n junction, the conditions of thermal equilibrium will be disturbed. We need to consider the effect of both forward and reverse bias potentials. For forward bias conditions, the applied potential will reduce the electrostatic potential across the depletion zone, which will in turn diminish the width of the depletion zone. As a consequence, the drift current will be reduced and the diffusion current will be enhanced, injecting minority carriers from either side of the junction towards the opposite side. In the case of reverse bias, the opposite will be true and the width and electrostatic potential will increase. The effect will be to remove minority carriers from the regions adjacent to the depletion zone on both n and p sides. These changes are illustrated in Figure 9.11.

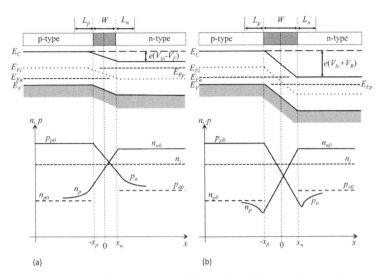

FIGURE 9.11: Depletion regions for (a) forward and (b) reverse bias conditions. Also shown in both cases are the band diagrams and charge carrier distributions across the p-n junction.

We will here consider the ideal current - voltage $(I - V)$ characteristics, which are based on the following assumptions: i) An abrupt depletion zone, ii) the densities of charge carriers at the boundaries are related to the electrostatic potential difference at the junction, iii) low-level injection, iv) there are no significant current due to the generation and recombination of charge carriers in the depletion zone.

At thermal equilibrium, the majority carrier concentration is essentially given by the doping level, and for zero applied voltage we have a built-in potential of:

$$V_{bi} = \frac{k_B T}{e} \ln\left(\frac{p_{p0} n_{n0}}{n_i^2}\right) = \frac{k_B T}{e} \ln\left(\frac{n_{n0}}{n_{p0}}\right) \qquad (9.87)$$

where we have used the law of mass action in the form: $n_i^2 = p_{p0} n_{p0}$. We can re-write this as:

$$n_{n0} = n_{p0} e^{eV_{bi}/k_B T} \qquad (9.88)$$

Similarly, we can establish the relation:

$$p_{p0} = p_{n0} e^{eV_{bi}/k_B T} \qquad (9.89)$$

Indeed, these expressions form part of the boundary conditions for the depletion zone, relating the carrier densities with the built-in potential at thermal equilibrium. It is a simple matter to extend this principle to form the boundary conditions for the ideal $I - V$ characteristics when an applied voltage exists between the p and n sides of the junction:

$$n_n = n_p e^{e(V_{bi}-V)/k_BT} \tag{9.90}$$

being the relevant boundary condition for non-equilibrium when a potential is applied. For conditions of low-level injection, the density of minority carriers injected is much less than the density of majority carriers, therefore we have: $n_n \simeq n_{n0}$. Applying this, along with Equation (9.90), we obtain the boundary condition for the electron density at the p-side of the depletion region ($x = -x_p$):

$$n_p = n_{p0} e^{eV/k_BT} \quad \text{or} \quad n_p - n_{p0} = n_{p0}(e^{eV/k_BT} - 1) \tag{9.91}$$

In an analogous manner we obtain the following for the boundary condition for hole injection on the n-side (at $x = x_n$):

$$p_n = p_{n0} e^{eV/k_BT} \quad \text{or} \quad p_n - p_{n0} = p_{n0}(e^{eV/k_BT} - 1) \tag{9.92}$$

To obtain the central governing equation for the p-n junction we need to consider the currents that flow in the system, remembering that the only currents to exist are those coming from the neutral regions on the p- and n-sides. This means we can assess the current flow in the neutral zone, where there is no electric field. For the n-side, considering the stationary state of the system, the continuity equation can be expressed as:

$$\frac{dp_n^2}{dx^2} - \frac{p_n - p_{n0}}{D_p \tau_p} = 0 \tag{9.93}$$

The solution to this equation is arrived at by using the correct boundary conditions, as expressed by Equation (9.92) and $p_n(x = \infty) = p_{n0}$, to give:

$$p_n - p_{n0} = p_{n0}(e^{eV/k_BT} - 1)e^{-(x-x_n)/L_p} \tag{9.94}$$

where $L_p = \sqrt{D_p \tau_p}$ is the diffusion length of holes in the n-region. The hole current is then obtained from:

$$J_p(x_n) = -eD_p \left(\frac{dp_n}{dx}\right)_{x_n} = \frac{eD_p p_{n0}}{L_p}(e^{eV/k_BT} - 1) \qquad (9.95)$$

In similar fashion we can write the electron concentration and current in the p-region as:

$$n_p - n_{p0} = n_{p0}(e^{eV/k_BT} - 1)e^{(x+x_p)/L_n} \qquad (9.96)$$

and

$$J_n(-x_p) = eD_n \left(\frac{dn_p}{dx}\right)_{-x_p} = \frac{eD_n n_{p0}}{L_n}(e^{eV/k_BT} - 1) \qquad (9.97)$$

where $L_n = \sqrt{D_n \tau_n}$ is the diffusion length for electrons in the n-region. We can now join the expressions for the current density to obtain the total current density due to the contributions to both type of charge carrier:

$$J = J_p(x_n) + J_n(-x_p) = J_s(e^{eV/k_BT} - 1) \qquad (9.98)$$

where the saturation current, J_s is expressed as:

$$J_s = \frac{eD_p p_{n0}}{L_p} + \frac{eD_n n_{p0}}{L_n} \qquad (9.99)$$

Equation (9.98) is called the *ideal diode* equation. The form of this equation is an exponential increase in the forward bias direction and a low current in the reverse bias polarization, defined by the *saturation* or *leakage current*, J_s. These characteristics are shown for Ge and Si p-n junction diodes in Figure 9.12. We note the different scales on the different portions of the graph. The leakage currents are of the order of μ A, while breakdown occurs at elevated voltages in the reverse bias configuration.

The p-n junction serves as the base for many semiconductor devices. In addition to the rectifying properties of the p-n junction diode, applications of the p-n interface are numerous. These include

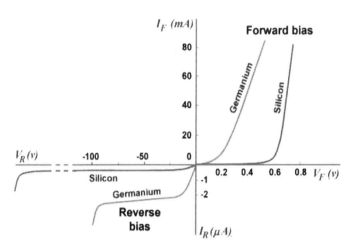

FIGURE 9.12: Current - voltage characteristics of p-n junction diodes made from Si and Ge. Note the difference in the forward and reverse bias voltage and current scales.

the LED (light emitting diode), the photovoltaic or solar cell and the semiconductor laser. Furthermore, the p-n junction forms the basic element in more complex structures, such as the bipolar transistor, the thyristor, and in field effect transistors (FET). The detailed description of these devices goes beyond the scope of the book. The interested reader is encouraged to consult specialized books, for example Sze (1985) is a good starting point.

9.6 HETEROSTRUCTURES AND QUANTUM WELLS

With the development of advanced preparation techniques for the deposition of thin films of semiconductors, enormous progress was made in the research and development in semiconductor devices. Of the available techniques, molecular beam epitaxy (MBE) and metal-organic chemical vapor deposition (MOCVD) are the principal methods that are employed for the preparation of thin and ultrathin films. MBE is essentially a research tool that allows for the preparation of high quality single crystal (epitaxial) films and multlilayered structures with excellent control over film thicknesses and preparation conditions. It generally employs a number of in-situ analytical techniques to monitor the structure and chemical purity of the depositing layers. Such tools are usually the reflection high-energy electron diffraction (RHEED) and Auger electron spectroscopy (AES) techniques. The former uses a high energy electron beam at low angles of incidence on the film surface to produce diffraction patterns, which permit a detailed study of the sample crystallinity and structural properties, see Chapter 2. The latter technique also employs an electron beam, though in this case we are interested in inelastic processes and the excitation of the atoms of the surface, which via the Auger process emits electrons of energies specific to the atomic species at the sample surface, see Chapter 13. In this way it is possible to chemically characterize the sample.

In the Introduction, we discussed the various types of semiconductor materials and their band structures. It is possible to modify the band structure and importantly the band gap itself by varying the composition of the semiconductor. For example, we can introduce Al into the GaAs semiconductor in controlled quantities and alter the properties of the semiconductor. The study of these properties led to the so-called *band structure engineering* of semiconductors, where we can manipulate the composition to obtain the desired band gap and hence properties to perform specific device functions. As an example, we may require a semiconductor device (LED) to emit light of a certain wavelength, we would then wish to adjust the band gap of the semiconductor to a certain energy which corresponds to the desired wavelength; $\lambda = hc/\Delta$. In Figure 9.13, we illustrate the relationship between the band gap energy of various semiconductors with their lattice parameter. It should be noted that the lines between the binary alloys corresponds to the ternary compounds with varying composition. It is clear that the lattice parameter is a function of the composition of the compound and thus affects the band structure and band gap correspondingly. The line between the binary points for GaAs and AlAs correspond to the various compositions of the $Al_xGa_{1-x}As$ compound. We note here that for $x \geq 0.35$ the semiconductor goes from a direct to an indirect band gap material. Importantly, there is virtually no change in the lattice parameter

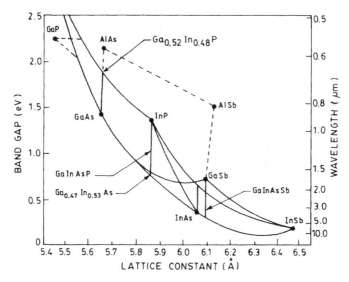

FIGURE 9.13: Variation of semiconductor band gap energy with lattice parameter for some III-V and II-VI compounds. Solid lines correspond to semiconductors with direct band gaps, while the dashed lines indicate indirect band gap materials. Wavelengths for the band gap transition are indicated on the right hand side of the graph.

between the extremal points (corresponding to GaAs and AlAs), meaning that it is very suitable for fabricating heterostructures with little or no lattice strain.

It is clear that from Figure 9.13 that we can simply choose the band gap we require and deposit the material with the correct composition. Interesting things begin to happen when we deposit various different layers with different compositions. Such a multilayer structure is referred to as a *heterostructure*. The MBE technique is especially adapted to the preparation of such types of structures, since we can control the deposition to a high degree of precision. In a simple case we can join three layers to make a *quantum well* (QW). The quantum well has a number of specific properties which depend on both intrinsic and extrinsic characteristics of the heterostructure.

The QW structure can be characterized in terms of the various energies of the band gaps and the thickness of the well layer itself, see Figure 9.14. The difference in band edge energies are important in defining the depth of the quantum well and depending on

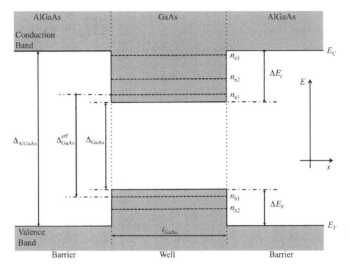

FIGURE 9.14: Band diagram of a quantum well (QW) structure, where we show the band edges as a function of the deposition thickness in the x - direction. We Indicate AlGaAs and GaAs as the layers, where $\Delta_{AlGaAs} > \Delta_{GaAs}$ gives a region in which the electrons and holes can be spatially confined within the thickness of the well (GaAs) layer, t_{GaAs}.

the doping (and hence Fermi energies) the well depths for holes in the valence band, ΔE_V, and electrons in the conduction band, ΔE_C can be different. It is possible using simple quantum mechanics to define the eigenstates of these quantum wells. If we approximate them as infinite wells we can obtain the analytical expressions. These can be expressed as:

$$E_e = \frac{\hbar^2}{2m_e^*}\left(\frac{n_e^2 \pi^2}{t_{GaAs}^2} + k_y^2 + k_z^2\right) \quad (9.100)$$

for electrons, and

$$E_h = \frac{\hbar^2}{2m_h^*}\left(\frac{n_h^2 \pi^2}{t_{GaAs}^2} + k_y^2 + k_z^2\right) \quad (9.101)$$

for holes, where n_e and n_h are integers, with values 1, 2, 3, etc, for the electrons and holes in the QW conduction and valence bands, respectively. One of the effects of quantum confinement is that the energy states become discrete in the direction of the spatial confinement, which in our example is in the x-direction. We note that the energies of these confined states which occur in the well are referred to as *bound states*. There will now be no energy state available at the GaAs valence and conduction band edges since the numbers n_e and n_h cannot be zero. This means that the effective band gap in the well (GaAs) layer, Δ_{GaAs}^{eff}, is generally larger than its bulk value, Δ_{GaAs}, and can be expressed in the relation:

$$\Delta_{GaAs}^{eff} = \Delta_{GaAs} + \frac{\hbar^2 \pi^2}{2 t_{GaAs}^2} \left(\frac{1}{m_e^*} + \frac{1}{m_h^*} \right) \quad (9.102)$$

In terms of the expressions given for the electrons and holes, Equations (9.100) and (9.101), each value of n_e and n_h produce their own *sub-band* in the conduction and valence bands, respectively. These are indicated in Figure 9.14 as n_{e1}, n_{e2} and n_{h1}, n_{h2} etc, for the conduction and valence bands. We note that the effective masses of the charge carriers will affect the size of the effective band gap, but it is via the thickness of the well layer that we can really manipulate the properties of the QW structure, since we can finely control this parameter during the growth process. It is important to stress here that while there is electron confinement in the x - direction, which is the cause of the quantization effects, no such restrictions exist in the y and z - directions, and electrons are free to move in the plane of the layer. In fact, in these directions, they are still subject to the usual dispersion relations, which are indicated as parabolic in Equations (9.100) and (9.101).

Clearly there is enormous scope for tailoring bands for specific energy transitions, and this is frequently used in device architecture, particularly for optoelectronic applications, such as lasers and optical detectors. In fact, often many more layers are used in the QW structures and form what are termed *superlattices*, due to the formation of periodic structures with periodicities greater than the crystal lattice parameter. There are several types of superlattice, which are related to their band structures. Due to the relative energies involved, these quantum effect become more noticeable for thicknesses below a few tens of nm. This really takes us into the realm of *nanotechnologies* and we will have more to say on this subject in the final chapter.

9.7 SUMMARY

Semiconductors form an important class of materials, having very adaptable electronic properties which make them extremely manipulable for electronic and optical device applications. At very low temperatures, intrinsic semiconductors have no free charge carriers for electrical conduction, and thermal excitation is require to liberate them from bound states on the semiconductor atoms. The concentrations of electrons and holes in an intrinsic semiconductor can be evaluated using the Fermi distribution function and the density of states. The introduction of controlled concentrations of impurity atoms offers another method of controlling the electronic properties of semiconductors. The impurities can either furnish additional electrons or remove electrons from bound states to create hole states. In either case the impurities are used to create specific concentrations of charge carriers, which are then essentially free to contribute to electrical conduction. Materials which have donor impurities are called n-type because they have electrons (with negative charge) as the majority charge carriers. On the other hand, semiconductors with acceptor impurities are p-type, as the majority carriers are holes (having positive charge). The equilibrium statistics allows the determination of carrier concentrations as a function of temperature as well as to determine the variation of the Fermi level in the solid.

Specific applications of potentials and/or the incidence of electromagnetic radiation on the semiconductor can introduce excess charge carriers in a process called injection. The level of injection will depend on the doping concentrations in the semiconductor, where the situation is one of non-equilibrium. The excitation process, called generation, can be achieved by the application of light of an energy equal to or greater than the band gap energy. This energy represents a binding energy of electrons, in the valence band, to be liberated from localized states on atoms. Such a process will produce excesses of both electrons and holes. The relaxation process, called recombination, occurs when an electron "descends" from the higher energy of the conduction band back to a bound state in the valence band, thus annihilating an electron and hole pair. Such a process will have a specific lifetime and is generally material dependent. The process can occur directly in a direct band gap material or via trapping (impurity) states in the band gap in indirect semiconductors.

We can take into account the various process of generation, recombination, diffusion, and drift processes in the continuity equations. These are versatile equations that allow us to assess specific conditions in the semiconductor and determine how the charge carriers will behave. These can be applied in many useful applications that are frequently found in devices.

One of the most important structures in semiconductor device physics is the p-n junction, which forms the basis of many applications, such as diodes, LEDs, lasers, as well as being the essential component in transistors. The interface between the n and p-type semiconductors is an important region and controls the motion of charge carriers in the device as a whole. The p-n junction is a rectifier, allowing current to flow more freely in one direction, while hindering motion in the other. The interface region, called the depletion zone, usually extends in the region of a few microns, and in thermal equilibrium presents a variation of space-charge. This fixed space charge occurs due to the exchange of electrons and holes from the n and p-sides on the junction and produces an electric field or built-in potential at the interface. It is this which prevents the flow of charge in the so-called reverse direction. We used the continuity equations as well as the Poisson equation to analyze the p-n junction. This allowed us to determine the size of the built-in potential and extent of the depletion zone in terms of the doping concentrations in the n and p semiconductors. We also derived the diode equation, which shows the current - voltage characteristics of the device.

It is possible, using modern deposition techniques to create layered structures of different compositions of semiconductor, to produce artificial properties and manipulate the band structures in semiconductor devices. The modification of lattice parameter can be used to adjust the band gap of a semiconductor and can be tailored to a specific energy and thus working wavelength for optoelectronic devices. Additionally, the growth of thin and ultrathin layers of different semiconductor materials with different band gap energies can be used to confine the motion of charge carriers.

Such heterostructures are called quantum wells. The are the basis of many modern devices, such as lasers, LEDs as well as transistors and optical detectors.

REFERENCES AND FURTHER READING

Basic Texts

- S. M. Sze, *Semiconductor Devices: Physics and Technology*, J. Wiley and Sons, (1985)
- J. S. Blakemore, *Solid State Physics*, Cambridge University Press, Cambridge (1985)
- H. P. Myers, *Introductory Solid State Physics*, Taylor and Francis, London (1998)

Advanced Texts

- M. Balkanski and R. F. Wallis, *Semiconductor Physics and Applications*, Oxford University Press, Oxford (2000)
- Ming-Fu Li, *Modern Semiconductor Quantum Physics*, World Scientific, Singapore (1994)
- N. W. Ashcroft and N. D. Mermin, *Solid State Physics*, Saunders College, Philadelphia (1976)
- C. M. Wolfe, J. Holonyak and G. E. Stillman, *Physical Properties of Semiconductors*, Prentice - Hall (1989)

EXERCISES

Q1. Evaluate the probability that a state with energy 0.01 eV below the Fermi energy is unoccupied at temperatures of 300, 500, and 700 K.

Q2. Consider the case of an ideal insulator in which the concentrations of electrons in the conduction band and holes in the valence band take the form:

$$n_c = \frac{g}{e^{(E_c - E_F)/k_B T} + 1} \qquad (9.103)$$

and

$$n_v = \frac{g}{e^{(E_F - E_v)/k_B T} + 1} \qquad (9.104)$$

in which g represents the number of states in each band per unit volume. Show: (a) that the Fermi level sits exactly in the middle of the band gap. (b) that the density of electrons in the conduction band can be approximated as:

$$n_c \simeq g e^{-\Delta/2 k_B T} \qquad (9.105)$$

Q3. Derive expressions for the Fermi level in an intrinsic semiconductor in terms of the effective masses, m_c^* and m_v^*, of the charge carriers in the conduction and valence bands.

Q4. Use the law of mass action to show that the intrinsic carrier concentration takes the general form:

$$n_i(T) = \sqrt{N_c(T) N_v(T)} e^{-\Delta/2 k_B T} \qquad (9.106)$$

Q5. Find expressions for the Fermi energy for extrinsic semiconductors in terms of the doping concentrations, N_D and N_A, for n and p-type materials.

Q6. Calculate the position of the intrinsic fermi level for silicon at the following temperatures: i) Liquid nitrogen ii) Room temperature and iii) 100°C. Use the following values of the effective masses: $m_c^* = 0.3 m_e$ and $m_v^* = 0.5 m_e$. Comment on whether it is reasonable to assume that the intrinsic Fermi level lies in the middle of the band gap. N.B. $\Delta_{Si} = 1.12$ eV.

Q7. Calculate the concentration of charge carriers and the Fermi level in a sample of silicon at room temperature for the case of doping with 10^{16} atoms of As (cm^{-3}) and 3×10^{16} atoms of B (cm^{-3}). State whether the sample is n or p-type.

Q8. Using Equation (9.1) show that the intrinsic conductivity for a semiconductor can be expressed as:

$$\sigma_i(T) = n_i(T)e\mu_p(b+1) \qquad (9.107)$$

where $b = \mu_n/\mu_p$. Further show that in the extrinsic case we can write the conductivity as:

$$\sigma = \frac{\sigma_i}{n_i}\left(\frac{nb+p}{b+1}\right) \qquad (9.108)$$

Q9. Prove Equation (9.54).

Q10. A sample of pure silicon, with dimensions: $10 \times 5 \times 1$ mm, is heated at one of its extremities (short end) to a temperature of 600 K, while the other end (10 mm away) is maintained at room temperature. Assuming that the temperature varies linearly with distance in the sample, evaluate the variation of the following (graphically or at specific points along the sample):

a) Concentration of charge carriers, n_i.

b) The charge carrier concentration gradient, dn_i/dx.

c) The diffusion current.

Use the following constants for Si: $\Delta_{Si} = 1.12$ eV; $m_n^* = 0.3 m_e$; $m_p^* = 0.5 m_e$; $(\mu_n)_{Si} = 0.145 \text{m}^2\text{V}^{-1}\text{s}^{-1}$; $(\mu_p)_{Si} = 0.045 \text{m}^2\text{V}^{-1}\text{s}^{-1}$.

Q11. A Si pn junction is formed using the following doping concentrations; $N_A = 10^{19}\text{cm}^{-3}$ and $N_D = 10^{16}\text{cm}^{-3}$. Calculate the changes in the depletion zone thickness with an applied voltage of 0.6V (forward bias) and -0.9V (reverse bias). Also evaluate the effect of these potentials on the maximum electric field in the pn junction. For all calculations use room temperature conditions and assume an intrinsic carrier concentration of $n_i = 1.45 \times 10^{10} \text{cm}^{-3}$ and a dielectric constant of $\epsilon_r = 11.7$ for silicon.

Q12. Derive the diode equation.

Q13. Derive Equation (9.102).

NOTES

[1] We should strictly use the chemical potential here, as we stated in Section 6.7. However, for the low temperature approximation we take $\mu \simeq E_F$.

[2] The integral has the form: $\int_0^\infty x^{1/2} e^{-x} \, dx = \pi^{1/2}/2$

[3] The existence of heavy and light holes derives from the gradients of the valence energy bands, where a steep curve will give rise to light effective masses, while a shallow curve produces a heavy-mass band. See Section 8.3 for further explanation.

CHAPTER 10

MAGNETIC MATERIALS AND PHENOMENA

"The fundamental laws necessary for the mathematical treatment of a large part of physics and the whole of chemistry are thus completely known, and the difficulty lies only in the fact that application of these laws leads to equations that are too complex to be solved."

—Paul A. M. Dirac

"Politics is the art of looking for trouble, finding it everywhere, diagnosing it incorrectly and applying the wrong remedies."

—Groucho Marx

10.1 INTRODUCTION

When a magnetic field is applied to a solid, it will react in a way that reflects the manner in which the electrons in that solid are ordered in the atoms and how they interact among themselves. The magnetic field interacts with the electrons in the atoms of the solid, and since all solids have atoms with electrons, all solids must respond in some form to the application of a magnetic field. It is the form of this interaction and reaction that distinguishes materials into the various categories. Strong responses arise from those materials whose atoms have a magnetic moment and these magnetic moments interact amongst themselves. In fact, materials are said to be magnetic only when there exists some form of magnetic order between the magnetic moments on the constituent atoms, where there is a strong response to a magnetic field. However, the situation isn't as simple as that. In many cases substances can have magnetically ordered phases, but there is still a weak response to a magnetic field. Indeed, it is one of the principal objectives of the present chapter to discuss magnetic ordering in solids and how this affects the bulk properties of the system.

In terms of the elements of the periodic table, there are very few natural ferromagnetic solids. The transition elements iron, cobalt and nickel are the only transition metals which display ferromagnetism at room temperature. Chromium has antiferromagnetic order at room temperature and gadolinium is just ferromagnetic at room temperature. At low temperature further rare-earth elements display magnetic ordering. The details of exactly what ferromagnetism and antiferromagnetism is will be discussed in later sections of this chapter. In addition to the elemental species, alloys and compounds can be found and fabricated which also exhibit different magnetic ordering. There are many examples and some fairly complex systems can be found. We will mention some examples of these as we discuss the various types of magnetic order.

Apart from their scientific interest, magnetic materials have a number of very important applications, some of which we find on a daily basis. Probably the most widely found applications of magnetic materials are in transformers, in motors and the hard disk drives found in most computers. While we will not go in to the details of these applications it is important that we are aware of the importance of this class of material.

In this chapter, we will firstly look at how atoms *acquire* their magnetic moment, which will be outlined in the following section. A vast majority of materials have very weak responses to an applied magnetic field. These can be separated into paramagnetic and diamagnetic materials, and these will form the topics of the subsequent sections. Ordering of the magnetic moments can only arise if there is an interaction between them, which is characterized as the *exchange* interaction, of which there are a number of mechanisms, which can be either direct or indirect. We will discuss this in Section 10.5 in more detail. The nature of the exchange interaction allows us to distinguish the magnetic ordering into the main categories of ferromagnetism, antiferromagnetism, and ferrimagnetism. These form the topics of the sections following on from the discussion on exchange interactions.

Related to the theme of interactions are the topics which follow. The fact that there exist interactions between the atomic magnetic moments in a cooperative manner implies some form of directionality of the moments with respect to the crystal lattice. This is called *spontaneous magnetization*. This has some fundamentally important implications for how we treat magnetic phenomena in solids. For example, if we apply a magnetic field in some orientation with respect to the natural direction of the magnetic moment (or magnetization vector), then the response will essentially be different. This dependence on orientation is termed, in general, *magnetic anisotropy* and can have a number of forms. We will outline the principal forms of magnetic anisotropy; Section 10.9.

As we discussed in the early chapters of the book, the crystalline structures of solids give rise to specific symmetry properties due to the periodic arrangement of the atoms. Therefore, the orientation of the spontaneous magnetization can have a number of equivalent directions. This, along with magnetic energy considerations leads us to the notion of magnetic domains. These are regions of spontaneous magnetization in the solid, which can now be separated into various zones with regions of transition between them called *domain walls*. The specific domain structure depends on a number of factors, such as the magnetic anisotropies and the strength of the exchange interaction aligning the magnetic moments of the atoms.

The magnetic order of a solid can be perturbed by excitations, such as thermal agitation, electromagnetic wave absorption or the application of time varying magnetic fields. Such excitations can be understood in terms of quantum excitations or quasi-particles, much in the same way as we described the existence of lattice oscillations or phonons. In the case of magnetic perturbations, we call these excitations *spin waves* or *magnons*. These are dynamic properties of the magnetic material and will form the subject of Section 10.11. Actually the dynamic response of a magnetic lattice can be observed in a number of phenomena, such as magnetic resonances and magnetization reversal processes.

The discovery of the giant magnetoresistance (GMR) in the late 1980s paved the way for the development of novel magnetic devices, which have important applications, especially in the magnetic recording industry. The importance of this discovery is attested to by the attribution of the 2007 Nobel Prize in Physics to Albert Fert and Peter Grünberg, who were jointly awarded the prize for the discovery of this effect. In fact, this discovery led to a new field of electronics based on the spin of the electron and not just its charge. The topic is now known as *spintronics,* and is a fast growing area of research and development. We will outline some of the basic physics of spintronics in Section 10.12.

10.2 THE ATOMIC MAGNETIC MOMENT

10.2.1 Orbital and Spin Angular Momenta

At the most fundamental level, all magnetic phenomena in solids depends on the magnetic moment of the atoms in the solid itself. The simplest way to consider the origin of a magnetic moment of an atom is to think of the electron motion in a classical orbit around the nucleus. If the orbit has a radius of r and orbital velocity, v, the orbital period is: $T = 2\pi r/v$, see Figure 10.1.

Now the motion of a charged particle, such as an electron, constitutes a current, defined as the quantity of charge passing a certain point per unit time. The current due to the motion of the single electron in its circular orbit will therefore be: $I = -e/T = -ev/2\pi r$. The negative sign comes from the convention for the electronic charge, and therefore the current direction is opposite to that of the electron motion. From Ampere's Law, such a current loop will create a magnetic dipole moment of strength:

$$\mu_l = Id\mathbf{A} \qquad (10.1)$$

where $d\mathbf{A} = \hat{\mathbf{n}}dA$, with dA being the area of the loop and $\hat{\mathbf{n}}$ the unit vector normal to the loop area. It is now possible to write the magnetic moment as:

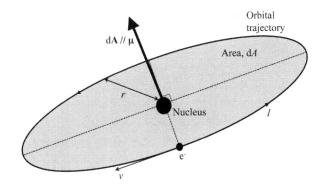

FIGURE 10.1: An electron in a circular orbit about the nucleus of an atom can be considered as an electrical current in a loop. Such a current loop will produce a magnetic moment, μ.

$$\mu_l = -\frac{ev}{2\pi r} d\mathbf{A} \qquad (10.2)$$

We can now substitute in for the area, πr^2, and the orbital angular momentum, $|\hbar \mathbf{l}| = \hat{\mathbf{n}} m_e v r$, to obtain:

$$\mu_l = -\frac{e\hbar}{2m_e} \mathbf{l} \qquad (10.3)$$

where we note that $\hbar$ is the natural unit of the orbital angular momentum for electrons in atoms. The pre-factor in the above equation is referred to as the *Bohr magneton*:

$$\mu_B = \frac{e\hbar}{2m_e} \qquad (10.4)$$

and has a numerical value of $9.271 \times 10^{-24} \mathrm{JT}^{-1}$. The above is valid for the quantum mechanical treatment provided that $\hbar \mathbf{l}$ is considered to be the operator for the angular momentum of the electron. In addition to the charge of the electron creating a magnetic moment due to the current flow described by the orbital motion above, the electron also has the property of *spin*. This is the rotational motion of the electron as a spinning object and produces a further contribution to the magnetic moment of the electron. This can be expressed as

$$\boldsymbol{\mu}_s = -\mu_B g_s \mathbf{s} \qquad (10.5)$$

in which the quantity $\hbar\mathbf{s}$ represents the intrinsic spin angular momentum of the electron. The constant g_s has a value of very close to 2. A comparison of Equations (10.3) and (10.5), therefore shows that the spins angular momentum has a value twice that of the orbital angular momentum. As we saw in Chapter 1, the eigenvalues of the spin (in the z- orientation) are $\hbar s_z = \pm\hbar/2$. This means that the value of the z-component of the moment will be $\pm\mu_B$.

Since atoms can have more than one electron, we evaluate the total magnetic moment of the atom as:

$$\mu_{TOT} = -\mu_B(\mathbf{L} + 2\mathbf{S}) \tag{10.6}$$

where we have introduced; $\hbar\mathbf{L} = \hbar\sum\mathbf{l}$ and $\hbar\mathbf{S} = \hbar\sum\mathbf{s}$. The summations being over all electrons in the atom, with $\hbar\mathbf{L}$ and $\hbar\mathbf{S}$ being the total orbital and spin angular momenta of the atom, respectively. In Chapter 1, we saw the general rules relating to the filling of electrons in shells. However, this does not help us in evaluation how the different quantum states are occupied in the shells. As electrons fill a particular shell, they must do so such as to minimize the energy of the atom. In addition to this we must also consider the total angular momentum, **J**, of the atom. The manner in which this occurs is via a vector sum of the orbital and spin angular momenta, though the specifics depend on the particular circumstances of the atom.

10.2.2 Hund's Rules and the Ground State

The most common scheme in isolated atoms with incomplete electron shells is via the *Russell - Saunders* or **L-S** coupling mechanism. In this case, the orbital angular moments, $\mathbf{l}_i$, are coupled together, such that:

$$\mathbf{L} = \sum_i \mathbf{l}_i \tag{10.7}$$

Similarly the spin angular momenta couple together into a resultant spin:

$$\mathbf{S} = \sum_i \mathbf{s}_i \tag{10.8}$$

The momenta then interact (spin-orbit interaction), to give a total angular momentum:

$$\mathbf{J} = \mathbf{L} + \mathbf{S} \tag{10.9}$$

According the this scheme, the stationary states of the shell have eigenstates $\mathbf{L}^2$, $\mathbf{S}^2$ and $\mathbf{J}^2$, with respective eigenvalues: $\sqrt{L(L+1)}$, $\sqrt{S(S+1)}$, $\sqrt{J(J+1)}$. The values for L, S and J, which correspond to the lowest energy state can be derived from the *Hund's rules*, which must be obeyed in the following order of priority:

1. S must take the maximum value possible allowed by the Pauli exclusion principle. This means that as many as possible of the electron spins must align parallel. This in effect minimizes the Coulomb interaction between electrons since they are kept as far apart as possible from one another.

2. L must take the maximum value, consistent with the first Hund's rule, thus aligning as much as possible the orbital angular momenta of the electrons in the atom. This can be envisaged as the electrons orbiting as much as possible in the same direction, hence they will interact less, and again reduce the Coulomb repulsion.

3. The final rule concerns the total angular momentum, J. This is given as: i) $J = |L - S|$, for a shell which is less than half full, and ii) $J = |L + S|$, for a shell which is more than half full.

This last rule is sometimes disobeyed, such as in the transition metal ions. In this case the spin-orbit interaction energies are weak compared to other energies, such as the crystal field energy. The third rule does work well for the rare earth ions. The Hund's rules leads to a specific ground state of the atom, but predicts nothing about the excited states. It therefore allows a useful estimate of the magnetic moment, assuming all atoms exits in their ground state.

For this situation we can express the magnetic moment in the form:

$$\mu_{TOT} = -\mu_B g \sqrt{J(J+1)} \tag{10.10}$$

where g is called the *Landé factor* or simply the *g-factor*, which can be expressed in terms of the angular momenta as:

$$g = \frac{3}{2} + \frac{S(S+1) - L(L+1)}{2J(J+1)} \quad (10.11)$$

It is instructive to show some examples of how the Hund's rules work. Let us consider the case of Fe^{2+}. This ion has 6 electrons in the 3D shell, where we have $n = 3$ and $l = 2$, see Table 10.1. (Note, we do not need to consider the other electron shells as they are full, and as such have no net angular momentum and hence no magnetic moment.) The situation can be expressed as:

TABLE 10.1: Electronic configuration of Fe^{2+}.

l	2	1	0	−1	−2
s	1/2	1/2	1/2	1/2	1/2
	↑	↑	↑	↑	↑
	↓				

We can now see that for this ion, we have: $\mathbf{L} = \sum_i \mathbf{l}_i = 2$ and $\mathbf{S} = \sum_i \mathbf{s}_i = 2$. Since the 3D shell is more than half full, we obtain: $J = |L + S| = 4$. Given that $L = S$, we have $g = 3/2$ and the magnetic moment of the ion is obtained from Equation (10.10) to be $\mu = 6.71 \mu_B$. In Figure 10.2, we show the 3D transition metal series, giving the results obtained from the Hund's rules.

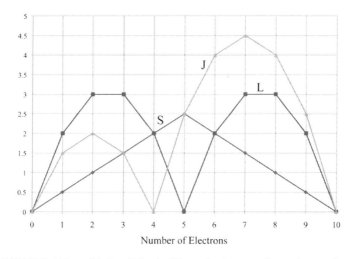

FIGURE 10.2: Values of L, S and J for the 3D metal series according to the Hund's rules.

In general, the electric forces that couple the individual angular momenta, l_i and s_i, into the single vectors **L** and **S**, respectively, are stronger than the magnetic spin - orbit forces, which couple **L** and **S** to form **J** in light atoms. Such forces tend to dominate even when a moderate magnetic field is applied[1]. For heavy atoms, the nuclear charge becomes sufficient to produce a spin-orbit interaction, which is comparable to the electric one and the **L-S** coupling scheme starts to break down. A similar break down occurs when a strong magnetic field is applied, producing the *Paschen-Back effect* in atomic spectra. Once this occurs the total angular momentum of the individual electron, J_i, add directly together to form the total angular momentum of the atom. Such a situation is called *jj coupling*. In this case we can write:

$$\mathbf{J} = \sum_i \mathbf{J}_i \tag{10.12}$$

where, $\mathbf{J}_i = \mathbf{L}_i + \mathbf{S}_i$ is the total angular momentum of each electron.

As we have stated above, the Hund's rules do not agree very well with the 3D transition metal ions. In Table 10.2, we show a comparison of the magnetic moments of various ions in this series, where we compare experimental and calculated values. Of particular note is the agreement between the experimental value and that obtained

TABLE 10.2: Ground states for the 3D ion series using the Hund's rules. The magnetic moments are calculated using J and S and are compared with experimental values. All moments are expressed in unites of μ_B. Notable is the agreement between the experimental values and those from $g\sqrt{S(S+1)}$.

Ion	3d shell	S	L	J	$\mu_{exp}(\mu_B)$	$g\sqrt{J(J+1)}$	$g\sqrt{S(S+1)}$
Ti^{3+}, V^{4+}	$3d^1$	1/2	2	3/2	1.70	1.55	1.73
V^{3+}	$3d^2$	1	3	2	2.61	1.63	2.83
Cr^{3+}, V^{2+}	$3d^3$	3/2	3	3/2	3.85	0.77	3.84
Mn^{3+}, Cr^{2+}	$3d^4$	2	2	0	4.82	0	4.90
Fe^{3+}, Mn^{2+}	$3d^5$	5/2	0	5/2	5.82	5.92	5.92
Fe^{2+}	$3d^6$	2	2	4	5.36	6.70	4.90
Co^{2+}	$3d^7$	3/2	3	9/2	4.90	6.63	3.87
Ni^{2+}	$3d^8$	1	3	4	3.12	5.59	2.83
Cu^{2+}	$3d^9$	1/2	2	5/2	1.83	3.55	1.73
Zn^{2+}	$3d^{10}$	0	0	0	0	0	0

when we use $\mu = g\mu_B\sqrt{S(S+1)}$. This agreement can be achieved by putting $L = 0$, an effect referred to as *orbital quenching*. This arises due to the crystal field interaction being much stronger than the spin - orbit interaction. The third Hund's rule is no longer obeyed in this situation.

10.2.3 Moments and Energies

When a magnetic moment is subject to a uniform magnetic field, there will be an energy of interaction between the two, this is commonly termed the *Zeeman energy*. Let us consider the magnetic moment, μ, in an applied magnetic field, **B**. The energy of interaction is expressed as:

$$E = -\mu \cdot \mathbf{B} \tag{10.13}$$

The energy will be a minimum when the magnetic moment aligns along the direction of the applied magnetic field. The presence of a magnetic field will produce a torque, **T**, on the magnetic moment, given by:

$$\mathbf{T} = \mu \times \mathbf{B} \tag{10.14}$$

This torque leads to a re-directioning of the magnetic moment, and since the torque is equal to the rate of change of the angular momentum, we can write:

$$\frac{d\mu}{dt} = \gamma(\mu \times \mathbf{B}) \tag{10.15}$$

where $\gamma = e/2m_e = \mu_B/\hbar$ is the magnetogyric (often referred to as *gyromagnetic*) ratio. The motion of the magnetic moment is a precession about the direction of the magnetic field, whose frequency is given by the *Larmor precessional frequency*; $\omega_L = \gamma B$. From the above, we see that there is a complex relation between the magnetic field and the magnetic moment, and the magnetic field does not simply align the magnetic moment, but can induce a number of subtle dynamic effects in spin systems. Dynamic phenomena in magnetic systems will be further discussed in Section 10.13.

Let us consider the Hamiltonian for an atom with Z electrons. In the simplest form this can be expressed as:

$$\hat{\mathcal{H}}_0 = \sum_{j=1}^{Z} \left(\frac{p_j^2}{2m_e} + V_j \right) \quad (10.16)$$

where $p_j^2/2m_e$ represents the kinetic energy and V_j the potential energy of the jth electron. In the presence of a magnetic field the the momentum can be expressed as: $\mathbf{p}_j \to \mathbf{p}_j + e\mathbf{A}(\mathbf{r}_j)$, where the vector potential and the magnetic field are related via: $\mathbf{B} = \nabla \times \mathbf{A}$. This means that the Hamiltonian will take the form:

$$\hat{\mathcal{H}} = \sum_{j=1}^{Z} \left\{ \frac{[\mathbf{p}_j + e\mathbf{A}(\mathbf{r}_j)]^2}{2m_e} + V_j \right\} + g\mu_B \mathbf{B} \cdot \mathbf{S} \quad (10.17)$$

Using the fact that $\mathbf{L} = \mathbf{r} \times \mathbf{p}$ and that the magnetic vector potential has the form: $\mathbf{A}(\mathbf{r}) = (\mathbf{B} \times \mathbf{r})/2$, we obtain:

$$\hat{\mathcal{H}} = \hat{\mathcal{H}}_0 + \mu_B(\mathbf{L} + g\mathbf{S}) + \frac{e^2}{8m_e} \sum_{j=1}^{Z} (\mathbf{B} \times \mathbf{r}_j)^2 \quad (10.18)$$

As we shall see shortly, the second and third terms are perturbations to the original Hamiltonian of the system, where the first is related to paramagnetic effects while the latter to diamagnetic properties of the atom.

10.3 DIAMAGNETISM

It is useful to outline some of the principal physical quantities that will be used in the following. Firstly, the magnetic field strength, $\mathbf{H}$, is related to the magnetic induction, $\mathbf{B}$, via the relation:

$$\mathbf{B} = \mu_0 \mathbf{H} \quad (10.19)$$

where $\mu_0 = 4\pi \times 10^{-7} \, \text{Hm}^{-1}(\text{TmA}^{-1})$ is the permeability of vacuum, or free space. In the case of a magnetic material, this expression is modified to:

$$\mathbf{B} = \mu_0(\mathbf{H} + \mathbf{M}) \quad (10.20)$$

where $\mathbf{M}$ is the magnetization, or density of magnetic dipole moments:

$$\mathbf{M} = \mu \frac{N}{V} \qquad (10.21)$$

Finally we can express the magnetic susceptibility as the ratio of the magnetization to the applied magnetic field:

$$\chi = \frac{\mathbf{M}}{\mathbf{H}} \qquad (10.22)$$

In Figure 10.3, we show the susceptibilities of the elements. We note that some are negative while others are positive. This provides a first classification to the response of materials to a magnetic field, and in the majority of cases they are either paramagnetic (weak positive response) or diamagnetic (weak negative response). In this and the following section, we will discuss these effects. The ferromagnetic elements have a spontaneous magnetization in zero field and are not included in this figure.

Virtually all materials have some form of diamagnetic response, which is a weak and negative magnetic susceptibility. The negative susceptibility is due to the induced magnetic moment aligning in

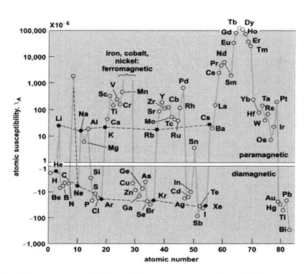

FIGURE 10.3: Atomic susceptibilities of the elements. Fe, Co, and Ni have a spontaneous magnetization in the absence of a applied field since they are ferromagnetic and are off the chart. The majority of the elements generally have a weakly positive or weakly negative susceptibility. The former are paramagnetic, while the latter diamagnetic.

the direction opposed to that of the applied magnetic field. For electronic shells that are full or empty, the orbital and spin angular momenta, and hence the total angular momentum, of the shell will be zero. This means that such a situation leads to zero contribution to the magnetic moment. Therefore any atom from Li and above in the periodic table will have atoms in which this situation occurs. Referring to the energy of the atom in a magnetic field, Equation (10.18), the only changes to the ground state term will occur via second order perturbations; i.e., the third term, in which the energy shift can be assessed via:

$$\delta E_0 = \frac{e^2 B^2}{8m_e} \sum_{j=1}^{Z} \langle \psi_0 |(x_j^2 + y_j^2)| \psi_0 \rangle = \frac{e^2 B^2}{12 m_e} \sum_{j=1}^{Z} \langle \psi_0 | r_j^2 | \psi_0 \rangle \quad (10.23)$$

where we have assumed that the magnetic field is aligned in the z-direction, such that $(\mathbf{B} \times \mathbf{r}_j)^2 = B^2(x_j^2 + y_j^2)$, and from the spherical symmetry of the system we have; $\langle x_j^2 \rangle = \langle y_j^2 \rangle = \langle r_j^2 \rangle / 3$. The magnetic susceptibility is now evaluated from:

$$\chi = -\frac{N}{V} \frac{\partial^2 [\delta E_0]}{\partial B^2} = -\frac{N e^2 \mu_0}{6 V m_e} \sum_{j=1}^{Z} \langle | r_j^2 | \rangle \quad (10.24)$$

This expression is known as the *Larmor diamagnetic susceptibility*. As we noted above, the susceptibility has the negative sign for diamagnetism. The diamagnetic response shows no temperature dependence, but there is a dependence on the atomic mass number, since the number of electrons is an important component in the susceptibility. This is borne out by experiment. We mentioned above that most materials have some form of diamagnetic response. In fact, this was amply demonstrated in a series of experiments in which magnetic levitation was used to demonstrate the diamagnetism of many common objects. The levitation of a live frog was probably the object that attracted the most attention. The levitation itself requires only that the object be placed in sufficient magnetic field so that the repulsive force, caused by the opposition to the applied field, overcomes the gravitational force on the object.

In addition to the diamagnetic contribution due to full electron shells, the motion of free electrons also has both paramagnetic and diamagnetic responses. These were discussed in Chapter 6, Section 6.9, these are referred to as Pauli paramagnetism and Landau diamagnetism, respectively.

10.4 PARAMAGNETISM

If there is no interaction between the atomic magnetic moments in a solid, i.e., they behave independently, then in the absence of a magnetic field the magnetic moments will be randomly oriented. A magnetic field acts to align these moments in the same direction as the field itself, and hence has a positive magnetic susceptibility. This is termed *paramagnetism*. Actually, we have already introduced paramagnetic effects when we discussed the properties of free electrons in metals. The response of free electrons in metals is referred to as Pauli paramagnetism, see Section 6.9. In this section, we will discuss the paramagnetic response of the atomic magnetic moments in solids.

10.4.1 Classical Treatment

The classical theory of paramagnetism is due to Langevin (1905) and assumes that the magnetic moment can align in any direction and has a probability of having an energy, E at a temperature, T, given by the Boltzmann factor:

$$P(E) = e^{-E/k_BT} = e^{\mu \cdot \mathbf{B}/k_BT} = e^{\mu B \cos\theta/k_BT} \qquad (10.25)$$

We can evaluate the number of magnetic moments in the orientation between θ and $\theta + d\theta$, see Figure 10.4, which can be expressed in terms of the area of the annulus on the surface of the sphere: $2\pi \sin\theta d\theta$, considering the sphere to have unit radius.

The average moment along the field, $\mathbf{B}$, takes the form:

$$\langle \mu \rangle = \frac{\mu \int_0^\pi \cos\theta \sin\theta e^{\mu B \cos\theta/k_BT} d\theta}{\int_0^\pi \sin\theta e^{\mu B \cos\theta/k_BT} d\theta} \qquad (10.26)$$

Solving can be achieved by substituting: $x = \cos\theta$, $dx = \sin\theta d\theta$ and $\alpha = \mu B/k_BT$. The magnetization can be expressed as:

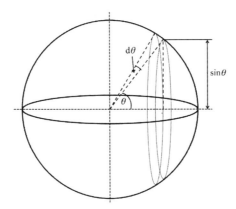

FIGURE 10.4: Geometry for the calculation of the number of magnetic moments between direction θ and $\theta + d\theta$.

$$M = N\mu \frac{\int_{-1}^{1} x e^{\alpha x}\, dx}{\int_{-1}^{1} e^{\alpha x}\, dx} \qquad (10.27)$$

where N is the total number density of moments. It can easily be shown that:

$$M = N\mu \left[\coth\left(\frac{\mu B}{k_B T}\right) - \frac{k_B T}{\mu B} \right] = n\mu L\left(\frac{k_B T}{\mu B}\right) \qquad (10.28)$$

The function $L(x)$ is called the *Langevin function*, and is illustrated in Figure 10.5.

For low values of the argument; $\mu B/k_B T \ll 1$, we find:

$$M = N\mu \left(\frac{\mu B}{3k_B T}\right) = \frac{n\mu^2 B}{3k_B T} = \frac{n\mu_0 \mu^2 H}{3k_B T} \qquad (10.29)$$

From this we find the corresponding paramagnetic susceptibility:

$$\chi = \frac{M}{H} = \frac{n\mu_0 \mu^2}{3k_B T} \qquad (10.30)$$

which is valid for small applied fields. This demonstrates that the susceptibility is inversely proportional to the inverse of the temperature, a variation known as the *Curie law*, after Pierre Curie (1895).

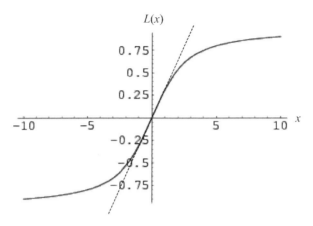

FIGURE 10.5: The classical paramagnetic magnetization follows the Langevin function, $L(x) = \coth x - 1/x$. For small values of x this is essentially linear: $L(x) \simeq x/3$, as shown by the dashed line.

10.4.2 Quantum Mechanical Treatment

In the quantum description of paramagnetism, we limit the orientation of the magnetic moment to those permitted by the quantization of the angular momentum, J. The energy of the system is of the form: $E = -g\mu_B M_J B$, where M_J can take values: $-J, -(J-1), \ldots, (J-1), J$. The partition function[2] can be written:

$$Z = \sum_{M_J=-J}^{J} e^{g\mu_B M_J B/k_B T} \tag{10.31}$$

Writing $x = g\mu_B B/k_B T$, we can express the average value of M_J as:

$$\langle M_J \rangle = \frac{\sum_{M_J=-J}^{J} M_J e^{xM_J}}{\sum_{M_J=-J}^{J} e^{xM_J}} = \frac{\sum_{M_J=-J}^{J} M_J e^{g\mu_B M_J B/k_B T}}{\sum_{M_J=-J}^{J} e^{g\mu_B M_J B/k_B T}} = \frac{1}{Z}\frac{\partial Z}{\partial x} \tag{10.32}$$

The magnetization can now be written in the form:

$$M = Ng\mu_B \langle M_J \rangle = Ng\mu_B \frac{d}{dx}\left[\ln\left(\sum_{M_J=-J}^{J} e^{xM_J}\right)\right] \tag{10.33}$$

The summation can be simplified by writing:

$$\sum_{M_J=-J}^{J} e^{M_J x} = \sum_{i=1}^{P} ar^{i-1} = a + ar + ar^2 + \ldots + arP - 1 = a\frac{1-r^P}{1-r} \quad (10.34)$$

Since P is the number of terms in the series, for a given value of J, it will have a value of $2J + 1$. We thus obtain:

$$\sum_{M_J=-J}^{J} e^{M_J x} = e^{-Jx}\frac{(1-e^{(2J+1)x})}{(1-e^x)} = \frac{\sinh[(2J+1)x/2]}{\sinh(x/2)} \quad (10.35)$$

This must be substituted into Equation (10.33). Developing the differential can be shown to give:

$$M = Ng\mu_B J \left\{ \left[\frac{(2J+1)}{2J}\right] \coth\left[\frac{(2J+1)}{2J}y\right] - \frac{1}{2J}\coth\left(\frac{y}{2J}\right) \right\} = M_0 B_J(y)$$

$$(10.36)$$

where we have: $y = xJ$, $M_0 = ng\mu_B J$ and $B_J(y)$ is known as the Brillouin function. The form of this function is illustrated in Figure 10.6. The subscript of the Brillouin function refers to the value of the angular momentum. In the classical limit we have $J = \infty$ and we therefore obtain $B_J(y) = B_\infty(y) \to L(y)$, i.e., we recover the classical result, showing that the Brillouin function is a more general form of the Langevin case.

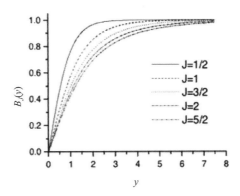

FIGURE 10.6: The Brillouin function, $B_J(y)$. Shown are various curves for different values of J. In the classical limit $J = \infty$ and $B_J(y) = B_\infty(y) \to L(y)$.

In the limit of low magnetic fields, we have: $B_J(y) \simeq (J+1)y/3J$, the magnetic susceptibility in this limit can be given by:

$$\chi = \frac{M}{H} = Ng\mu_B \frac{J(J+1)}{3J} \frac{g\mu_B J}{k_B T} \mu_0 = \frac{Ng^2\mu_B^2 J(J+1)}{3k_B T} \quad (10.37)$$

This has the form of the classical Curie law. In Figure 10.7, we show some experimental results for some paramagnetic salts along with the fits to various Brillouin functions, for the spin value given.

10.4.3 Van Vleck Paramagnetism

As we will see in the next section, the ground state for atoms or ions with full electrons shells has $\mathbf{J} = 0$. In this case there will be no resulting magnetic moment and thus no paramagnetic effect. However, this is only true to first-order perturbations and second-order effects give rise to the modification of the ground state. Any

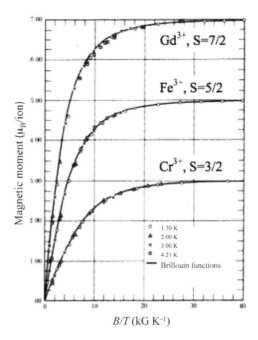

FIGURE 10.7: Magnetization curves for some paramagnetic salts. The atomic ions are indicated and the fit to the relevant form of the Brillouin function is shown. N.B. The fits are made according to the quenching of orbital angular momentum: $\mathbf{L} = 0 \rightarrow J = S$.

alteration of the ground state must involve a promotion of an electron from one of the closed shell occupied states to an unoccupied state in another shell and thus produces a change in the angular momentum state, and therefore $\mathbf{J} \neq 0$. Based on Equation (10.18), we can write the corresponding change in energy as:

$$\delta E_0 = \sum_n \frac{|\langle \psi_0 | (\mathbf{L} + g\mathbf{S}) \cdot \mathbf{B} | \psi_n \rangle|^2}{E_0 - E_n} + \frac{e^2}{8m_e} \sum_{j=1}^{Z} (\mathbf{B} \times \mathbf{r}_j)^2 \quad (10.38)$$

The second term comprises diamagnetic effects, see Section 10.3. The summation in the first term is over excited states of the atom or ion. Since the solid will contain N/V atoms/ions per unit volume, we can obtain the corresponding paramagnetic susceptibility as:

$$\chi = -\frac{N}{V} \frac{\partial^2 [\delta E_0]}{\partial B^2} = \frac{N}{V} \left[2\mu_B^2 \sum_n \frac{|\langle \psi_0 | (L_z + gS_z) \cdot \mathbf{B} | \psi_n \rangle|^2}{E_0 - E_n} - \frac{e^2 \mu_0}{6m_e} \sum_{j=1}^{Z} \langle |r_j^2| \rangle \right]$$
$$(10.39)$$

We note that the two terms have opposite signs. The second term is the Larmor diamagnetism that we discussed previously, while the first term is referred to as the *Van Vleck paramagnetism*.

10.5 INTERACTIONS, EXCHANGE, AND MAGNETIC ORDER

We now turn our attention to the collective effects that occur when the magnetic moments of the individual atoms are coupled together. There are a number of physical mechanisms by which this can occur and it depends very much on the nature of the materials involved which type of interaction predominates.

10.5.1 Dipolar Interaction

Before we consider the physics of the interaction between atomic magnetic moments, we will outline the purely dipolar energy between two magnetic dipoles, μ_1 and μ_2, which are separated by a position vector, **r**. The mathematical expression for the dipolar energy can be given in the form:

$$E_{dip} = \frac{\mu_0}{4\pi r^3}\left[\boldsymbol{\mu}_1 \cdot \boldsymbol{\mu}_2 - \frac{3}{r^2}(\boldsymbol{\mu}_1 \cdot \mathbf{r})(\boldsymbol{\mu}_2 \cdot \mathbf{r})\right] \qquad (10.40)$$

This energy depends not only on the physical separation of the magnetic moments, but also importantly, on their relative orientations. For example, the energy of interaction will be zero when the spins are oriented at right angles. For other orientations, the vector between the magnetic moments is also important in evaluating the energy of interaction. The dipolar energy can be simplified in a number of ways. For example, if the magnetic moments have the same magnitude (i.e., $|\boldsymbol{\mu}_1| = |\boldsymbol{\mu}_2| = \mu$). This can be expressed in terms of the various angles, as defined in Figure 10.8, in the form:

$$\begin{aligned}E_{dip} = \frac{\mu_0 \mu^2}{4\pi r^3}&\{[\sin\theta_1 \sin\theta_2 \cos(\phi_1-\phi_2)+\cos\theta_1\cos\theta_2]\\&-3[\sin\Theta\sin\theta_1\cos(\Phi-\phi_1)+\cos\Theta\cos\theta_1]\\&[\sin\Theta\sin\theta_2\cos(\Phi-\phi_2)+\cos\Theta\cos\theta_2]\} \qquad (10.41)\end{aligned}$$

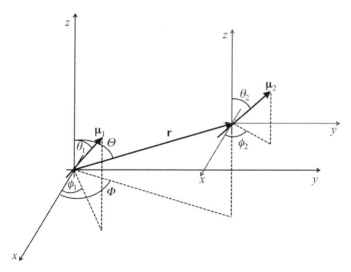

FIGURE 10.8: Definition of angles for the spherical coordinate system used to define the dipolar energy between two magnetic moments in Equation (10.41).

Furthermore, if we take the moments to be along the same direction ($\mu_1 // \mu_2$) we obtain:

$$E_{dip} = \frac{\mu_0 \mu^2}{4\pi r^3} \{1 - 3[\sin\Theta \sin\theta \cos(\Phi - \phi) + \cos\Theta \cos\theta]^2\} \quad (10.42)$$

Further simplification is possible by setting the vector between the moment along the y-axis; $\Theta = 90°$; $\Phi = 90°$, from which we find:

$$E_{dip} = \frac{\mu_0 \mu^2}{4\pi r^3}(1 - 3\sin^2\theta \sin^2\phi) \quad (10.43)$$

The lowest energy state occurs when the moments are aligned along the y-axis ($\theta = \phi = 90°$), where we obtain: $E_{dip} = -\mu_0\mu^2/2\pi r^3$. The maximum energy is for the spins to be aligned along the z-axis ($\theta = \phi = 0°$), where $E_{dip} = -\mu_0\mu^2/4\pi r^3$. At intermediate orientations there is a $(1 - 3\sin^2\theta)$ dependence. The other important limiting case to note is when $\phi = 0°$, which case there is no dependence on the θ angle and the two moments can rotate in the $x - z$-plane with no alteration in their energy state. The existence of dipolar fields has important consequences for magnetic anisotropies in magnetic objects, in fact it is the origin of *magnetostatic* or *shape anisotropy* in solids.

10.5.2 Exchange Interactions

The exchange interaction is used in a generic way to characterize the interactions between the fixed atomic magnetic moments in solids and is used to describe the long-range magnetic order in *magnetic materials*. The theory of exchange stems from quantum mechanical behavior of electrons and is related to the *indistinguishability* of these particles. This may seem a little abstract, but it is another example of the surprises that quantum mechanics has brought to our understanding of magnetism and physics in general. At its origin, the exchange interaction is of an electrostatic nature, not dissimilar to the Coulomb interaction. The exchange interaction, usually written as a constant, also referred to as the *exchange integral*, J_{ij}, can be expressed in terms of the wave-functions of electrons (i and j) in a system, and is determined by the energy it takes to exchange the electrons among themselves in the system[3].

The Hamiltonian of the exchange interaction between two electron spins can be expressed as:

$$\hat{\mathcal{H}}_{ij} = -2J_{ij}\mathbf{S}_i \cdot \mathbf{S}_j \tag{10.44}$$

For systems with more than two electrons and in solids, the situation is more complex. Limiting the interactions to nearest neighbors, we can express the Hamiltonian in the form:

$$\hat{\mathcal{H}} = -\sum_{ij} J_{ij}\mathbf{S}_i \cdot \mathbf{S}_j \tag{10.45}$$

This is called the *Heisenberg* Hamiltonian. In this apparently simple form, the exchange between neighboring spins is characterized via an analysis of the sign of the constant J_{ij}. For positive values, the lowest energy state will be that in which the spins are parallel, this is the case of *ferromagnetic order*. For negative signs of J_{ij}, the lowest energy will be for antiparallel alignment between neighboring spins; this corresponds to *antiferromagnetic order*. This, however, tells us nothing about the nature of the interaction itself. In fact, the complexity of exchange is reflected in the various mechanisms which can occur. We will here outline some of the more important of these.

As a first step, we can distinguish between *direct* and *indirect* exchange interactions. The former refers to the case where magnetic atoms interact between nearest neighbors, while the latter concerns interactions between magnetic atoms or ions via an intervening atom. While this may seem a simple picture, the physical reality is far from simple. The direct exchange may seem an obvious route for interactions via an overlap of wave-functions between neighboring spins, however, this is rarely the case. For example, in the 4f (rare earth) series, the electronic wave-functions are too localized for any significant overlap and it is unlikely that a direct interaction is responsible for the ferromagnetic effects, and indirect mechanisms probably via conduction electrons are required. Even for the case of the transition metal ferromagnets (Fe, Co, Ni), where the 3D electron orbitals extend further from the nucleus, there are important contributions via conduction electrons. The interaction via the conduction electrons is referred to as *itinerant exchange* and can lead to some rather complex behavior in metallic systems. One such mechanism was devised in the 1950s by Ruderman, Kittel, Kasuya and Yosida (commonly known as the *RKKY interaction*), whereby the interaction between magnetic ions in a metal produce a polarization of the conduction electrons which gives rise to spin density oscillations. This results in an indirect exchange mechanism which oscillates between a positive and negative coupling between ions (i.e., between ferromagnetic and antiferromagnetic), and depends on the separation of the ions and is related in a complex manner with the Fermi surface of the host metal, having an oscillatory wavelength of π/k_F; $J_{RKKY} \propto \cos(k_F r)/r^3$. This mechanism is one of the principal candidates for explaining the *giant magnetoresistance effect* in magnetic multilayers, see Section 10.12.

Another important indirect exchange mechanism is *superexchange* and is responsible for the antiferromagnetic coupling between magnetic moments in many oxides and flourides, e.g., MnO, CoO, MnF_2 etc. The interaction between the the magnetic moments, which reside on the transition metal (e.g.Mn^{2+}) ions, occurs via the intervening non-magnetic ion and is a consequence of the Pauli exclusion principle and the exchange of electrons between the constituent atoms to form the ionic solid.

In many magnetic oxides, there are complex and indirect exchange forces which give rise to a number of non-trivial magnetic structures. Such interactions, in addition to the superexchange mechanism, include *double exchange* and *Dzyaloshinski - Moriya* (DM) interactions. The double exchange mechanism occurs in oxide systems, such as Fe_3O_4, where mixed valency ions exist. In this example, Fe^{2+} and Fe^{3+} ions couple via the *hopping* of an electron between them, thus swapping the ionic states between 2+ and 3+. Such a mechanism is known to occur in many manganite systems which exhibit *colossal magnetoresistance* (CMR). The DM interaction is unusual in that it is anisotropic, being expressed via a Hamiltonian of the form:

$$\hat{\mathcal{H}}_{DM} = -\sum_{ij} \mathbf{D}_{ij} \mathbf{S}_i \times \mathbf{S}_j \tag{10.46}$$

here the vector, $\mathbf{D}_{ij} = -\mathbf{D}_{ji}$, vanishes when the crystal field has an inversion symmetry between the two magnetic ions. This type of interaction favors non-collinear spin structures, since parallel structures will result in $\hat{\mathcal{H}}_{DM} = 0$. Frequently the DM interaction causes a canting between the magnetic moments and produces *weak ferromagnetism*, such as in the following compounds: $\alpha - Fe_2O_3$, $MnCO_3$ and $CoCO_3$.

10.6 FERROMAGNETIC ORDER

When the word "magnet" is used, the image conjured up is that of iron filings, bar magnets and "fridge magnets". This is the public perception of what magnetism is. Of course this simplistic view is, for the most part, limited to the case of strong magnetic effects, especially in the mysterious forces which produce magnetic repulsion between bar magnets. Actually, as we have seen, magnetic phenomena are much more complex and varied than this. This popular view of magnetism is part of what captures the imagination of the public in general. The strong magnetic response of certain materials is mainly due to the spontaneous magnetization which occurs in ordered magnetic structures like ferromagnets. As was mentioned above, ferromagnetism is a specific type of order in which the atomic or ionic magnetic moments align in the same direction as their neighbors, which results from a positive exchange integral. In the case of ferromagnets in the presence of a magnetic field, the Hamiltonian takes the form:

$$\hat{\mathcal{H}} = -\sum_{ij(i \neq j)} J_{ij} \mathbf{S}_i \cdot \mathbf{S}_j + g\mu_B \sum_j \mathbf{S}_j \cdot \mathbf{B} \qquad (10.47)$$

The first term is the Heisenberg exchange and the second corresponds to the Zeeman energy.

10.6.1 Mean Field Theory

We will now introduce the *mean field* or *molecular field approximation*, which is due to P. Weiss (1906), and is also known as the *Weiss molecular field theory*. In this approximation we consider the effect on the ith spin of all the other spins in the solid, taken as the *mean effective field* of the ferromagnetic system. Using the above Hamiltonian, we can express the energy of for the ith spin as:

$$E_i = -2\sum_j J_{ij}\mathbf{S}_i \cdot \langle \mathbf{S}_j \rangle - g\mu_B \mathbf{S}_i \cdot \mathbf{B} = -\mathbf{S}_i \cdot \mathbf{B}_i^{eff} \quad (10.48)$$

where $\langle \mathbf{S}_j \rangle$ is the average value of the other spins in the system, and the effective field on spin i can be expressed as:

$$\mathbf{B}_i^{eff} = \mathbf{B} - \frac{2\sum_j J_{ij}\langle \mathbf{S}_j \rangle}{g\mu_B} = \mathbf{B} + \mathbf{B}_{mf} \quad (10.49)$$

due to the applied field and the interaction field from the other spins. We note that the factor of 2 arises from the summation of the interactions between pairs of spins. As we have written it here, the interactions thus appear as an effective magnetic field on spin i, where $\mathbf{B}_{mf}$ is the so-called *mean* or *molecular field*. Clearly the effective field is non-zero even when there is no applied field. It is worth noting that we can readily interchange between fields and energies via the relation:

$$\mathbf{B}_{eff} = -\frac{\partial E_{eff}}{\partial \mathbf{S}} \quad (10.50)$$

We can now write an effective form for the Hamiltonian as:

$$\hat{\mathcal{H}}_{eff} = g\mu_B \sum_i \mathbf{S}_i \cdot (\mathbf{B} + \mathbf{B}_{mf}) \quad (10.51)$$

This has the same form as the Hamiltonian for a paramagnetic solid in a field $\mathbf{B} + \mathbf{B}_{mf}$. Actually, we can now state that all magnetic moments in the solid are subject to the same internal field, $\mathbf{B}_{mf}$, which arises from the ordering of all the other magnetic moments in the solid. Since the ordered state will have a spontaneous magnetization, $\mathbf{M}$, we can equate the internal field and the magnetization as:

$$\mathbf{B}_{mf} = \lambda_{mf}\mathbf{M} \quad (10.52)$$

where λ_{mf} is a constant of proportionality, which is obviously positive for ferromagnetic materials. Ferromagnetic materials reduce their magnetization with increase of temperature. This occurs due to thermal fluctuations in the spin system, which interrupt the exchange interactions. Clearly at a critical temperature, designated as the *Curie temperature*, T_c, all magnetization is lost and the ferromagnetic undergoes a phase transition to the paramagnetic state. To account for this, we can solve the following equations:

$$\mathbf{M} = \mathbf{M}_s B_J(y) \tag{10.53}$$

$$y = \frac{g\mu_B J}{k_B T}(\mathbf{B} + \lambda_{mf}\mathbf{M}) \tag{10.54}$$

For the case of $\lambda_{mf}M = 0$, we have the same variation as for normal paramagnetism. At zero applied field we obtain: $M = y k_B T / g\mu_B J \lambda_{mf}$. Therefore, the simultaneous solution of equations is the intersections between the Brillouin function and a straight line, as illustrated in Figure 10.9.

We note that when $T < T_c$, there are three points of intersection between the curves, one at $M = 0$ and two for $M = \pm M(T)$, only these latter two are stable solutions, which indicate that for any temperature below the Curie temperature, there is a stable value of the magnetization. At $T \geq T_c$, there is only one point of cross

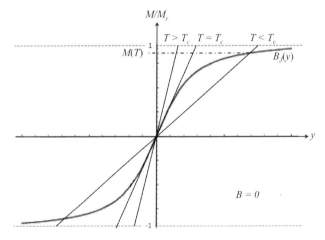

FIGURE 10.9: Simultaneous solutions of Equations (10.53) and (10.54) are given by the points of intersection between the straight line and the Brillouin function. Three cases are shown: (i) $T > T_c$, (ii) $T = T_c$ and (iii) $T < T_c$.

over between the curves, which corresponds to that at the origin where $y = 0$ and $M = 0$. It is easy to see from the model that as the temperature decreases (the gradient of the line also decreases), the spontaneous values of the magnetization will increase, even without an applied magnetic field.

We can evaluate the Curie temperature from the gradient of the straight line: $M = yk_BT/g\mu_BJ\lambda_{mf}$ and $M = M_sB_J(y)$ near the origin, where $B_J(y) \simeq (J+1)y/3J$, from which we find at $T = T_c$:

$$\frac{k_BT_c}{g\mu_BJ\lambda_{mf}} = M_s\frac{(J+1)}{3J} \qquad (10.55)$$

Thus we can write the Curie temperature as:

$$T_c = M_s\frac{g\mu_B(J+1)\lambda_{mf}}{3k_B} \qquad (10.56)$$

Substituting in the molecular field; $B_{mf} = \lambda_{mf}M_s$, we obtain:

$$B_{mf} = \frac{3k_BT_c}{g\mu_B(J+1)} \qquad (10.57)$$

We can now estimate the size of the Weiss molecular field, where using $J = 1/2$ and $T_c \sim 1000$ K we obtain $B_{mf} = k_BT/\mu_B = 1488$ Tesla! This huge effective magnetic field is a reflection of the strength of the exchange interaction in ferromagnets.

The temperature dependence of the magnetization can also be evaluated by the simultaneous solution of the Brillouin function, from which we can derive the $M(T)$ curves, as shown in Figure 10.10. By applying a magnetic field to the system, the simultaneous solution of Equations (10.53) and (10.54) will be modified; the straight lines of Equation (10.54) will be shifted along the x–axis, as shown in Figure 10.11. The effect of a magnetic field is to prolong the magnetized state even when the temperature is above the Curie temperature.

For the case where $T \simeq T_c$ the value of $y \ll 1$ and the approximation to the Brillouin function allows us to write:

$$\frac{M}{M_s} \sim \frac{g\mu_B(J+1)}{3k_B}\left(\frac{B+\lambda_{mf}M}{T}\right) = \frac{T_c}{T}\left(\frac{B+\lambda_{mf}M}{\lambda_{mf}M_s}\right) \qquad (10.58)$$

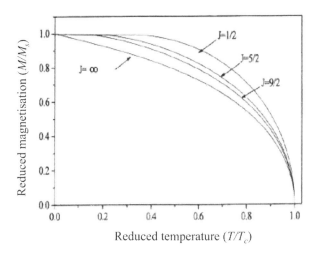

FIGURE 10.10: Temperature dependence of the magnetization in the mean-field model. The different curves correspond to different values of J in the Brillouin function of Equation (10.53).

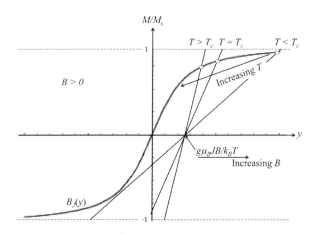

FIGURE 10.11: Simultaneous solution of Equations (10.53) and (10.54) allows us to assess the magnetization in the presence of a magnetic field, B.

Re-arranging we find that the magnetic susceptibility for low fields can be expressed as:

$$\chi = \frac{C}{T - T_c} \tag{10.59}$$

This result is known as the *Curie - Weiss* law. We note that in the case where T_c vanishes, i.e., in a paramagnetic state, we recover the Curie law; $\chi \propto C/T$. Assuming that the exchange interaction only extends as far as nearest neighbor spins, it is possible show that the Curie temperature is related to the exchange integral via:

$$T_c = \frac{2zJ_{ij}}{3k_B}(J+1) \tag{10.60}$$

Here z refers to the atomic coordination number. This means that a measurement of the Curie temperature allows us to assess the strength of the exchange interaction. This is analogous to the relationship between the melting temperature of a solid and the bond strength between atoms. In Table 10.3, we show some typical values of the principal magnetic constants for some ferromagnets.

TABLE 10.3: Magnetic properties of some common ferromagnets.

Material	T_c (K)	$M(\times 10^6 Am^{-1})$	μ (μB/atom)
Fe	1043	1.71	2.22
Co	1394	1.42	1.715
Ni	631	0.48	0.605
Gd	289		7.5

10.6.2 Itinerant Ferromagnetism

It seems clear from the previous table (Table 10.3), where the 3D ferromagnets have non-integer moments per atom, that there is a problem with the theory of localized moments. This points to deficiencies in the model and is used as evidence for the theory of *band ferromagnetism* or *itinerant ferromagnetism*. In this model, the magnetization is considered as arising from a disequilibrium in the spin-split bands. So what do we mean by this? In Section 6.9, we introduced the idea of spin-split bands, where the electron densities of the spin-up and spin-down electrons separately. The magnetization of the solid is then taken as being proportional to the difference in the electron spin densities times the Bohr magneton, see Equation (6.55). As we have seen in the chapter on band structures, Chapter 7, the specific form depends strongly on the crystalline structure, but also on the type of atoms present. In the case of ferromagnetic metals, the existence of localized atomic moments causes a polarization in the free electron gas, and thus there will be a contribution to the magnetization from the free electrons in the solid due to the Weiss molecular field. The internal effective magnetic field will then naturally cause a spin density imbalance between spin-up and spin-down electrons, as envisaged in Figure 6.4.

Let us consider the spin-splitting in terms of the densities of states and the energies involved. In terms of the imbalance in the total number of electrons in the spin-up and spin-down bands, we need to consider the density of states, and given that, the number in each spin sub-band. As for the case where we considered the Pauli paramagnetism, we can write the magnetization in terms of this imbalance:

$$M = \mu_B [n_\uparrow(E) - n_\downarrow(E)] \qquad (10.61)$$

The difference here is clearly that we are now dealing with electrons that are subject to an effective exchange energy, J_{ij}^{eff}, or an internal effective (Weiss) field, $\lambda_{mf} M$. Whichever way we choose to look at this, the result is that there will be an energy difference involved for the ferromagnetic and a non-ferromagnetic states. A consideration of this energy involves the evaluation of the difference in electron populations in the spin sub-bands, $n_\uparrow(E) - n_\downarrow(E)$, where

$$n_{\uparrow(\downarrow)}(E) = \frac{1}{2} \int_0^\infty f_{FD}(E) g_{\uparrow(\downarrow)}(E \pm \mu_B B \pm \delta E) dE \qquad (10.62)$$

The total energy change between the two states, in the absence of an applied magnetic field, can be expressed as:

$$\Delta E = \frac{1}{2} g(E_F)(\delta E)^2 [1 - U g(E_F)] \qquad (10.63)$$

the energy U is given by: $U = \mu_0 \mu_B^2 \lambda_{mf}$. The above equation is interpreted by stating that the existence of ferromagnetism will occur when $\Delta E < 0$, which implies that we have:

$$U g(E_F) \geq 1 \qquad (10.64)$$

Equation (10.64) is known as the *Stoner criterion* for ferromagnetism. The existence of ferromagnetism requires a large density of states at the Fermi energy, without this $g(E_F)$ will tend to zero.

We also require that λ_{mf} be appreciable. To evaluate the magnetic susceptibility we add the magnetic energy to the energy difference equation, since this will also alter the spin populations, and we obtain:

$$\Delta E = \frac{1}{2}g(E_F)(\delta E)^2[1-Ug(E_F)] - \mu_0 MH = \frac{M^2}{2\mu_B^2 g(E_F)}[1-Ug(E_F)] - \mu_0 MH \quad (10.65)$$

Minimizing this energy, we can obtain the magnetic susceptibility as:

$$\chi = \frac{M}{H} = \frac{\mu_0 \mu_B^2 g(E_F)}{[1-Ug(E_F)]} \quad (10.66)$$

Comparing this result with the Pauli paramagnetic susceptibility we find:

$$\chi = \frac{\chi_p}{[1-Ug(E_F)]} \quad (10.67)$$

Therefore, in the case where the Stoner criterion is satisfied, and we have ferromagnetism, the susceptibility will be larger than χ_p by the factor $[1 - Ug(E_F)]^{-1}$. This is referred to as *Stoner enhancement*. While the ferromagnetic transition metals meet the criterion for ferromagnetism, other metals come close, but do not reach the threshold. One such example is the case of Pd, which has a Stoner parameter of $Ug(E_F) \simeq 0.84$. Pd is considered to be on the verge of ferromagnetism, but with insufficient energy to create spontaneous ordering of its moments. In Table 10.4, below we give some calculated values for the Stoner parameter for selected metals.

TABLE 10.4: Stoner parameters for some common metals. [Data adapted from M. M. Sigalas and D. A. Papaconstantopoulos, Phys. Rev. B. **50**, 7255, (1994).]

Material	$Ug(E_F)$ bcc	$Ug(E_F)$ fcc
Fe	1.302	0.541
Co	1.353	0.913
Ni	0.875	2.145
Pd	0.40	0.849

10.7 ANTIFERROMAGNETIC ORDER

In the case of a negative exchange integral, the lowest energy between neighboring spins will occur when they are aligned in an anti-parallel order. This is called *antiferromagnetism*. Since the alternating spins have opposite sign, in zero field the magnetization will be zero. Antiferromagnetic order requires that the magnetic moments have the same magnitude. The ordering temperature in the case of antiferromagnetism is called the *Néel temperature*, T_N. Above this, the system becomes disordered, or paramagnetic. It is common to consider the antiferromagnetic state as two sublattices (we will call the A and B) with opposite magnetizations, such that: $\mathbf{M}_A = -\mathbf{M}_B$, $|\mathbf{M}_A| = |\mathbf{M}_B| = M$ and $\mathbf{M} = \mathbf{M}_A + \mathbf{M}_B = 0$. Building on the Weiss theory, Néel considered the molecular fields of one sublattice to be proportional to the magnetization of the other sublattice, such that:

$$B_A = -|\lambda_{mf}|M_B$$
$$B_B = -|\lambda_{mf}|M_A \qquad (10.68)$$

Each sublattice follows the Brillouin form of the variation of magnetization:

$$\mathbf{M}_{A(B)} = \mathbf{M}_s B_J\left(-\frac{g\mu_B J |\lambda_{mf}| M_{B(A)}}{k_B T}\right) \qquad (10.69)$$

Following the same arguments used in the case of ferromagnetism we can evaluate the Néel temperature as:

$$T_N = \frac{g\mu_B |\lambda_{mf}| M_s}{3k_B}(J+1) \qquad (10.70)$$

The above analysis is based on considering only nearest neighbor interactions. Considering next nearest neighbor interactions leads to:

$$B_A = -|\lambda_{AB}|M_B + |\lambda_{AA}|M_A$$
$$B_B = -|\lambda_{AB}|M_A + |\lambda_{BB}|M_B \qquad (10.71)$$

This can also be expressed in abbreviated form as: $B_{A(B)} = (|\lambda_{AA(BB)}| - |\lambda_{AB}|)M$. The Néel temperature is thus given by

$$T_N = \frac{g\mu_B |(\lambda_{AA} - \lambda_{AB})| M_s}{3k_B}(J+1) \qquad (10.72)$$

where we have assumed $|\lambda_{AA}| = |\lambda_{BB}|$.

When a field is applied to the antiferromagnet, usually the strong exchange interactions only produce small changes in the magnetization: $M = |M_A - M_B|$. The direction of the field also has an important effect on how the magnetization of the sample varies, see Figure 10.12. For the perpendicular configuration, the magnetic field is taken along the direction that is orthogonal to both magnetic sublattices. The resultant magnetization under the magnetic field causes a gradual rotation of the two magnetic sublattices, giving rise to an increase in M. When the field is applied along the direction of one of the sublattices, then only very small changes in magnetization are observed for low fields. However, once a critical field is reached, $B \geq B_{cr}$, the spin system will relax to a new state, called a *spin-flop* state, as illustrated in Figure 10.12 (b). This is usually favored because the magnetic sublattices have a negative exchange integral

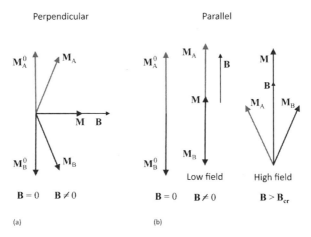

FIGURE 10.12: Schematic representation of the magnetization in an antiferromagnet when a magnetic field is applied in (a) a direction perpendicular to the magnetic sublattice orientations and (b) when the field is applied parallel to one of the magnetic sublattices. For low fields, the magnetization gradually grows along the direction of the field, however, at a critical field, B_{cr}, the spin system undergoes a *spin-flop* transition.

and do not want to align. Further increase of the field brings about a gradual alignment of the sublattice magnetic moments. In rare cases, a single transition can occur from the antiparallel to the parallel alignment. This is distinguished from the previous case and is termed a *spin-flip* transition. This will occur when the antiferromagnetic exchange strength is very strong.

For zero temperature, we have $\mathbf{M}_A(0) = -\mathbf{M}_B(0)$ and for a field applied in the parallel configuration, there will be no change in magnetization, and hence we expect $\chi_\parallel = 0$ at $T = 0$ K. For the perpendicular case, the magnetic sublattices can be expected to suffer some rotation and therefore $\chi_\perp \neq 0$ at $T = 0$ K. The basic form of the variation of the magnetic susceptibility for the two configurations as a function of sample temperature are illustrated in Figure 10.13.

Above the Néel temperature, the material becomes paramagnetic. For a magnetic field applied at some intermediate orientation, ϕ, we can write:

$$\chi_\phi = \chi_\parallel \cos^2 \phi + \chi_\perp \sin^2 \phi \tag{10.73}$$

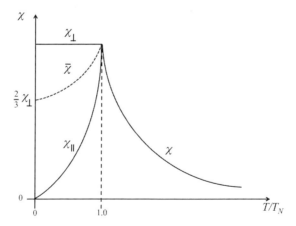

FIGURE 10.13: Temperature dependence of the different magnetic susceptibility contributions, $\chi_\parallel$ - parallel and $\chi_\perp$ - perpendicular. In the case of a polycrystalline materials with a random orientation of grains, the antiferromagnetic susceptibility will have a temperature dependence that is a mixture of both $\chi_\parallel$ and $\chi_\perp$, and expressed as $\overline{\chi}$, the average susceptibility, see text.

For a polycrystalline sample, there will be grains of random orientation and the average susceptibility takes the form:

$$\bar{\chi} = \int_0^{\frac{\pi}{2}} \chi_\phi \sin\phi \, d\phi = \chi_\| \int_0^{\frac{\pi}{2}} \cos^2\phi \sin\phi \, d\phi + \chi_\perp \int_0^{\frac{\pi}{2}} \sin^3\phi \, d\phi$$

$$= \frac{1}{3}\chi_\| + \frac{2}{3}\chi_\perp \qquad (10.74)$$

The parallel susceptibility can be evaluated at low fields from

$$\chi_\| = \frac{\partial M_A}{\partial H} + \frac{\partial M_B}{\partial H} \qquad (10.75)$$

Using $M_{A(B)} = M_s B_J(y_{A(B)})$ it is possible to show:

$$\chi_\| = \frac{2\mu_0 M_s B'_J(y)(g\mu_B J / k_B T)}{\{1 + M_s B'_J(y)(g\mu_B J / k_B T)[\lambda_{AB} - \lambda_{AA}]\}} \propto \frac{1}{T + T_N} \qquad (10.76)$$

This takes the form of the Curie - Weiss law:

$$\chi \propto \frac{1}{T + \Theta} \qquad (10.77)$$

Here Θ is called the Weiss temperature. We can now distinguish experimentally between the principle types of magnetic response, where $\Theta > 0$ for ferromagnetism ($\Theta = T_c$); $\Theta < 0$ for antiferromagnetism ($\Theta = -T_N$) and $\Theta = 0$ for paramagnetism, see Figure 10.14.

TABLE 10.5: Properties of some common antiferromagnetic materials.

Material	T_N (K)	Θ (K)	J
FeO	198	−570	2
CoO	292	−330	$\frac{3}{2}$
NiO	524	−1310	1
MnO	116	−610	$\frac{5}{2}$
$\alpha - Fe_2O_3$	950	−2000	$\frac{5}{2}$

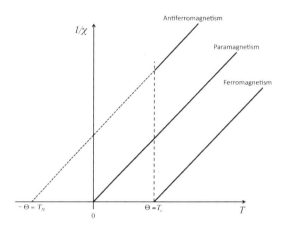

FIGURE 10.14: Generalized Curie - Weiss laws for antiferromagnetism ($\Theta = -T_N$), paramagnetism ($\Theta = 0$) and ferromagnetism ($\Theta = T_c$).

10.8 FERRIMAGNETIC ORDER

Ferrimagnetism is a special case of antiferromagnetism and can be thought of in a very similar way, with the principal difference being that the magnitude of the sublattice magnetizations being different. The main consequence here is that the magnetizations will not cancel and we are left with a global or net magnetization. Experimentally, ferrimagnetism looks very much like ferromagnetism, though with some important differences. One major difference arises because the molecular fields of the sublattices are different and have different temperature dependences. This can lead to some unexpected variations of the magnetization with temperature.

We can treat the ferrimagnet as an inequivalent two sublattice system, much in the same way as we did for the case of antiferromagnentism. The magnetizations of the sublattices can thus be expressed in the same form using the Brillouin function:

$$\mathbf{M}_{A(B)} = \mathbf{M}_s^{A(B)} B_J \left(\frac{g\mu_B J}{k_B T} [\lambda_{2(3)} M_{B(A)} - \lambda_1 M_{A(B)} + B] \right) \quad (10.78)$$

This can be approximated as:

$$\mathbf{M}_{A(B)} \simeq \mathbf{M}_s^{A(B)} \frac{J+1}{3J} \left(\frac{g\mu_B J}{k_B T} \right) [\lambda_{2(3)} M_{B(A)} - \lambda_1 M_{A(B)} + B]$$

$$= \frac{C_{A(B)}}{T} [\lambda_{2(3)} M_{B(A)} - \lambda_1 M_{A(B)} + B] \quad (10.79)$$

We note that the full sublattice system has more constants than in the case for antiferromagnetism. Since $\mathbf{M}_A \neq \mathbf{M}_B$ there will always be a residual net magnetization, which vanishes at the order temperature, T_c. The overall magnetization dependence will take the form:

$$\mathbf{M}(T) = \mathbf{M}_A(T) - \mathbf{M}_B(T) \quad (10.80)$$

The expressions for $\mathbf{M}_A$ and $\mathbf{M}_B$ can be re-written in the form:

$$M_{A(B)} = \frac{B[C_{A(B)}(T - C_{B(A)}\lambda_{3(2)}) - C_A C_B \lambda_1]}{(T - C_A \lambda_2)(T - C_B \lambda_3) - C_A C_B \lambda_1^2} \quad (10.81)$$

We can now express the inverse susceptibility as:

$$\frac{1}{\chi} = \frac{B}{\mu_0 (M_A + M_B)} = \frac{[T^2 - T(C_A \lambda_2 + C_B \lambda_3) + C_A C_B (\lambda_2 \lambda_3 - \lambda_1^2)]}{T(C_A + C_B) - C_A C_B (2\lambda_1 + \lambda_2 + \lambda_3)} \quad (10.82)$$

The basic form of this equation can be written:

$$\frac{1}{\chi} = \frac{T}{C} - \frac{\sigma}{T - \Theta} + \frac{1}{\chi_0} \quad (10.83)$$

where the following substitutions have been used:

$$C = \frac{C_A + C_B}{T}$$

$$\Theta = \frac{C_A C_B}{C \mu_0} (2\lambda_1 + \lambda_2 + \lambda_3)$$

$$\sigma = \frac{C_A C_B}{C \mu_0} (\lambda_1^2 - \lambda_2 \lambda_3) - \frac{\Theta}{\mu_0 \chi_0}$$

$$\frac{1}{\chi_0} = \frac{\Theta}{\mu_0 C} - \frac{1}{\mu_0 C} (C_A \lambda_2 + C_B \lambda_3)$$

For high temperatures Equation (10.83) takes the form:

$$\frac{1}{\chi} \simeq \frac{T}{C} + \frac{1}{\chi_0} \tag{10.84}$$

These relations for the inverse magnetic susceptibility are illustrated in Figure 10.15. We can see how the curves for the inverse magnetic susceptibility differ for ferrimagnetic and ferromagnetic materials (compare Figure 10.14). We note that the curves for Equations (10.83) and (10.84) diverge from $T \sim \Theta$.

From the fact that at $T = T_c$, $1/\chi = 0$, we can evaluate the Curie temperature as:

$$T_c = \frac{(C_A \lambda_2 + C_B \lambda_3)}{2} + \frac{1}{2}[(C_A \lambda_2 - C_B \lambda_3)^2 + 4 C_A C_B \lambda_1^2]^{1/2} \tag{10.85}$$

We recover the antiferromagnetic order temperature for $|M_A| = |M_B| = M$, $C_A = C_B = C$ and $\lambda_2 = \lambda_3$:

$$T_c = \frac{g\mu_B(J+1)M}{3k_B}(\lambda_2 \pm \lambda_1) \tag{10.86}$$

compare with Equation (10.72), where $(\lambda_2 \pm \lambda_1) = (\lambda_{AA} - \lambda_{AB})$. The temperature dependence of the net magnetization, $|M_A(T) + M_B(T)|$, is illustrated in Figure 10.16 for the general cases typically

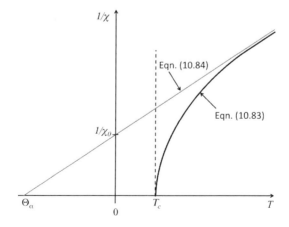

FIGURE 10.15: Inverse magnetic susceptibility for a generalized ferrimagnet. We note that $\Theta_\alpha = -C/\chi_0$.

found in many ferrites. In the case of Figure 10.16(a), the temperature dependent net moment always points in the same direction, where $|M_A| > |M_A|$. When there is a larger variation in the relative magnitudes of the sublattice magnetizations, the net magnetization can flip orientation from say the A - sublattice to the direction of the B - sublattice. Therefore, at a temperature designated as the *compensation temperature*, T_{comp}, there will be a net magnetization of zero, as illustrated in Figure 10.16(b).

There are many classes of ferrimagnet, the most important of which are: spinel ferrites, inverse spinels, hexaferrites, rare earth iron garnets, and in particular yttrium iron garnet (or YIG). The spinels have a general molecular formula: Fe_2MO_4, where M is a transition metal, such as Cu, Pb, Co, or Ni. For the case where M is Fe we have the compound *magnetite*; Fe_3O_4. In this type of compound, the oxygen atoms are doubly ionized as O^{2-}, by removing electrons from the metals. As such the metal ions have a positive ionized state; in general: Fe^{3+} and M^{2+}. For the normal spinel structure, the M^{2+} ions are located in tetrahedral (A) sites (with four O neighbors), while the Fe^{3+} ions are positioned at octahedral (B) sites (with six oxygen neighbors). There are twice as many B sites as A sites in the spinel structure. These sites are crystallographically different, and contain ions with different magnetic moments. In the case of the inverse spinel structure the M^{2+} ions are located on half of the B sites, while rest and the A sites are occupied by occupied by the Fe^{3+} ions. This case is particularly important, since given the two sites form the

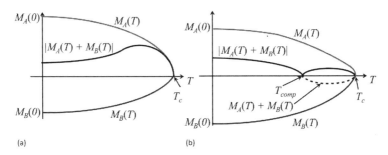

FIGURE 10.16: Schematic representation of the temperature dependence of the magnetic sublattices, $M_A(T)$ and $M_B(T)$ as well as the overall net magnetization, $|M_A(T) + M_B(T)|$ for (a) $\lambda M_A < \lambda M_B$ and (b) $\lambda M_A > \lambda M_B$.

magnetic sublattices of the ferrimagnet, the resultant magnetization is given only by the M^{2+} ions, since the Fe^{3+} ions are located in equal numbers on the A and B sites.

The iron garnets have the molecular formula: $R_3Fe_5O_{12}$, where R is a rare earth or yttrium. The crystal structure is rather complex, with 8 chemical formula units per unit cell, which thus has a total of 160 atoms. There are again octahedral and tetrahedral sites within the unit cell, both being occupied by Fe^{3+} ions, with the ratio 3:2. The R^{3+} ions are located on dodecahedral sites. In the case of YIG, the Y^{3+} ions have no magnetic moment (being $4d^0$), so the ferrimagnetic state is derived from the number imbalance of the Fe^{3+} ions on octahedral and tetrahedral sites. Since Fe^{3+} has a magnetic moment of $5\mu_B$, this will be the net moment per formula unit.

The hexaferrites, with chemical formula $MFe_{12}O_{19}$, have a hexagonal crystalline structure, with M typically being occupied by one of the following ions: Ba^{2+}, Pb^{2+} or Sr^{2+}. The ferrimagnetic structure has eight Fe^{3+} ions with moments opposed to the other four, as the net moment per formula unit will be $20\mu_B$. The hexagonal crystall typically exhibits a rather strong uniaxial magnetic anisotropy. In Table 10.6, we give some of the properties of some of the more important ferrites.

TABLE 10.6: Properties of some common ferrimagnetic materials.

Material	Type	T_c (K)	T_{comp} (K)	Magnetic moment (μ_B/formula unit)
Fe_3O_4	inverse spinel	860	-	4.1
$CoFe_2O_4$	inverse spinel	790	-	3.7
$NiFe_2O_4$	inverse spinel	865	-	2.3
$CuFe_2O_4$	inverse spinel	728	-	1.3
$Y_3Fe_5O_{12}$	garnet	560	-	5.0
$Gd_3Fe_5O_{12}$	garnet	564	290	16.0
$Dy_3Fe_5O_{12}$	garnet	563	220	18.2
$Ho_3Fe_5O_{12}$	garnet	567	137	15.2
$BaFe_{12}O_{19}$	hexaferrite	740	-	19.9
$SrFe_{12}O_{19}$	hexaferrite	746	-	20.2
$PbFe_{12}O_{19}$	hexaferrite	725	-	19.6

10.9 MAGNETIC ANISOTROPIES

From the consideration of ordered magnetic structures, it should be fairly clear that if we have an ordering of the magnetic moments in a solid, this implies a certain directionality of the moments and thus magnetization, with respect to the principal axes of the crystalline structure. Such directionality is referred to in magnetism as anisotropy. The principal origins of magnetic anisotropy are in the spin - orbit coupling and crystal field interaction, which gives rise to the so-called *magnetocrystalline anisotropy*. Magnetostatic effects due to the dipolar interaction gives rise to the *shape anisotropy* in non-spherical magnetic samples, as outlined in Section 10.5.1. While there can be other forms of magnetic anisotropy, such as surface and field induced anisotropies, we will limit our discussion to the two principal sources of magnetic anisotropy.

10.9.1 Shape Anisotropy

Let us consider a magnetic body, with a saturation magnetization of M_s. In the presence of a magnetic field, the magnetostatic energy is described by:

$$E = -\mu_0 \int \mathbf{M} \cdot \mathbf{H} \; dV \qquad (10.87)$$

where the integral is over the volume of the magnetic body. Since a uniformly magnetized body[4] produces a magnetostatic or demagnetizing field; $\mathbf{H}_D = -\mathcal{N}\mathbf{M}$, which depends of the specific shape of the body, we can express this as a demagnetizing energy of the form:

$$E_{ms} = -\mu_0 \int \mathbf{M} \cdot \mathbf{H}_D \; dV = \mu_0 \mathcal{N} \int M^2 \, dV = \frac{\mu_0}{2} \mathcal{N} M^2 \qquad (10.88)$$

In the presence of an applied magnetic field, $\mathbf{H}$, the effective magnetic field can be expressed as:

$$\mathbf{H}_{eff} = \mathbf{H} - \mathbf{H}_D = \mathbf{H} - \mathcal{N}\mathbf{M} \qquad (10.89)$$

see Figure 10.17. The factor $\mathcal{N}$ is called the demagnetization tensor, which for ellipsoids of rotations requires just the three leading

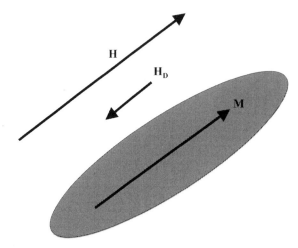

FIGURE 10.17: A spontaneously magnetized single domain ferromagnetic body produces a demagnetizing field, H_D, which is dependent on the shape of the body.

terms. In such cases we can express the magnetostatic energy in component form as:

$$E_{ms} = \frac{\mu_0}{2}(\mathcal{N}_x M_x^2 + \mathcal{N}_y M_y^2 + \mathcal{N}_z M_z^2) \tag{10.90}$$

We can express this using a spherical coordinate system as:

$$E_{ms} = \frac{\mu_0}{2} M^2 [(\mathcal{N}_x \cos^2\phi + \mathcal{N}_y \sin^2\phi)\sin^2\theta + \mathcal{N}_z \cos^2\theta] \tag{10.91}$$

where we have used: $M_x = M\cos\phi\sin\theta$; $M_y = M\sin\phi\sin\theta$; $M_z = M\cos\theta$. We note that the demagnetization factors are restricted by the relation: $\mathcal{N}_x + \mathcal{N}_y + \mathcal{N}_z = 1$. The form of Equation (10.91) provides a more useful expression for specific calculations. The principal demagnetizing factors can be evaluated analytically in terms of the dimensions of the axes for ellipsoids of rotation[5]. It is thus possible to obtain the specific form of the magnetostatic energy: For a sphere we have $\mathcal{N}_x = \mathcal{N}_y = \mathcal{N}_z = 1/3; \Rightarrow E_{ms} = \mu_0 M^2/6$. For a long cylindrical ferromagnet we obtain: $\mathcal{N}_x = \mathcal{N}_y = 1/2; \mathcal{N}_z = 0; \Rightarrow E_{ms} = \mu_0 M^2 \sin^2\theta/4$. In the case of a film or disc shaped sample we can write: $\mathcal{N}_x = \mathcal{N}_y = 0; \mathcal{N}_z = 1; \Rightarrow E_{ms} = \mu_0 M^2 \cos^2\theta/2$. The spherical sample clearly has no anisotropy, while the disc and cylinder are characterized by a uniaxial type anisotropy.

10.9.2 Magnetocrystalline Anisotropy

As the name suggests, magnetocrystalline anisotropy is strongly related to the type of crystalline structure of the material in question. As an example, we can show the variation of the sample magnetization with the application of a magnetic field in different directions with respect to the crystalline structure. In Figure 10.18, we show the initial magnetization curves for Fe and Ni, where we note that in the case of Fe the sample magnetizes more readily in the [100], being the so-called *easy-axis*, while in the [110] orientation the saturated state requires much higher magnetic fields to reach saturation. This is called the *hard-axis* of magnetization. In Ni the easy-axis corresponds to the [111] direction, while the hard-axis is along the [100] orientation. The other transition metal ferromagnet, Co, has an easy axis aligned along the [0001] axis of the hcp structure, where the basal plane is a hard direction for magnetization.

Despite the origins of magnetocrystalline anisotropy being related to the crystal field and spin - orbit interactions at the atomic level, the description is generally based on a phenomenological analysis of these symmetry effects. The simplest form we can give to the magnetocrystalline anisotropy is that for a uniaxial anisotropy, which means that the equivalent easy axes of magnetization are aligned at 180 from each other, i.e., in opposite directions along a single

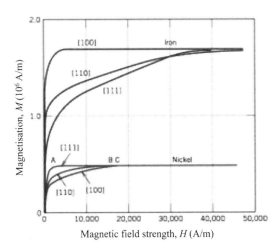

FIGURE 10.18: Magnetization curves for single crystals of Fe and Ni. Magnetic fields were applied along the principal crystalline axes; [100], [110] and [111].

axis, e.g. as in the case for Co. There are several ways to express this mathematically, one of the most general ways of doing this is to include several orders of a series, such as:

$$E_{Ku} = \sum_n K_{un} \sin^{2n}\theta = K_{u0} + K_{u1}\sin^2\theta + K_{u2}\sin^4\theta + K_{u3}\sin^6\theta + \ldots$$
(10.92)

where the constants, K_{un}, characterize the form of the energy landscape. For example, if $K_{u1} > 0$, and dominates the other constants, the the energy surface will be an oblate spheroid, with easy directions along the $\pm z$ directions. However, when $K_{u1} < 0$, the energy surface is a prolate spheroid and the easy axes are anywhere in the xy-plane. Cobalt has room temperature anisotropy constants of $K_{u1} = 4.1 \times 10^5$ Jm^{-1} and $K_{u2} = 1.5 \times 10^5$ Jm^{-1}.

For cubic structures, such as Fe and Ni, the situation is more complex from a mathematical descriptive point of view. As a first step, we can express a cubic form of anisotropy energy with the use of the directional cosines in the form:

$$E_{Ku} = K_0 + K_1(\alpha_1^2 \alpha_2^2 + \alpha_1 2^2 \alpha_3^2 + \alpha_1^2 \alpha_3^2) + K_2(\alpha_1^2 \alpha_2^2 \alpha_3^2) \quad (10.93)$$

where we have:

$$\alpha_1 = \cos\theta_1 = \cos\phi\sin\theta$$
$$\alpha_2 = \cos\theta_2 = \sin\phi\sin\theta$$
$$\alpha_3 = \cos\theta_3 = \cos\theta$$

here θ_n is the angle between the principal axes ($x = 1, y = 2$ and $z = 3$), with the magnetization vector in some arbitrary orientation. Substituting in for the directional cosines, Equation (10.93) can be expressed as:

$$E_K(\theta,\phi) = K_0 + \frac{K_1}{64}[(1-\cos 4\phi)(3-4\cos 2\theta + \cos 4\theta) + 8(1-\cos 4\theta)]$$
$$+ \frac{K_2}{128}(1-\cos 4\phi)(1-\cos 2\theta - \cos 4\theta + \cos 4\theta \cos 2\theta)$$
(10.94)

Based on room temperature experimental measurements, the principal anisotropy constants for Fe and Ni are given by: $K_1 = 4.8 \times 10^4$ Jm^{-1} and $K_2 = 0.5 \times 10^4$ Jm^{-1} for Fe and $K_1 = -4.5 \times 10^3$ Jm^{-1} and $K_2 = 2.3 \times 10^3$ Jm^{-1} for Ni.

Due to the nature of the physical origin of magnetocrystalline anisotropy, the anisotropy constants are very sensitive functions of temperature. For example, in transition metals, the principal anisotropy constant follows the relation:

$$K_1(T) = K_1(0)\left[1 - \frac{T}{T_c}\right] \quad (10.95)$$

See for example O'Handley (2000), Herpin (1968) for more details.

10.10 MAGNETIC DOMAINS, DOMAIN WALLS, AND HYSTERESIS

The magnetostatic energy of a spontaneously magnetized sample can be significantly reduced or eliminated by the division of the magnetized state between various domains, in which the direction of the spontaneous magnetization aligns along equivalent crystallographic orientations for the easy direction of magnetization. In such a situation, the sample can reach a state with a net zero magnetization and thus removes the magnetostatic energy. This does, however, come at a cost. In the formation of these domains, there must be regions of transition of the magnetization direction from one magnetic domain to the next. These regions of transition are called *domain walls*. In considering the existence of magnetic domains, we must take into account the equivalent directions of the magnetization, as defined by the type of magnetocrystalline anisotropy. For example in the case of single crystal cobalt, we can only expect to find two types of magnetic domain, with orientations [0001] and [000$\bar{1}$]. For Fe the situation is more complex since there are six equivalent easy axes in its (bcc) cubic structure: [100], [010], [001], [$\bar{1}$00], [0$\bar{1}$0], and [00$\bar{1}$]. There are two types of domain wall, which are referred to as *Néel walls* and *Bloch walls*. In the former, the rotation of the magnetic moment occurs in the plane of the direction of the magnetic domains, while in the latter the rotation is out of the plane. The type of domain wall will depend on the specific energy configuration, being influenced by such factors as the relative magnetic anisotropies and the exchange energy, etc.

Since the transition regions require that the direction of the moment is shifted from the easy direction as well as the fact that the exchange energy is minimized when the spins are maintained parallel means that the existence of domain walls comes at an energy cost. The fine balance of how domains are formed is then dependent on how much magnetostatic energy is saved at the cost of domain wall energy. In the evaluation of the domain wall energy, the principal considerations, as we stated above, will come from the mangeto-crystalline anisotropy and exchange. We can demonstrate the way to calculate this if we consider the transition of the magnetic moments for a chain of spins, which cut across from one domain to the next through the domain wall. The exchange contribution takes the form:

$$E_{exch} = -\mu_0 z J_{ij} \boldsymbol{\mu}_i \cdot \boldsymbol{\mu}_j = -\mu_0 z J_{ij} \mu^2 \cos\eta \qquad (10.96)$$

where z refers to the coordination number, $|\boldsymbol{\mu}_i| = |\boldsymbol{\mu}_j| = \mu$ and η is the angle between adjacent spin orientations. Since we can expect this angle to be small, we can make the following approximation: $\cos\eta \simeq 1 - \eta^2/2$. Thus we obtain:

$$E_{exch} = -\mu_0 z J_{ij} \mu^2 (1 - \eta^2/2) = \mu_0 J_{ij} \mu^2 (\eta^2 - 2) \qquad (10.97)$$

We can now expect that the additional energy due to a domain wall is the sum of the individual interactions in the wall, such that:

$$E_{exch} = \mu_0 J_{ij} \mu^2 \eta^2 n \qquad (10.98)$$

where n is the number of spins in the domain wall. In the case of a 180° wall, we can divide the spin angles such that we have: $\eta = \pi/n$. If the separation between spins is equal to the lattice constant a, we can write the exchange energy per unit area of the wall as:

$$\varepsilon_{exch} = \frac{\mu_0 J_{ij} \mu^2 \pi^2}{na^2} \tag{10.99}$$

From this expression we can see that the interaction energy will be minimized when the angle between spins is very small, which would be satisfied for very broad domain walls. We now consider the anisotropy contribution, which we will consider as a first order anisotropy energy of the form: $E_K(\theta) = K_1 \sin^2 \theta$. Since there are n spins in the direction perpendicular to the domain wall, the energy will be a sum of all the moments in the wall, such that the anisotropy energy per unit area of the wall is:

$$\varepsilon_K = \frac{nK_1 a}{2} \tag{10.100}$$

From this form of the energy, we would expect that domain walls should be as narrow as possible to reduce this energy contribution. We now have a total energy of the form:

$$\varepsilon_{BW} = \varepsilon_{exch} + \varepsilon_K = \frac{\mu_0 J_{ij} \mu^2 \pi^2}{na^2} + \frac{nK_1 a}{2}$$

$$= \frac{\mu_0 J_{ij} \mu^2 \pi^2}{\delta_{BW} a} + \frac{\delta_{BW} K_1}{2} \tag{10.101}$$

where we have written the domain wall thickness, for the Bloch type wall as: $\delta_{BW} = na$. We now minimize the wall energy with respect to the thickness: $\partial \varepsilon_{BW}/\partial \delta_{BW} = 0$, from which we obtain the 180° Bloch domain wall thickness as:

$$\delta_{BW} = \pi \mu \sqrt{\frac{2\mu_0 J_{ij}}{aK_1}} = \pi \sqrt{\frac{A}{K_1}} \tag{10.102}$$

where we have introduced the *exchange stiffness constant*, A. Substituting back into the wall energy we now obtain:

$$\varepsilon_{BW} = \pi \sqrt{AK_1} \tag{10.103}$$

From Equation (10.102), we note that the competition between the exchange and anisotropy contributions becomes clear: Large exchange will produce broad domain walls, while large anisotropies will act to diminish the wall thickness. From the magnetic properties for Fe, Co, and Ni, we obtain the domain wall parameters given in Table 10.7.

In the absence of a magnetic field a magnetic sample will persist in a state defined by the minimization of the various energy contributions, exchange, magnetostatic, and magnetocrystalline energies. As noted above, the magnetic domains will form to reduce the overall energy of the magnetic system. When the domain pattern has no free poles at the surface, it is said to have *closure domains*, as illustrated in Figure 10.19.

TABLE 10.7: Magnetic domain wall parameters for selected ferromagnets.

Material	Domain Wall Thickness (nm)	Domain Wall Thickness (number of lattice parameters)	Domain Wall Energy (Jm^{-2})
Fe	40	138	3.0×10^{-3}
Co	15	36	8.0×10^{-3}
Ni	100	285	1.0×10^{-3}

Let us now consider what happens when a magnetic field is applied. As always we can calculate the energy of the system by adding the Zeeman energy term to the total energy and minimize

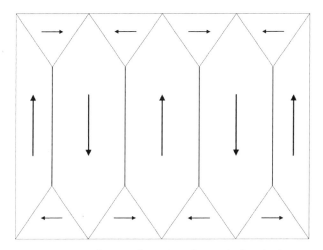

FIGURE 10.19: Schematic illustration of magnetic domain closure.

in the usual way. Of course the specific domain structure is more difficult to assess from a simple calculation of the type and we would have to define the size and shape of the sample. In recent years, such calculations have been performed using micromagnetic simulations, in which the system is divided up into cells of a certain size and energy minimization can be performed for specific cases. One such package is called the Object Oriented MicroMagnetic Framework (OOMMF)[6] and many examples can be found in the research literature. Of particular importance in the consideration of the changes incurred upon application of a magnetic field is the direction in which it is applied. The initial state of the sample, when the global magnetization is zero due to closure domains or an averaging to zero of the magnetic domains, is called the *virgin state*. If we imagine that the direction of an applied field is arbitrary with respect to the easy axes of the system, then typically what will happen is that magnetic domains which have an orientation close to the direction of the applied field will grow at the expense of those more opposed to the direction of the field. This process of *domain growth* will continue until the sample reaches a state of being a single domain. Once this occurs, further increase of the external magnetic field will now bring about a *rotation* of the magnetization until it is aligned with the direction of the applied field. Once this state is reached the magnetization in the direction of the applied field cannot grow any further and the sample is said to be at *magnetic saturation*, designated as M_s. The removal of the magnetic field to zero will allow the magnetization to relax back to a direction of local energy minimum, i.e., along the direction of an easy axis. This state will have a *remanent magnetization*, M_r. Reversing the direction of the magnetic field will allow the sample to over come the energy minimum of the local anisotropy and some domain walls can begin to form. This will reduce the magnetization. The point at which the magnetization crosses the field axis is called the *coercive field*, H_c. Further increase of the magnetization in the reverse direction will eventually lead to the sample saturation. This process can be completed by reducing the field to zero and then increasing it in the original direction. Usually the resulting magnetization - field curve is different for the ascending and descending portions, and the loop illustrates the *hysteresis* inherent in the magnetic system. The hysteresis loop is

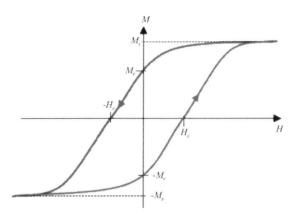

FIGURE 10.20: Schematic illustration of a magnetic hysteresis loop, showing the saturation magnetization, M_s, the remanent magnetization, M_r and the coercive field, H_c.

illustrated in Figure 10.20. The area of a hysteresis loop is a measure the magnetic energy dissipated in the system and can vary quite strongly, especially in magnetic materials with large anisotropies. The size of the coercive field distinguished hard magnetic materials from soft ones. In the former, the large coercive field means that they are difficult to reverse magnetically. These materials have very different and specific applications, such as magnetic recording media and motors. The use of soft magnetic materials is favorable for transformers since they will produce smaller loops and dissipate less energy. The heating that we note in transformers is directly related to the energy dissipated in going around the hysteresis loop. The origin of this dissipation in energy is related to the domain wall movements in the sample and produces the irreversibility due to the pinning of the domain wall at defects in the sample. This process is also related to the discontinuous motion of the domain walls as the applied field forces the domain wall to progress, causing jumps in the magnetization. This process is known as the *Barkhausen effect*.

There are a number of experimental methods that can be used to observe and measure magnetic domains. The *magneto-optic Kerr effect* (MOKE) is based on the rotation of the axis of polarization of incident light upon reflection at a magnetic surface. This provides a simple techniques for the observation of the magnetic state of a sample and allows domains to be visualized, using incident polarized light. For smaller magnetic structures, the use of electron beams in a transmission electron microscope (TEM), also permits the observation of magnetic domains. This relies on the Lorentz force which acts on the electrons as they pass through the magnetic material, thus deflecting them from their normal path. Since different domains will deflect the electrons in different directions, this provides a good method of the visualization of magnetic domains in small structures. This techniques is called *Lorentz microscopy*. Another technique frequently used in the observation of magnetic domains is *magnetic force microscopy* (MFM), and is an adaptation of the *atomic force microscopy* (AFM) technique, where a magnetic coating is applied to a normal AFM tip. Here the very fine tip is brought into close proximity with the surface of a sample, and deflections of the tip are produced by its interaction with the sample, which in the case of MFM is due to attractive and repulsive forces between the sample and tip magnetization. The domain images are produced by mapping the tip deflection as it is scanned over the surface of the sample.

10.11 SPIN WAVES

A ferromagnetic material is only in its ground state, i.e., with all magnetic spin pointing rigidly in the same direction, at zero temperature, (obviously ignoring and zero-point fluctuations). At non-zero temperatures thermal excitations will interrupt this perfect alignment. The process is not dissimilar to the excitation of the crystal lattice in the production of phonons. In the case of magnetic structures, these excitations are called *spin waves* or for the quantum description *magnons*.

The excitations of a magnetic spin system can be described by the consideration of the Heisenberg Hamiltonian, Equation (10.45). For a simple chain of spins, we can express the Hamiltonian as:

$$\hat{\mathcal{H}} = -2\mathcal{J}\sum_i \hat{\mathbf{S}}_i \cdot \hat{\mathbf{S}}_{i+1} = -2\mathcal{J}\sum_i \left[\hat{S}_i^z \hat{S}_{i+1}^z + \frac{1}{2}(\hat{S}_i^+ \hat{S}_{i+1}^- + \hat{S}_i^- \hat{S}_{i+1}^+) \right]$$
(10.104)

where $\hat{S}^{\pm} = \hat{S}_x \pm \hat{S}_y$ are called the raising and lowering operators. As we stated above, the absolute ground state can be considered as the set of all parallel spins, along the z-direction, which we can express as: $|\Phi_0\rangle = |\uparrow\uparrow\uparrow \ldots \uparrow\rangle$. The eigenvalue of this state can be obtained from:

$$\hat{\mathcal{H}}|\Phi_0\rangle = -NS^2 \mathcal{J}|\Phi_0\rangle \tag{10.105}$$

where N is the number of spins in the chain. The first excited state can be expressed by the flipping of a single spin, say at position k in the chain. We can write this as: $|\Phi_k\rangle = \hat{S}_k^- |\Phi_0\rangle$. Here the operator: $\hat{S}_k^-$ acts only on the spin at site k in the chain and flips it from, say, the up to the down state. This will produce and energy change of: $\frac{1}{2} - (-\frac{1}{2}) = 1$. The integer value of the spin change indicates that this excitation corresponds to a boson. Applying the Hamiltonian to the state yields:

$$\hat{\mathcal{H}}|\Phi_k\rangle = 2\left[(-NS^2\mathcal{J} + 2S\mathcal{J})|\Phi_k\rangle - S\mathcal{J}|\Phi_{k+1}\rangle - S\mathcal{J}|\Phi_{k-1}\rangle\right] \tag{10.106}$$

This is not an eigenstate of the spin chain since it is not a constant multiplied by $|\Phi_k\rangle$. Clearly a localized spin flip of this type would involve a large exchange energy between spin k with it neighbors, at $k + 1$ and $k - 1$. However, if we spread out the energy in a gradual way by delocalizing the excitation, we can represent the state as plane wave of the form:

$$|\Phi_q\rangle = \frac{1}{\sqrt{N}} \sum_n e^{i\mathbf{q} \cdot \mathbf{R}_k} |\Phi_k\rangle \tag{10.107}$$

where $\mathbf{q}$ is the magnon wave-vector. We can visualize the spin wave as shown in Figure 10.21. We can evaluate the eigenvalue of this magnonic state from:

FIGURE 10.21: Illustration of a spin wave as a variation of the spin orientation with position.

$$\hat{\mathcal{H}}|\Phi_{\mathbf{q}}\rangle = E(\mathbf{q})|\Phi_{\mathbf{q}}\rangle \qquad (10.108)$$

where

$$E(\mathbf{q}) = -NS^2\mathcal{J} + 4S\mathcal{J}[1-\cos(qa)] \qquad (10.109)$$

The excitation energy takes the form: $\hbar\omega_q = 4S\mathcal{J}[1-\cos(qa)]$. In a 3D solid, this result takes the form:

$$\hbar\omega_q = \mathcal{J}S\left[z - \sum_m \cos(\mathbf{q}\cdot\mathbf{a}_m)\right] \qquad (10.110)$$

the summation is over the nearest neighbor vectors of $\mathbf{a}_m$ and z is the coordination number. For the case of small $\mathbf{q}$, we can approximate this result as:

$$\hbar\omega_q \simeq \mathcal{J}S\left[z - \sum_m \left(1 - \frac{(\mathbf{q}\cdot\mathbf{a}_m)^2}{2} + \ldots\right)\right] = \frac{\mathcal{J}Sa^2q^2}{2}\sum_m \cos^2\theta_m \qquad (10.111)$$

where θ_m is the angle between $\mathbf{q}$ and $\mathbf{a}_m$. Since this angle is small, we again approximate to obtain:

$$\hbar\omega_q \simeq \frac{\mathcal{J}Sa^2}{2}q^2 = Dq^2 \qquad (10.112)$$

In the presence of a magnetic field this dispersion relation takes the form:

$$\hbar\omega_q = g\mu_B\mu_0 H + Dq^2 \qquad (10.113)$$

These dispersion relations are shown in Figure 10.22.

We can use spin wave theory to predict the variation of the magnetization with temperature. In an intuitive way we can see that the zero temperature magnetization, $M_s(0)$ will represent a maximum value. The effect of temperature in the excitation of the spin system will thus be to reduce the magnetization from this maximum value. To evaluate the trend we can start by considering the density of magnon states. Since magnons are bosons, they follow a similar trend to phonons and hence we can write:

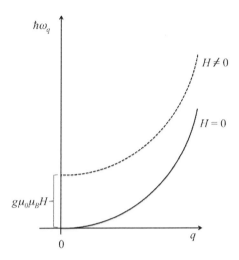

FIGURE 10.22: Spin wave dispersion relation for small wave vectors, q, with and without a magnetic field.

$$g(q)dq \propto q^2 dq$$
$$\Rightarrow g(\omega)d\omega \propto \omega^{1/2}d\omega \qquad (10.114)$$

Spin waves are quantized in the same way as phonons. The number of magnons excited at a specific temperature, T, can be calculated by integrating the magnon density of states over all frequencies, which we can now express as:

$$n_{magn} = \int_0^\infty \frac{g(\omega_q)d\omega_q}{e^{\hbar\omega_q/k_B T} - 1} \qquad (10.115)$$

Writing $x = \hbar\omega_q/k_B T$, we obtain:

$$n_{magn} = \left(\frac{k_B T}{\hbar}\right)^{3/2} \int_0^\infty \frac{x^{1/2}dx}{e^x - 1} = \left(\frac{k_B T}{\hbar}\right)^{3/2} \frac{\sqrt{\pi}}{2}\zeta(3/2) \qquad (10.116)$$

where $\zeta(3/2)$ is the Reimann zeta function, which being independent of temperature can be considered as a constant for our purposes. Since the magnon number is proportional to $T^{3/2}$, we obtain the result that the magnetization reduces from its zero temperature value according to:

$$\frac{M(0)-M(T)}{M(0)} \propto T^{3/2} \qquad (10.117)$$

This variation of the magnetization is referred to as the *Bloch $T^{3/2}$ law* and shows a good agreement with experiment in the low temperature range and up to around $T \simeq T_c/2$. In the high temperature regime, the magnetization, $M(T)$ follows a variation of the form $(T_c - T)^\beta$, where β is called the *critical exponent*, which is usually obtained by fitting to experimental data.

In addition to the excitation by spin waves, another form of excitation can also reduce the magnetization of a ferromagnet. This is called a *Stoner excitation* and occurs in metals when electrons flip from one spin state to the other. In itinerant ferromagnets, see Section 10.6.2, the conduction electrons are split in to spin-up and spin down bands. At a certain energy it is possible to flip the electron from a filled state $\mathbf{k}$ in one spin sub-band to an empty state $\mathbf{k} - \mathbf{q}$ in the other spin sub-band. The excitation energy conforms to the relation:

$$\hbar\omega_q = E_{\mathbf{k}} - E_{\mathbf{k-q}} + \Delta_{ex} \qquad (10.118)$$

where $E_{\mathbf{k}} = \hbar^2 k^2 / 2m_e^*$, and Δ_{ex} is the exchange splitting between the spin-up and spin down states, and corresponds to the energy necessary to flip a spin. In Figure 10.23, we show the energy spectrum for

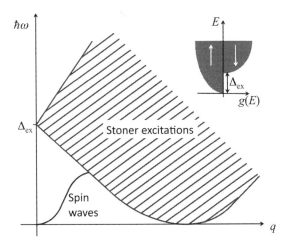

FIGURE 10.23: Excitations in the electron gas of metals. The spin wave branch is illustrated along with the continuum of Stoner excitations. Inset show the spin split bands separated by the exchange splitting energy, Δ_{ex}.

Stoner and spin wave excitations. The hatched region is referred to as the *Stoner continuum*, since there is a broad range of excitation energies in the spectrum of Stoner excited states.

10.12 GIANT MAGNETORESISTANCE AND SPINTRONICS

We have already seen (see Section 8.8.3) that magnetoresistance is a transport phenomenon that is fairly common in metals. In ferromagnetic transition metals the effects can be enhanced due the narrow d bands, which are spin split and provide a large $g(E_F)$ and hence satisfy the Stoner criterion. The variation of the magnetoresistance is due to s - d transitions, which are promoted by the applied magnetic field, and since the effective masses of the electrons in the different bands are different, this provides a mechanism for negative magnetoresistive effects, as are observed in Ni, for example. A model of magnetoresistance can be provided by Mott's *two-current model* (1936), in which the scattering of spin-up and spin-down electrons are considered as independent, providing two separate conduction channels. The spin-split bands in a ferromagnet means that the electrons undergoing spin-flip scattering will do so with different probabilities for the two spin states. The orientation of the magnetization with respect to the applied field can be important and produces *anisotropic magnetoresistance*.

By far the largest magnetoresistive effect in metals is that found in magnetic multilayers. In particular, due to the coupling between adjacent ferromagnetic layers having an oscillatory character as a function of a non-magnetic metallic interlayer thickness, it is possible to fabricate multilayer systems in which the magnetization in alternate layers are aligned in anti-parallel, this is called *antiferromagnetic coupling*. The RKKY interaction is one of the credible coupling mechanisms, which can bring about this variation. The application of a magnetic field can be used to invert one of the magnetic layers such that they become parallel. The difference in the resistance of the parallel and antiparallel alignments gives rise to the large change of electrical resistance, referred to as *giant magnetoresistance* (GMR). This effect was first observed in 1986 and led to the 2007 Nobel Prize in Physics for Fert and Grünberg. The two current model can then be used to consider the current paths for spin-up and spin-down electrons. This simple model can be explained as follows:

Let us consider the parallel and antiparallel states, as illustrated in Figure 10.24. In the case where the layers are initially antiparallel, the two-current resistance can be expressed as:

$$R_{AP} = \frac{R+r}{2} \qquad (10.119)$$

where R and r represent the resistance for electrons of a specific spin state passing through a magnetic layer whose magnetization (majority spin band) is opposed to its spin direction and parallel to it, respectively. The high resistance state occurs because the electrons can be expected to scatter more in a layer with spin opposing the magnetization. In this case the two electrons spin states experience a similar resistance, having to pass through layers with parallel and antiparallel spin bands. However, when we consider the case where the magnetization of the two ferromagnetic layers are parallel, then the situation is different. Referring to Figure 10.24, the up-state electrons cross two magnetic layers with parallel majority spin bands and feel a low resistance channel ($2r$), while down-spin electrons

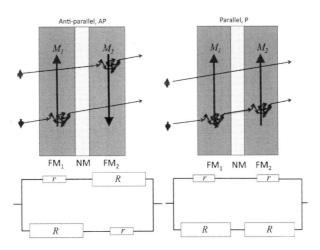

FIGURE 10.24: Two ferromagnetic layers; FM_1 and FM_2, are separated by a non-magnetic layer (NM). We consider the two current model for magnetic bilayer system: when the magnetization vectors are aligned antiparallel, both spin-up and spin-down electrons experience the same effective resistance ($R+r$) (left). For the case where the magnetization vectors of the two ferromagnetic layers are parallel the resistance can be thought of as two channels with resistances $2r$ and $2R$ for the spin-up and spin-down electrons, respectively (right).

encounter two opposite majority spin band layers, and hence pass through a high resistance channel ($2R$). The effective resistance for this parallel state can be expressed as:

$$R_P = \frac{2Rr}{R+r} \qquad (10.120)$$

It is now a simple matter to evaluate the maximum magnetoresistance produced when a magnetic field is applied to invert one of the magnetization states from the antiparallel to the parallel configuration. This can be expressed as:

$$GMR = \frac{R_{AP} - R_P}{R_P} = \frac{(R+r)^2}{2Rr} \qquad (10.121)$$

It should now be clear that the applied field acts to bring the ferromagnetic layers parallel and thus reduce the overall resistance. Using multiple repeats of the layers enhances further the magnetoresistance, as shown in Figure 10.25.

The discovery of spin dependent effects in transport properties has led to a large number of device applications, including the

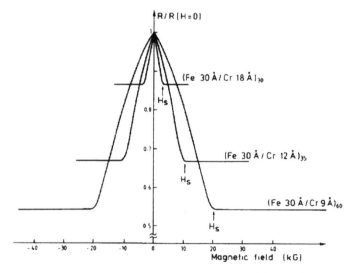

FIGURE 10.25: Giant magneto resistance in Fe/Cr/Fe multilayers. The ultimate magnetoresistance is seen to increase with the number of bilayers in the stack. (Reprinted figure with permission from: Baibich et al., Phys. Rev. Lett. **61**, 2472-2475 (1988). Copyright 2007 by the American Physical Society.)

spin-valve, which is based directly on the GMR between layers coupled to an antiferromagnetic layer which acts to pin one of the ferromagnetic layers via what is know as the *exchange bias* effect. This device is used in read-write heads of hard disks. In fact, so important was the discovery of GMR, that it has given rise to a new field of study based on the spin of the electron and derives from the fact that the spin state can be manipulated by passing an electrical current through a magnetic layer which can produce a spin filtering effect. This is popularly termed *spin electronics* or *spintronics* and is a major area of research and development. Progress in spintronic devices and effects has advanced well and many of the traditional semiconductor devices have their spintronic counterparts. For example designs for the spin FET have been advanced. There have been spin LEDs produced, which because of the excitations involved between spin polarized states, the resulting light is also left or right circular polarized and is due to the transfer of angular momentum from electron to photon being conserved.

10.13 SPIN DYNAMICS

The study of spin dynamics generally refers to any process in which the spin configuration and magnetization varies in time. The most common forms of spin dynamics will be via precessional motion of the magnetization as in a magnetic resonance experiment. In the case of a ferromagnetic material, this is referred to as *ferromagnetic resonance* or FMR. Other processes include standing spin wave resonance and magnetization reversal processes. Magnetic systems can undergo various types of excitation, which produce changes to the magnetization. Ultrafast demagnetization can occur to a spin system when it is excited by a high power laser, for example. The demagnetization in this case is produced by a transfer of energy from the photons to the spin system and through the heating of the electron gas to temperatures above the Curie temperature. In certain case the demagnetization can occur in the order of 10s of femtoseconds (1 fs = 10^{-15} s). Precessional dynamics typically occur at a much slower pace, with oscillation periods in the picosecond range (10^{-12} s). In this section, we will give a brief overview of some of the main principles of magnetization dynamics and magnetic resonance phenomena.

The description of the time variation of the magnetization vector can be expressed in a phenomenological formulation. The precessional motion derives from the interaction of the magnetization with the applied field. This produces a magnetic torque, which can be described by the vector product of these quantities. The response of the magnetization to an excitation depends very much on how the excitation is produced in terms of its strength, direction and speed, or if it is a continuous excitation, the frequency. Once a magnetic system is pushed out of its equilibrium state, the system will try to return to an energy minimum, which can be different from the one in which it started from equilibrium. This process is generally referred to as relaxation. There have been several different approaches to the description of such processes. The main forms of the phenomenological description are given below:

1) Landau - Lifshitz (LL) equation:

$$\frac{\partial \mathbf{M}}{\partial t} = -\gamma(\mathbf{M} \times \mathbf{H}_{eff}) - \frac{\lambda}{M_s^2}[\mathbf{M} \times (\mathbf{M} \times \mathbf{H}_{eff})] \quad (10.122)$$

2) Landau - Lifshitz - Gilbert (LLG) equation:

$$\frac{\partial \mathbf{M}}{\partial t} = -\gamma(\mathbf{M} \times \mathbf{H}_{eff}) + \frac{\alpha}{M_s}\left[\mathbf{M} \times \frac{\partial \mathbf{M}}{\partial t}\right] \quad (10.123)$$

3) Bloch - Bloembergen (BB) equations:

$$\frac{\partial M_x}{\partial t} = -\gamma(\mathbf{M} \times \mathbf{H}_{eff})_x - \frac{M_x(t)}{T_2} \quad (10.124)$$

$$\frac{\partial M_y}{\partial t} = -\gamma(\mathbf{M} \times \mathbf{H}_{eff})_y - \frac{M_y(t)}{T_2} \quad (10.125)$$

and

$$\frac{\partial M_z}{\partial t} = -\gamma(\mathbf{M} \times \mathbf{H}_{eff})_z - \frac{M_z(t) - M_0}{T_2} \quad (10.126)$$

The first point to note about these equations is that the left hand side are just the time evolutions of the magnetization, while

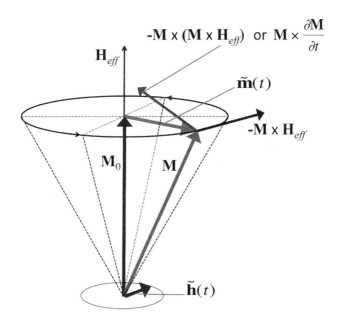

FIGURE 10.26: Vector construction of the Landau - Lifshitz and Landau - Lifshitz - Gilbert equations.

the first terms on the right hand side are all equal and correspond to the description of the precessional motion of the magnetization about the effective magnetic field, H_{eff}, and is the vector product of this with the magnetization. The vector construction for Equations (10.122) and (10.123) is shown in Figure 10.26.

The last terms can now be seen to describe the relaxation processes, being vectors which point the magnetization back to the equilibrium configuration of the system. These are also illustrated for Equations (10.122) and (10.123) in Figure 10.26. The constant γ is the magnetogyric ratio. For the LL and LLG equations the prefactors will determine how quickly this occurs, being defined by the constants, λ and α, the Gilbert damping parameter, respectively. In these two equations, the combination of precession and damping lead to a spiral motion of the magnetization vector, where the magnetization is described by a classical vector which maintains its magnitude. For the case of the BB equation the damping is described by two parameters, T_1 and T_2, which are the longitudinal (spin - lattice) and transversal (spin - spin) relaxation times, respectively. In this case the relaxation process can be described by the following relations:

$$M_{x,y}(t) = M_{x,y}(0)e^{-t/T_2} \tag{10.127}$$

and

$$M_z(t) = M_{z,eq} - [M_{z,eq} - M_z(0)]e^{-t/T_1} \tag{10.128}$$

The time varying z-component of the magnetization can thus change length in this formulation. The BB formulation is usually reserved for the description of nuclear magnetic resonance (NMR), with most descriptions (of FMR) being performed with the LL or LLG approaches. The magnetization and magnetic field vectors, as shown in Figure 10.26, has static and dynamic components in general, which can be expressed as: $\mathbf{M}(t) = \mathbf{M}_0 + \tilde{\mathbf{m}}(t)$ and $\mathbf{H}_{eff}(t) = \mathbf{H}_{eff}^0 + \tilde{\mathbf{h}}(t)$, where the static components are generally considered to be much larger than the dynamic ones, which are usually taken to be small perturbations, $|\mathbf{M}_0| \gg |\tilde{\mathbf{m}}|$. In a ferromagnetic system, the effective field is made up of various components which comprise the external fields: static magnetic field, $\mathbf{H}_0$, dynamic (or rf) field, $\tilde{\mathbf{h}}$, and internal fields which arise from the effective fields of the shape (or demagnetizing) field, $\mathbf{H}_{dem}$ and magnetocrystalline anisotropies, $\mathbf{H}_K$ as well as the exchange field, $\mathbf{H}_{exch} = (2A/M_s^2)\nabla^2\mathbf{M}$.

It is possible to manipulate the LL and LLG equations of motion by recognizing that a magnetic torque, $\tau = \mathbf{M} \times \mathbf{H}$, can be replaced by the classical vector $\tau = -\mathbf{r} \times \nabla E$, where E is the magnetic free energy of the system. From this we can obtain the resonance condition in terms of the second derivatives of the free energy:

$$\left(\frac{\omega}{\gamma}\right)^2 = \frac{1}{M_s^2 \sin^2\theta}\left[\frac{\partial^2 E}{\partial \theta^2}\frac{\partial^2 E}{\partial \phi^2} - \left(\frac{\partial^2 E}{\partial \theta \partial \phi}\right)^2\right] \tag{10.129}$$

where ω is the angular frequency of the precession. This equation is known as the *Smit - Beljers* or *Smit - Suhl* equation. This equation must be used in conjunction with the equilibrium conditions as expressed by:

$$\frac{\partial E}{\partial \theta} = 0 \quad \text{and} \quad \frac{\partial E}{\partial \phi} = 0 \tag{10.130}$$

These equilibrium conditions are extremely important, since the magnetic system will be disturbed from this static state and, once the perturbation has ceased, will relax back to this condition, providing the perturbation isn't too strong. In the case where there are magnetic boundaries in the direction of the applied static magnetic field, a consideration of the exchange field and boundary conditions allows a more general condition to be obtained:

$$\left(\frac{\omega}{\gamma}\right)^2 = \left(\frac{2A}{M_s}\right)^2 k^4 + \left(\frac{1}{\sin^2\theta}\frac{\partial^2 E}{\partial \phi^2} + \frac{\partial^2 E}{\partial \theta^2}\right)\frac{2A}{M_s^2}k^2 \\ + \frac{1}{M_s^2 \sin^2\theta}\left[\frac{\partial^2 E}{\partial \theta^2}\frac{\partial^2 E}{\partial \phi^2} - \left(\frac{\partial^2 E}{\partial \theta \partial \phi}\right)^2\right] \quad (10.131)$$

It should be noted that this equation is a quadratic equation in terms of k^2, where k is a wave vector for standing spin waves. Standing spin wave can be supported in confined magnetic structures such as thin films, with an applied magnetic field applied in the perpendicular direction to the film plane. The boundary conditions, which depend on the surface anisotropies of the film, mean that a number of excitation modes can be observed. Such spectra are referred to as the *spin wave resonance* (SWR) spectra. We can now note that when there are no interfaces, the wavelength of the spin waves will be infinite (i.e., $k = 0$) and we will recover to Equation (10.129) for the case of ferromagnetic resonance.

To use the resonance equations we require expressions for the free energy of the system under study. This will have various components and is related to the effective field via the relation:

$$\mathbf{H}_{eff} = -\frac{\nabla_{\hat{u}} E}{M_s} \quad (10.132)$$

where $\hat{u} = \mathbf{M}/M_s$ is the versor of the magnetization. In Figure 10.27, we show the spin wave resonance spectra for a thin film of GaMn at low temperature. We note that the spectra exhibit multiple peaks, which correspond to different spin wave modes. Spectra are also shown as a function of the angle of the magnetic field away from the perpendicular direction, where shifts occur due to the variation of the internal field arising from shape and other anisotropies, as

384 • Solid State Physics

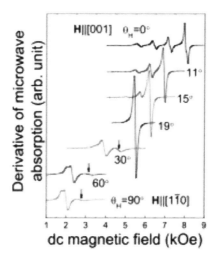

FIGURE 10.27: Spin wave resonance spectra for a 120 nm GaMn thin film at low temperature (4K). Various spin wave spectra are shown for different orientations of the applied magnetic field with respect to the film normal. (Reprinted figure with permission from: X. Liu, Y. Y. Zhou, and J. K. Furdyna, Phys. Rev. B, **75**, 195220 (2007). Copyright 2007 by the American Physical Society.)

well as the pinning (boundary) conditions. The distribution of peaks can provide valuable information on the exchange constants. For the case of a thin magnetic film with uniaxial magnetocrystalline anisotropy and an applied magnetic field along the film normal, the resonance equation can be expressed as:

$$\frac{\omega}{\gamma} = H_n - H_K - \mu_0 M_s + D k_n^2 \quad (10.133)$$

where n denotes the order of the spin standing wave mode, and thus H_n is the field of that particular resonance mode, $D = 2A/M_s$ is the spin wave constant and k_n is the wave vector of the nth mode. For the case where there is perfect pinning, which corresponds to the spins being completely pinned at the film surfaces, it is a simple matter to obtain the relation, $k_n = n\pi/t$, where t is the film thickness. Equation (10.132) is known as the *Kittel equation* and the above describes the basics of the *Kittel model*. We can compare Equations (10.113) and (10.132), where closer inspection should show that they both represent the spin wave spectra of a magnetic sample.

The general resonance equation for a sample will depend on the specific magnetic anisotropies of the sample in question. For example, a thin film with uniaxial anisotropy can be shown to have a resonance equation of the form:

$$\left(\frac{\omega}{\gamma}\right)^2 = \left[H\cos(\theta_H - \theta) + \left(\frac{2K_u}{\mu_0 M_s} - \mu_0 M_s\right)\cos 2\theta\right]$$
$$\left[H\cos(\theta_H - \theta) + \left(\frac{2K_u}{\mu_0 M_s} - \mu_0 M_s\right)\cos^2\theta\right] \quad (10.134)$$

where θ_H and θ are the in-plane orientations of the applied magnetic field and magnetization, respectively. This equation will give the resonance condition in the plane of the film.

Ferromagnetic resonance and spin wave resonance are important experimental techniques which can provide a lot of information on the magnetic properties of materials, including bulk samples, thin films, magnetic multilayers, and nanostructures (see also Chapter 13). There are several experimental techniques for the observation of spin dynamics. Traditionally, the microwave spectrometer (EPR electron paramagnetic resonance) has been used to measure FMR and SWR. However, modern techniques include vector network analyzers and ultrafast optical methods for observing magnetization dynamics in real time via pump and probe type measurements. One of the main advantages of these latter techniques is that they provide a sensitive measure of the dispersion relations for magnetic excitations. This corresponds to the variation of the resonance field with frequency, or alternatively the resonance frequency with field. Such a plot is illustrated in Figure 10.28, which shows some of the principal features that can be observed. There are two lines for each sample, where the shape of the line indicates the type of effective magnetic anisotropy, uniaxial in this case. The minimum along the field axis gives the value of the uniaxial anisotropy field, $H_K = 2K/\mu_0 M_s$, while the intersection along the frequency axis gives the resonance frequency in the anisotropy field of the sample itself, $\omega_K = \gamma H_K$. For a paramagnetic sample, the $\omega - H$ curve is a straight line given by the Larmor frequency; $\omega = \gamma H$.

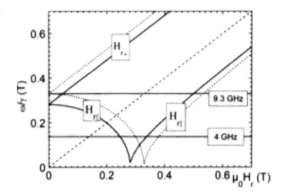

FIGURE 10.28: Frequency - field characteristics for magnetic samples. The solid and dotted lines show the variation for uniaxial anisotropies of 1.7×10^5 Jm^{-3} and 1.8×10^5 Jm^{-3}, respectively. The cusp along the field axis, for the field along the hard axis, has a minimum at he anisotropy field of $H_k = 2K/\mu_0 M_s$. The intersections at the origin show the positions of self-resonance in the anisotropy field, $\omega_K = \gamma H_K$. The dashed line shows the case for a paramagnetic sample. Intersections of the lines at 4 and 9 GHz show where the resonance fields occur for fixed frequency spectrometer measurements. (M. Farle, Rep. Prog. Phys., **61**, 755-826 (1998). ©IOP Publishing. Reproduced with permission. All rights reserved.)

10.14 SUMMARY

Magnetic effects in solids can be observed by applying a magnetic field to them. The response to this field depends greatly on the internal atomic and crystalline structure of the sample. In the case where there are only atoms with closed electron shells, this response will be weak and will produce as small magnetization which opposes the direction of the applied field. This is called diamagnetism and will occur in a majority of solids by virtue of the fact that most atoms contain at least some closed electron shells. When there are unpaired electrons in a solid whose magnetic moments are weakly or uncoupled, the reaction to a magnetic field will be to align their moments in the direction of the applied field. Such materials are said to be paramagnetic.

Once some form of interaction, called exchange, is present, then the response to an applied magnetic field will lead to a much stronger response to a magnetic applied field. In fact, a broad range of phenomena come into play that make magnetic materials rather exotic and of great technological interest. These magnetic phenomena are related to the specific magnetic internal structures of the materials, which in the simplest case is when all spins align in the same direction to form the ferromagnetic ordering. For antiferromagnetic order, neighboring spins align in antiparallel orientations. The magnetization of such materials will be zero and rather large magnetic fields are required to bring about any re-orientation in the spin structure. The antiferromagnetic state is generally modelled in terms of two antiparallel sublattice magnetizations. When the sublattice magnetizations are unequal, a ferrimagnetic state results, with experimental properties that are similar to ferromagnetic materials, though with some important differences. More complex magnetic structures can exist in some exotic alloys, such as helimagnetism, in which the ground state is characterized by a positional variation of the magnetic moments and can arise due to competing exchange interactions.

The existence of a long range magnetic order implies some form of orientational preference for the spin system. This directionality is referred to as magnetic anisotropy. The shape of a ferromagnetic or ferrimagnetic sample can have a very strong influence on the direction of its magnetizations due to magnetostatic energy considerations. For example, a long cylindrical sample will, in the absence of strong competing magnetic anisotropies, align its magnetization along its long axis since this reduces the free poles at the sample surface and minimizes the stray field produced by the magnetization. In terms of the shape, demagnetizing factors can be evaluated to account for the sample shape and provides a useful tool for evaluation the effective anisotropy. Crystal field and spin orbit interactions inside the crystal act to align the magnetic moments in a ferromagnet and give rise to the magnetocrysalline anisotropy. The crystal structure therefore can play a decisive role in the orientation of the magnetization in the absence of applied magnetic fields. Such situations can be modelled phenomenologically and can account for the orientational dependence of the magnetization in an arbitrarily aligned magnetic field. This process is usually done by minimizing the total energy of the system, while taking into account the various magnetic energy contributions in the solid.

The magnetocrystalline energy means that specific orientations of the magnetic moments with respect to the crystalline axes will be preferred. However, a fully saturated magnetization comes at the cost of magnetostatic energy. It is possible to reduce this energy by dividing the magnetization up into small regions of spontaneous magnetization called domains. The formation of domains, however, also has energy costs. Only when this energy is less than the magnetostatic energy will domains form. Otherwise the sample will exist in a single domain state. The existence of domains will form regions of transition between the different domains, which are called domain walls. The energy of a domain wall and it thickness depend directly on magnetic anisotropies and the exchange energy between neighboring spins.

The ground state of a ferromagnet can be described as the perfect alignment of all spins along a direction on easy magnetization. This can strictly only occur when no fluctuations of the spin system occur. Thermal energy produces an excitation of the spin system in a similar way to the production of lattice vibrations (phonons) in solids. The excitation of the magnetic spin systems produces a variation of the spin orientations called spin waves. These can be quantized (in quantum theory) as magnons. The existence of such magnons act to reduce the overall magnetization of the solid and for low temperatures spin wave theory provides a good model for the variation of the magnetization with temperature. Spin waves can be characterized as standing or travelling waves.

The motion of the magnetization under stimulation by a fast moving magnetic field provides a sensitive measure of magnetic properties and behavior. Precessional processes, such as ferromagnetic resonance and spin wave resonance have been used for many decades as a tool for the measurement of magnetic properties of materials. Indeed, they can furnish information on the principal magnetic properties, such as magnetization, g-factors, magnetic anisotropies, and exchange constants. Furthermore, in coupled magnetic systems, such as magnetic multilayers, it is possible to measure the interaction strength between the layers. Modern ultrafast optics techniques can be used to study real-time dynamics, such as sample demagnetization and reversal processes in addition to the slower precessional dynamics.

The interactions of fixed magnetic moments in a solid and the free electrons can produce variations in the electrical resistivity with an applied magnetic field. This is called magnetoresistance. In artificially produced layered structures a giant magnetoresistance can occur due to the spin scattering of electrons. Such processes are at the root of spin-tronics, which is a major new area of study and has produced significant progress in recent years with the existence of a number of spintronic devices. Much research is dedicated to the study of low dimensional effects in magnetism, such as in magnetic nanoparticles and nanostructures. We will describe some of these topics in the final chapter of the book.

REFERENCES AND FURTHER READING

Basic Texts

- S. J. Blundell, *Magnetism in Condensed Matter*, Oxford University Press, Oxford (2001)

- H. P. Myers, *Introductory Solid State Physics*, Taylor and Francis, London (1998)

- J. M. D. Coey, *Magnetism and Magnetic Materials*, Cambridge University Press, Cambridge (2010)

Advanced Texts

- R. Skomski, *Simple Models of Magnetism*, Oxford University Press, Oxford (2008)

- N. W. Ashcroft and N. D. Mermin, *Solid State Physics*, Saunders College, Philadelphia (1976)

- R. C. O'Handley, *Modern Magnetic Materials: Principles and Applications*, Wiley Interscience, New York (2000)

- A. Herpin, *Théorie du Magnétisme*, Presses Universitaires de France, (1968)

- D. C. Mattis, *The Theory of Magnetism Made Simple: An Introduction to Physical Concepts and to Some Useful Mathematical Methods*, World Scientific, Singapore, (2006)

- N. Majlis, *The Quantum Theory of Magnetism*, (2e), World Scientific, Singapore, (2007)

- B. D. Cullity and C. D: Graham, *Introduction to Magnetic Materials*, (2e), J. Wiley and Sons, New Jersey, (2009)

- R. M. White, *Quantum Theory of Magnetism*, (3e), Springer, Berlin, (2007)

EXERCISES

Q1. Derive the Langevin function, given in Equation (10.28).

Q2. Referring to Equation (10.15), show that the frequency of precession is given by the Larmor frequency; $\omega_L = \gamma B$.

Q3. Calculate the spin, orbital and total angular momenta for the following ions: Sm^{3+}, Fe^{2+} and V^{4+}. What are the magnetic moments for these ions?

Q4. Show that the Hunds rules can be expressed as:

$$S = \frac{1}{2}[(2l+1) - |2l+1-n_e|] \quad (10.135)$$

$$L = S|2l+1-n_e| \quad (10.136)$$

$$J = S|2l-n_e| \quad (10.137)$$

Q5. If a particle of charge 3e moves in a circular orbit of 2 Å with a frequency of 5×10^{16} Hz, what is the resulting magnetic moment? Give your answer in units of Bohr magnetons.

Q6. Show that the Heisenberg Hamiltonian can be expressed as:

$$\mathcal{H}_{ij} = \sum_{ij} J_{ij}(S_i^+ S_j^- + S_i^- S_j^+ + 2S_i^z S_j^z) \qquad (10.138)$$

where $S^\pm = S^x \pm iS^y$.

Q7. Determine the symmetry of the following spin wave-functions:

$$\chi_\alpha(\sigma_1)\chi_\alpha(\sigma_2)$$
$$\chi_\alpha(\sigma_1)\chi_\beta(\sigma_2)$$
$$\chi_\alpha(\sigma_1)\chi_\beta(\sigma_2) - \chi_\beta(\sigma_1)\chi_\alpha(\sigma_2)$$

Q8. Prove that the domain wall width for a Bloch type wall is given by Equation (10.102).

Q9. Given the magnetic energy:

$$E = -\mu_0 \mathbf{H} \cdot \mathbf{M} + K_u \sin^2\theta \qquad (10.139)$$

show that the magnetization direction will be defined by the condition:

$$\frac{\sin(\Theta_H - \theta)}{\sin\theta \cos\theta} = \frac{2K_u}{\mu_0 HM} \qquad (10.140)$$

where Θ_H defines the polar angle of the applied magnetic field and θ that of the magnetization.

Q10. Consider the case for GMR where the high-resistance spin channel is twice that of the low-resistance spin channel. What will the GMR % be in this case?

Q11. A magnetic field is applied in the basal plane of a Co crystal, whose easy axis is aligned along the c-direction if its hcp structure. Given the uniaxial anisotropy constant $K_u = 4.1 \times 10^5 \mathrm{Jm^{-3}}$ and magnetization $M = 1.42 \times 10^6 \mathrm{Am^{-1}}$ for Co, find the field necessary to align the magnetization in (a) the basal plane; and (b) at an angle of 45° to the easy axis.

Q12. A ferromagnetic sample has cubic anisotropy with an easy axis aligned in the (100) directions. Show that, when a magnetic field is applied along [110], the free energy of the system can be expressed as:

$$E = -\mu_0 HM \cos(\pi/4 - \phi) + \frac{K_1}{4} \sin^2 2\phi \qquad (10.141)$$

Find the equilibrium condition for this magnetic system.

Q13. Show that the Landau - Lifshitz equation, in the absence of damping, can be expressed as:

$$\frac{1}{\mu_0 \gamma} \frac{\partial \mathbf{M}(t)}{\partial t} = \hat{\mathbf{e}}_\theta \frac{1}{\sin\theta} \frac{\partial E}{\partial \phi} - \hat{\mathbf{e}}_\phi \frac{\partial E}{\partial \theta} \qquad (10.142)$$

Q14. Prove Equation (10.133).

NOTES

[1] In this case, the precession of **J** around the magnetic field **B** is slower than the precession of **L** and **S** about **J**
[2] The partition function is a "sum over states", which reflects how the energy states are divided up in a system
[3] The exchange integral is usually expressed in the form: $J_{ij} = \int \psi_i(\mathbf{r}_a)^* \psi_j(\mathbf{r}_b)^* \frac{e2}{|\mathbf{r}_a - \mathbf{r}_b|} \psi_i(\mathbf{r}_b) \psi_j(\mathbf{r}_a) d^3\mathbf{r}$. Comparing this with the Coulomb energy: $K = \int \psi_i(\mathbf{r}_a)^* \psi_j(\mathbf{r}_b)^* \frac{e2}{|\mathbf{r}_a - \mathbf{r}_b|} \psi_i(\mathbf{r}_a) \psi_j(\mathbf{r}_b) d^3\mathbf{r}$, we see where the name for exchange arises.
[4] We note that this will only strictly occur if the magnetic body has the form of an ellipsoid of rotation.
[5] See for example J. A. Osborn Phys. Rev. **67**, 351 (1945)
[6] This software is available free from: *http://math.nist.gov/oommf/*

CHAPTER 11

SUPERCONDUCTIVITY

"The greatness of a nation can be judged by the way its animals are treated."
—Mahatma Gandhi

"I think, at a child's birth, if a mother could ask a fairy godmother to endow it with the most useful gift, that gift should be curiosity."
—Eleanor Roosevelt

"True friends stab you in the front."
—Oscar Wilde

11.1   INTRODUCTION

The phenomenon of superconductivity was discovered in 1911 by H. Kamerlingh-Onnes while studying the low temperature resistivity properties of Hg. This first glimpse of the extraordinary behavior termed *superconductivity* was initially seen as the complete disappearance of electrical resistivity. Actually, the removal of electrical resistance from a body has a number of other physical implications, which are of fundamental importance in the study of solids. Onnes continued to study the low temperature transport properties of metals for a number of years. To his surprise even the addition of impurities in the solids did not bring about an increase in the residual resistivity, as was expected, and he concluded that mercury undergoes a transition into a new phase, which he termed the *superconductive state*. The temperature at which a transition to the superconductive state occurs is called the *critical temperature*, T_c. Above this temperature the material exhibits normal resistive behavior and is said to be in the *normal state*. The transition from normal to superconducting state is very abrupt, as illustrated in Figure 11.1. In Table 11.1, we give some of the superconducting transition temperatures for selected metals.

The discovery of the phenomenon of superconductivity lead to the survey of many elements and compounds to see if this state exists at more readily available temperatures.

It was found that mosts elements which do exhibit superconductivity, only do so at very low temperature. Niobium being the element with the highest T_c, of 9.2 K. Certain compounds and alloys were later found to have higher transition temperatures, see Table 11.1. A major advance was made in 1986 with the discovery of *high-temperature superconductors*, which were based on oxide ceramic cuprates: In fact, so important was this development considered, that the discoverers, J. G. Bednorz and K. A. Müller were awarded the 1987 Nobel Prize in Physics. Much frenetic activity in this research

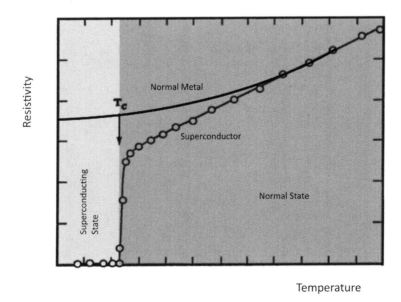

FIGURE 11.1: The transition from the normal state to the superconducting state.

TABLE 11.1: Superconducting transition temperatures for selected metals and compounds.

Element	Critical Temperature T_c (K)	Compound	Critical Temperature T_c (K)
Niobium, Nb	9.2	Nb_3Sn	18.0
Lead, Pb	7.2	Nb_3Ge	23.2
Tantalum, Ta	4.48	$La_{1.8}Sr_{0.2}CuO_4$	35
Mercury, Hg	4.15	YBa_2CuO_7	95
Aluminium, Al	1.2	$Tl_2Ba_2Ca_2Cu_2O_{10}$	125
Iridium, Ir	0.14	$HgBa_2Ca_2Cu_3O_8$	133

field followed this discovery with the promise of room temperature superconductors. Despite this, at the time of writing (2013), the highest temperature at which superconductivity has been found is 133 K for the compound $HgBa_2Ca_2Cu_3O_8$. This transition temperature can be raised to about 164 K under a pressure of around 30 GPa

Superconductivity is related to the phenomenon of superfluidity, both of which are manifestations of the *Bose - Einstein condensation*(BEC). This form of transition gives rise to rather strange physical properties, which are direct consequences of quantum mechanics, and occur at low temperatures, where thermal energy effects are kept under control. The fact that most quantum effects are observed in very small objects, like atoms, and as we shall see in the final chapter, in nanosized objects, makes superconductivity rather unusual in that it can be observed in bulk materials. The Bose - Einstein condensation essentially relates to the coupling of fermions to produce bosonic quasi-particles. The very long time between the discovery of superconductivity and the formulation of a satisfactory theory is evidence of the conceptual problems encountered in adequately describing this physical phenomenon. The Bardeen Cooper Schrieffer (BCS) theory was published in 1957 and was a significant step forward in the understanding of low temperature superconductivity. Bardeen, Cooper, and Schrieffer were awarded the 1972 Nobel Prize for their work. Despite this success, the normal BCS theory cannot be applied to high-temperature superconductors and this remains one of the important problems yet to be solved in solid state physics.

In this chapter, we aim to outline some of the main properties exhibited by superconducting materials. We will also give a brief introduction to some of the principal theoretical concepts necessary to understand this physical phenomenon. Finally, we shall describe some of the more important applications of superconductivity and give a brief overview of some of the recent developments.

11.2 PHENOMENA RELATED TO SUPERCONDUCTIVITY

As stated in the Introduction, the disappearance of electrical resistivity also comes with a number of physical consequences. We will give a brief introduction into the principal phenomena which are characteristic of superconductors.

11.2.1 Zero-Resistivity/Infinite Conductivity and Persistent Currents

We have already discussed the fact that the electrical resistance or resistivity disappears when a materials enters the superconducting state, as illustrated in Figure 11.1. This means that the conductivity should become infinite below the critical temperature. However, to be consistent with the current density: $j = \rho \varepsilon$, we must always have a zero electric field, $\varepsilon = 0$, throughout the superconductor. In this way we are not troubled by an infinite current density and we have no current flow without an electric field.

To guarantee that the resistance is truly zero, the preferred measurement techniques uses a four-point measurement, see Figure 11.2. In the two-terminal measurement, contact resistance will be included in the measurement. The four-point method is the preferred method for the measurement of low resistances, in which the four contacts connect directly to the sample, where only the current in the sample is of importance when the voltage drop is measured between two positions in the sample (when the sample is resistive). The resistivity is then assessed from Ohm's law ($V = IR$) and taking the sample geometry into account, since $R = \rho L / A$.

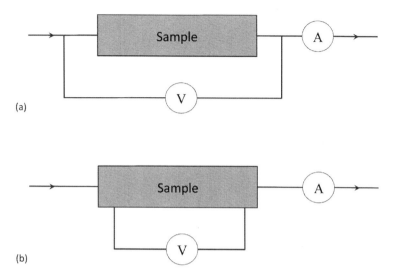

FIGURE 11.2: The evaluation of resistivity requires an experimental measurement of current and voltage. This can be performed using a two-terminal (a) or a four-terminal (b) technique.

The currents in a superconductor flow with no discernible dissipation of energy. However, the *supercurrent* can be readily destroyed if a sufficiently large magnetic field is applied to the sample (see below). Furthermore, should the current density exceed a certain value, called the *critical current*, the superconducting state will also be destroyed. This is called the *Silsbee effect*. The size of this critical current will depend on the nature and geometry of the of the sample. The existence of *persistent currents* is probably one of the more convincing methods to demonstrate superconductivity. We can show this by applying the magnetic flux, Φ, along with the Maxwell equation:

$$\nabla \times \boldsymbol{\varepsilon} = -\frac{\partial \mathbf{B}}{\partial t} \qquad (11.1)$$

and Stoke's theorem:

$$\int (\nabla \times \boldsymbol{\varepsilon}) d\mathbf{S} = -\oint \boldsymbol{\varepsilon} \cdot d\mathbf{r} \qquad (11.2)$$

to obtain:

$$\frac{d\Phi}{dt} = -\oint \boldsymbol{\varepsilon} \cdot d\mathbf{r} \qquad (11.3)$$

Since we can take the line integral around a closed path within the superconductor, where we have $\boldsymbol{\varepsilon} = 0$, we thus have

$$\frac{d\Phi}{dt} = 0 \qquad (11.4)$$

which means that the magnetic flux passing through the loop remains constant over time. The persistent current is now set up by applying a magnetic field, **B**, to the materials in the normal state. Since it is in the normal state the magnetic field can penetrate the material. The sample is then cooled to below the transition temperature to the superconducting state. The initial flux in the sample, which we take to be a ring, is $\Phi = \int \mathbf{B} \cdot d\mathbf{S}$. The external magnetic field is then removed and since Equation (11.4) must still hold, the superconductor must generate an internal flux to maintain the flux constant. This is done by the circulation of a current, I, inside the superconductor. Now since Φ must be constant in time, so must the current that generates it. Therefore, a circulating persistent current will be set up inside the ring of the superconductor. Since there is no dissipation of the current through electrical resistance, the currents should remain. Such persistent currents have been observed to remain constant over a period of years.

11.2.2 Meissner-Ochsenfeld Effect

While the zero-resistivity is an important property of the superconducting state, it is not generally held to be a proof of superconductivity. The demonstration of the *Meissner - Ochsenfeld effect* (which dates from 1933) is also a requirement. This effect arises from the fact that a superconductor expels a weak magnetic field. This is illustrated in Figure 11.3.

If we start with a sample above the critical temperature, and cool below the critical temperature, the sample will enter the superconducting state. If we apply an external magnetic field, the field must

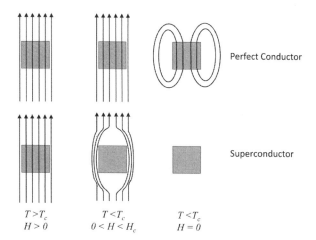

FIGURE 11.3: Illustration of the Meissner - Ochsenfeld effect and the difference between an imaginary perfect conductor and a superconductor. Above the critical temperature, magnetic flux can penetrate both materials. Reducing the temperature below T_c, the superconductor expels the magnetic flux from its interior. This is the Meissner - Ochsenfeld effect.

remain zero inside the superconductor, since $\varepsilon = 0$ and from Equation (11.1), we have:

$$\frac{\partial \mathbf{B}}{\partial t} = 0 \tag{11.5}$$

We can approach the situation from another point of view. Again, we start from the normal state with $T > T_c$, then apply a magnetic field, the flux lines can penetrate the interior of the sample. If we now cool the sample to $T < T_c$. The Meissner - Ochsenfeld effect is the expulsion of the magnetic flux from the sample and differs from the case of a material which happens to be a perfect conductor. The important point is that the expulsion of the magnetic flux is independent of the samples' history, i.e., the sequence of events leading up to the expulsion of the magnetic flux.

The Meissner - Ochsenfeld effect is a preferred method of demonstrating the existence of superconductivity since it is a property of thermal equilibrium, while transport is a non-equilibrium situation.

11.2.3 Perfect Diamagnetism

It turns out that the expulsion of magnetic flux from a superconductor is only possible if there are screening currents set up around the edges of the sample. These will produce a magnetic field which is equal and opposite to that of the external field and thus cancelling out the external field in the interior of the superconductor. Using the Maxwell equation:

$$\nabla \times \mathbf{B} = \mu_0 \mathbf{j} + \mu_0 \epsilon_0 \frac{\partial \mathcal{E}}{\partial t} \qquad (11.6)$$

we can show that the currents produced by the external field and internal magnetization, due to the applied current in the coils of a solenoid, and the screening currents in the sample, can be expressed as:

$$\nabla \times \mathbf{H} = \mathbf{j}_{ext} \qquad (11.7)$$

and

$$\nabla \times \mathbf{M} = \mathbf{j}_{int} \qquad (11.8)$$

respectively. Now since these currents must cancel out; $\mathbf{j}_{int} = -\mathbf{j}_{ext}$, we must have:

$$\mathbf{M} = -\mathbf{H} \qquad (11.9)$$

(Of course this condition can be seen from the expression: $\mathbf{B} = \mu_0(\mathbf{H} + \mathbf{M})$ and applying the Meissner - Ochsenfeld condition, $\mathbf{B} = 0$.) It is now a simple matter to express the magnetic susceptibility as:

$$\chi = \frac{M}{H} = -1 \qquad (11.10)$$

As we saw in the last chapter, a negative magnetic susceptibility occurs for diamagnetic materials. Diamagnets screen out part of the magnetic field applied to them, and thus become magnetized in the direction opposing the field. In the superconductor, the screening is complete and hence we refer to the superconductor as a *perfect diamagnet*. Therefore, a good way of detecting if a material is a superconductor is to measure its susceptibility. If the susceptibility drops abruptly to -1 at some temperature, this is a good confirmation of its status as a superconductor. This is generally regarded as a more reliable form of evidence for superconductivity than a simple measurement of zero resistance.

11.2.4 Critical Fields and Critical Current

Our description for the perfect diamagnetism in superconductors was made with the restriction of the applied magnetic field to low values. For higher values there are two possible outcomes, which depend on the materials itself.

In the first, simpler case, as the magnetic field increases, the **B**-field inside the sample remains zero, with a proportional increase in the sample magnetization. Then at a *critical field*, H_c, the superconductivity is completely destroyed. Such a material is called a *type I superconductor*. This situation is illustrated in Figure 11.4 (a), where the magnetization follows the trend $M = -H$ up to the point where $H = H_c$, at which point the superconductivity is lost and therefore the susceptibility abruptly changes (vanishes).

A more complex situation is observed in many superconductors, which are referred to as *type II superconductors*. In this case the initial behavior is the same as in a type I material, then at a critical field, H_{c1}, the magnetic susceptibility begins to fall off gradually from -1 and eventually disappears at a second critical field, denoted as H_{c2}. This situation is illustrated in Figure 11.4 (b).

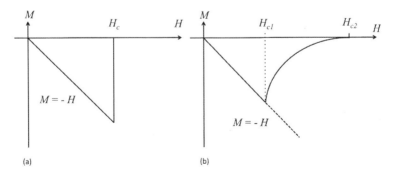

FIGURE 11.4: Magnetization - field curves for (a) type I and (b) type II superconductors. The perfectly diamagnetic state persists up to a critical field of H_c and then vanishes, in a type I material. In the type II superconductor, there is a gradual monotonic decrease of the susceptibility above the first critical field, H_{c1}, where it stops being a perfect diamagnet, and vanishes at a second critical field, H_{c2}.

Evidently, if the magnetic susceptibility falls below a value of -1, then some magnetic flux will penetrate the superconductor (type II) and will continue increasingly to do so until the sample is completely returned to the normal state. In the region $H_{c1} < H < H_{c2}$, the magnetic field penetrates the type II superconductor in the form of so-called *vortices*. Each vortex is a region of circulating supercurrent around the region of the superconductor which, being penetrated by the magnetic flux, has reverted to the normal state. Therefore, in this field range of fields, the material is in a mixed state, with regions which are superconducting and regions that are normal. The supercurrents circulate at the interface between the two and serve to screen the magnetic field from the superconducting regions. This mixed or vortex state is illustrated in Figure 11.5. The existence of the normal state core regions thus provide zone in which the magnetic flux can penetrate the sample and reduces the overall energy of the system. The postulation of the vortex state was originally proposed by A. A. Abrikosov (1952). Abrikosov was one of a trio awarded the Nobel Prize in Physics (2003), along with V. Ginzburg and A. J. Leggett, for their "pioneering contributions to the theory of superconductors and superfluids".

The critical fields also vary as a function of temperature; reducing gradually up to the critical temperature. This is schematically

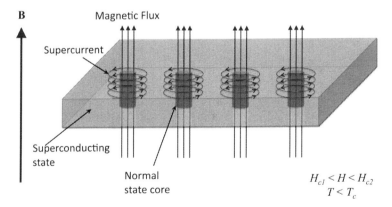

FIGURE 11.5: In type II superconductors at fields $H_{c1} < H < H_{c2}$, partial penetration of magnetic flux is concentrated in channels, called vortices. Inside the vortex core the material is in the normal state and is separated from the superconducting state by a supercurrent vortex. In this field regime the sample is said to be in the mixed or vortex state.

illustrated in Figure 11.6 for type I and type II superconductors. The temperature dependence of the critical fields can be expressed in the form:

$$H_c = H_{c0}\left[1 - \left(\frac{T}{T_c}\right)^2\right] \quad (11.11)$$

where H_{c0} is the value of the critical field at absolute zero of temperature. This is sometimes referred to as *Tuyn's law*, and can be also expressed using the magnetic induction; $B_c = \mu_0 H_c$.

In addition to the critical fields, there is also a critical current density, J_c, that will induce the critical field at the surface of the superconductor and drive it into the normal state. This relation between the critical field and the critical current takes the form:

$$H_c(T) = \lambda(T) J_c(T) \quad (11.12)$$

where we indicate that all three parameters are temperature dependent. The factor λ, is the thickness of the layer at the surface, also called the *penetration depth* of the superconductor where the current will flow. In fact, the current density varies as a function of position from the surface into the superconductor, decaying with

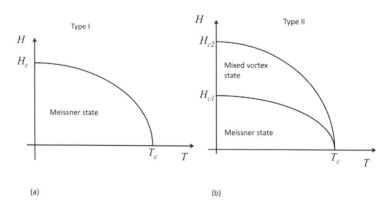

FIGURE 11.6: Magnetic field - temperature phase diagrams for type I (a) and type II (b) superconductors. Below the $H_c - T_c$ line in type I and the $H_{c1} - T_c$ line in type II superconductors the material is in the full Meissner state, being a perfect diamagnet. For the type II super conductor, in the region between the $H_{c1} - H_{c2}$ lines the material is in the mixed vortex state, also referred to as the Abrikosov - state.

distance from the surface. We will discuss this effect in Section 11.4. The temperature dependences of the penetration depth and the critical current can be expressed as:

$$\lambda = \lambda_0 \left[1 - \left(\frac{T}{T_c}\right)^4\right]^{-1/2} \qquad (11.13)$$

and

$$J_c = J_{c0} \left[1 - \left(\frac{T}{T_c}\right)^2\right]\left[1 - \left(\frac{T}{T_c}\right)^4\right]^{1/2} \qquad (11.14)$$

The critical behavior of superconductors can be described in terms of a three-dimensional surface, bounded by the critical parameters; T_c, B_c and J_c, as illustrated in Figure 11.7.

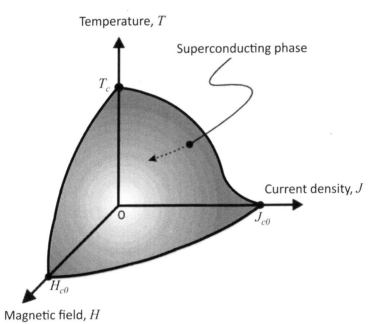

FIGURE 11.7: The critical surface of a superconductor. This is derived from the temperature dependence of the critical field and the fact that a current will produce a magnetic field, and this will also be temperature dependent.

11.3 THERMODYNAMICS OF THE SUPERCONDUCTING TRANSITION

11.3.1 Phase Stability of the Superconducting State

We can define the work done on a magnetic material by a magnetic field from the consideration of the change of flux density inside the material. In increasing the magnetic induction from **B** to **B** + d**B** in the presence of a field H, the work done on the sample can be expressed as: $\mathbf{H} \cdot d\mathbf{B} = \mu_0 \mathbf{H} \cdot d\mathbf{M}$, per unit volume. The first law of thermodynamics for the magnetic material can thus be expressed as:

$$dU = TdS - PdV + \mu_0 \mathbf{H} \cdot d\mathbf{M} \qquad (11.15)$$

The first term represents the heat energy (where T is the temperature and S is the entropy), the second term is the work done on a volume of gas (here we will take $dV = 0$) and the final term is the work done on a magnetic material.

The Gibbs free energy of a system is defined as: $G = U - TS - PV - \mathbf{B} \cdot \mathbf{M}$. For our purposes, we can consider the Gibbs free energy to be a function of temperature and field, where differentiation gives:

$$dG = dU - TdS - SdT - \mathbf{B} \cdot d\mathbf{M} - \mathbf{M} \cdot d\mathbf{B} \qquad (11.16)$$

Using Equation (11.15) we can write:

$$dG = -SdT - \mathbf{M} \cdot d\mathbf{B} = -SdT - \mu_0 \mathbf{M} \cdot d\mathbf{H} \qquad (11.17)$$

where we have considered P and V to be constants.

In the normal state, i.e., for a normal metal, where the magnetic susceptibility is essentially zero ($\chi \simeq 0$), there is virtually no interaction with an applied magnetic field. Therefore, we can affirm that there is no contribution to the free energy due to an external magnetic field. We can take this to be our state of zero energy. The phase stability is governed by the Gibbs free energy, $G(\mathbf{H}, T)$. For the normal (N) state we can write: $G_N(\mathbf{H}, T) = G_N(0, T)$, since the field does not affect the sample. We can now consider the case for a superconductor, where the diamagnetic properties give:

$$G_S(\mathbf{H},T) - G_S(0,T) = \int dG = -\mu_0 \int \mathbf{M} \cdot d\mathbf{H} \quad (11.18)$$

In the case of a type I superconductor at $T < T_c$, we have $\mathbf{M} = -\mathbf{H}$, and thus we have:

$$G_S(\mathbf{H},T) - G_S(0,T) = \frac{\mu_0 H^2}{2} \quad (11.19)$$

At the critical field $H = H_c$ we can write:

$$G_S(\mathbf{H}_c,T) = G_S(0,T) + \frac{\mu_0 H_c^2}{2} = G_N(\mathbf{H}_c,T) = G_N(0,T) \quad (11.20)$$

Re-arranging gives:

$$G_S(0,T) - G_N(0,T) = -\frac{\mu_0 H_c^2}{2} \quad (11.21)$$

Therefore, at $T = 0$,

$$G_S(0,0) - G_N(0,0) = -\frac{\mu_0 H_c^2}{2} \quad (11.22)$$

and since $G_S(0, 0) < G_N(0, 0)$, the superconducting state is the more stable phase for $H < H_c$. Thus we see that the critical field provides us with a measure of the stability of the superconducting state. From the above, it follows that:

$$G_S(\mathbf{H},T) = G_N(0,T) - \frac{\mu_0(H_c^2 - H^2)}{2} \quad (11.23)$$

which is valid for $H < H_c$, and shows the stability of the superconducting state at all fields below the critical field. We see clearly from this expression that the stability gradually reduces as the field increases, and finally vanishes at $H = H_c$.

11.3.2 Heat Capacity of a Superconductor

Equation (11.23) allows us to express the entropy, $(S = -(\partial G/\partial T))$, difference between the normal and superconducting states, which we can express as:

$$\Delta S = S_S(\mathbf{H},T) - S_N(0,T) = \mu_0 \frac{d}{dT}\left(\frac{H_c^2 - H^2}{2}\right) = \mu_0 H_c \frac{dH_c}{dT} \quad (11.24)$$

The heat capacity of the solid is given by:

$$C = T\frac{dS}{dT} \quad (11.25)$$

It is now a simple matter to write:

$$\begin{aligned} C_S - C_N &= T\frac{d}{dT}\mu_0\left(H_c\frac{dH_c}{dT}\right) \\ &= \mu_0 T\left[\left(\frac{dH_c}{dT}\right)^2 + H_c\frac{d^2H_c}{dT^2}\right] \end{aligned} \quad (11.26)$$

It should be clear that the entropy difference will disappear at the critical temperature, since at this point $H_c = 0$. However, the difference in the heat capacity will be finite and $(\Delta C)_{T_C} = \mu_0 T_c (dH_c/dT)^2_{T=T_c} > 0$. A discontinuity of the specific heat is observed at $T = T_c$, see Figure 11.8. The low temperature specific heat in a metal was discussed in Chapter 6, where a plot of C/T vs. T^2 is a straight line. In the inset of Figure 11.8, we see the experimental data for the metal vanadium. In the normal state, which is induced at $T < T_c$ by placing it in a field greater than its critical value, shows this linear behavior. In the superconducting state, measured with $H = 0$, we observe the expected discontinuity. It is worth noting that for $0 < T < T_C$, the derivative, dH_c/dT, is negative, such that $\Delta S < 0$. This implies that the superconducting state is more ordered than the normal state in this temperature range. Since we are only concerned with the electronic contributions to the specific heat, at $T = T_c$, we can write

$$\frac{(C_{eS} - C_{eN})}{\gamma T_c} = \frac{\mu_0}{\gamma}\left(\frac{dH_c}{dT}\right)^2_{T=T_c} \quad (11.27)$$

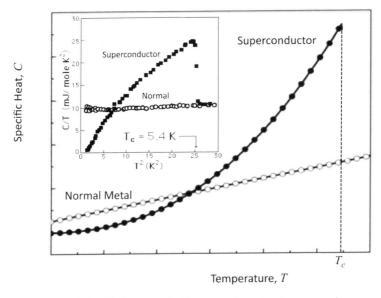

FIGURE 11.8: Specific heat capacity for a normal metal and a superconductor. The discontinuity of the specific heat in the superconductor marks the transition to the normal state. The inset shows the low temperature specific heat for Vanadium in the superconducting state and in the normal state. This latter is imposed by maintaining the metal in a magnetic field above the critical value.

where $\gamma = 2\pi^2 k_B^2 N(0)/3$. The jump in the specific heat at T_c is only related to the electronic contribution to the specific heat. Below the critical temperature C_{eS} decreases exponentially, roughly as $e^{-\Delta_0/k_B T}$. We will return later to discuss the significance of this term, see Section 11.6.

11.4 THE LONDON EQUATIONS

The equation of motion for electrons in a normal metal was given in Chapter 6, Equation (6.63). However, due to the nature of superconductivity, early researchers considered a modified form of the electron dynamics, which was different from that of normal metals. In 1935, the brothers Fritz and Heinz London considered that due to the vanishing resistivity, the classical form of the equation of motion should be modified and by neglecting the "friction" term $(m_e v_D / \tau)$, we can write:

$$m_e \frac{d\mathbf{v}}{dt} = -e\boldsymbol{\mathcal{E}} \quad (11.28)$$

It is further considered that the electrons in the system consists of normal electrons and superconducting electrons, in the postulated *two-fluid model*. The total number of electrons is thus: $n = n_N + n_S$. We can now write the current density of the superconductive electrons as: $\mathbf{j}_S = -n_S e \mathbf{v}$, from Equation (11.28) we obtain:

$$\frac{d\mathbf{j}_S}{dt} = -e n_S \frac{d\mathbf{v}}{dt} = \frac{n_S e^2}{m_e} \boldsymbol{\mathcal{E}} \quad (11.29)$$

This is known as the *first London equation*. We now take the curl of this equation:

$$\nabla \times \frac{d\mathbf{j}_S}{dt} = \left(\frac{n_S e^2}{m_e}\right) \nabla \times \boldsymbol{\mathcal{E}} \quad (11.30)$$

We now apply the Maxwell equation $(\nabla \times \boldsymbol{\mathcal{E}} = -\partial \mathbf{B}/\partial t)$ to obtain:

$$\nabla \times \frac{d\mathbf{j}_S}{dt} = -\left(\frac{n_S e^2}{m_e}\right) \frac{\partial \mathbf{B}}{\partial t} \quad (11.31)$$

Integrating with respect to time (and choosing a constant of integration of zero, in accord with the Meissner effect), we find:

$$\nabla \times \mathbf{j}_S = -\left(\frac{n_S e^2}{m_e}\right) \mathbf{B} \quad (11.32)$$

This is called the *second London equation*. We now apply the Maxwell equation $\nabla \times \mathbf{B} = \mu_0 \mathbf{j}_S$ and taking the curl of this we have:

$$\nabla \times (\nabla \times \mathbf{B}) = \nabla(\nabla \cdot \mathbf{B}) - \nabla^2 \mathbf{B} = \mu_0 \nabla \times \mathbf{j}_S \quad (11.33)$$

Since $\nabla \cdot \mathbf{B} = 0$, we can write:

$$\nabla^2 \mathbf{B} = \left(\frac{\mu_0 n_S e^2}{m_e}\right) \mathbf{B} = \frac{1}{\lambda^2} \mathbf{B} \quad (11.34)$$

where $\lambda = \sqrt{m_e / \mu_0 n_s e^2}$ is the *London penetration depth*. The solution of this equation takes the form:

$$\mathbf{B}(x) = \mathbf{B}(0) e^{-x/\lambda} \qquad (11.35)$$

This illustrated in Figure 11.9. The field does not penetrate very deeply into the superconductor and is just a surface effect, within depth λ. We can estimate the value of the penetration depth by considering that at $T = 0$ all electrons will be in the superconducting state. A reasonable estimation would be of the order of $n = n_S = 10^{29} \mathrm{m}^{-3}$, this gives a London penetration depth of $\lambda(0) \simeq 170 \text{Å}$. This is the same penetration depth we mentioned in Section 11.2.4, Equations (11.12) and (11.13). In fact, using this latter equation, we can evaluate the variation of the density of super electrons, n_S, as:

$$n_S = n_{S0} \left[1 - \left(\frac{T}{T_c} \right)^4 \right] \qquad (11.36)$$

From Equations (11.29) and (11.32), we can establish the relations that determine the microscopic electric and magnetic fields:

$$\varepsilon = \frac{\partial}{\partial t}(\Lambda \mathbf{j}_s) \qquad (11.37)$$

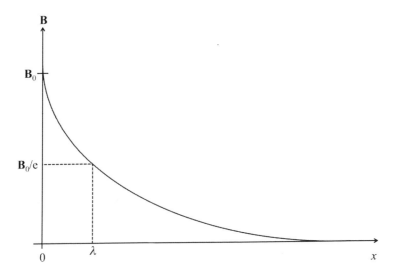

FIGURE 11.9: Decay of the magnetic field penetrating a superconductor in the space for $x \geq 0$.

and

$$\mathbf{B} = -\nabla \times (\Lambda \mathbf{j}_s) \quad (11.38)$$

where we have $\Lambda = m_e/n_s e^2$. Experimental measurements have been found to give a discrepancy to the value of this parameter, which is due to the local nature of the London equations. Pippard (1953) introduced the idea of a *coherence length*, ξ_0, for the wavefunction of superconducting carriers. The coherence length is a measure of the region of influence in which the carriers forming the current depend on the values of the vector potential $\mathbf{A}$ (a region of distance ξ_0).

An estimation of the coherence length can be obtained by the consideration that only electrons with an energy range of the order of $k_B T_c$ can be responsible for phenomena occurring at $T < T_c$. Given that only electrons with the Fermi velocity, v_F, can be involved, the range of values of the momentum must be of the order of $k_B T_c / v_F$. Using the uncertainty principle allows us then to estimate the coherence length:

$$\delta_x \geq \frac{\hbar}{\delta p} \simeq \frac{\hbar v_F}{k_B T_c} \quad (11.39)$$

For metals this corresponds to around $0.1 - 1 \mu$m. The coherence length is usually quoted as:

$$\xi_0 = a \left(\frac{\hbar v_F}{k_B T_c} \right) \quad (11.40)$$

where a is a constant to be determined by experiment, where for metals it has a value of about 0.15. Experiments on alloys show that the coherence length also depends on scattering, such that:

$$\frac{1}{\xi} = \frac{1}{\xi_0} + \frac{1}{\lambda_{mfp}} \quad (11.41)$$

thus depending on the mean free path of the material in the normal phase. Pippard suggested the modification of the London equations to incorporate this effect, which can be achieved by writing $\lambda' = \lambda \sqrt{\xi_0/\xi}$. Later, when the BCS theory was established, it was found that the coherence length also plays an important role and is related to the Δ_0 parameter or band gap as:

$$\xi_0 = \frac{\hbar v_F}{\pi \Delta_0} \qquad (11.42)$$

this being interpreted as the physical size of the Cooper pair bound state and Δ_0 being the binding energy. This will be discussed in more detail in Section 11.6.

We can define a dimensionless ratio; $\kappa = \lambda/\xi_0$ and this determines whether a superconductor is a type I or a type II material. Also, the mean free path is important, when $\lambda \gg \xi_0$, the superconductor is said to be in the *clean limit*. While for $\lambda < \xi_0$, it is said to be in the *dirty limit*. Surprisingly, most superconductors remain as such even when there are large concentrations of impurities, which makes λ very short. We note that for $\kappa \ll 1$ (i.e., $\lambda \ll \xi$), we have a type I superconductor, while superconductors with $\kappa \gg 1$ (i.e., $\lambda \gg \xi$) are of the type II variety.

11.5 GINZBURG - LANDAU MODEL

There are two principal theoretical approaches to superconductivity; i) a phenomenological theory which uses the London equations and ii) the microscopic theory known as the *BCS theory*. We will discuss this latter theory in the following section. A somewhat different perspective is given by the *Ginzburg - Landau (GL) theory*. The GL model is based on a very general approach to phase transitions and can be applied to phenomena related to other areas of physics, such as magnetism and liquid crystals. It is a *mean-field theory* applied to the thermodynamical state of a system. The model was originally proposed in the 1930s and relates to the change in symmetry of the system as it undergoes a phase change. For example, in the phase change that occurs when a ferromagnetic material passes the Curie temperature, the local moments align in a specific orientation. Such a phase transition is then characterized by an *order parameter*, which would be zero in the disordered state. So in the magnetic case, the magnetization could be used as the order parameter, which becomes zero at the Curie point of the ferromagnet.

In the case of a superconducting material the GL model presumes an order parameter, the complex pseudo-wave-function, which conforms to the following:

$$\psi = \begin{cases} 0 & T > T_c \\ \psi(T) \neq 0 & T < T_c \end{cases} \qquad (11.43)$$

In the GL theory, the order parameter is related to the density of superconducting electrons, such that; $|\psi|^2 = n_S$. (Later, Gorkov (1959) showed that a correlation with the BCS theory is possible near the critical temperature, T_c, where ψ is directly proportional to the *gap parameter*, Δ_0.)

The order parameter should be a smooth function of temperature and related to the free energy of the superconductor. Making a Taylor expansion of the free energy in powers of the order parameter, we can write:

$$F_S(T,V,n_S) = F_S(T,V,0) + \alpha(T)|\psi|^2 + \frac{1}{2}\beta(T)|\psi|^4 + \ldots \qquad (11.44)$$

which can be equivalently written as:

$$F_S(T,V,n_S) = F_N(T,V,0) + \alpha(T)n_s + \frac{1}{2}\beta(T)n_S^2 + \ldots \qquad (11.45)$$

From the equilibrium condition; $\partial F_S / \partial n_S = 0$, we obtain:

$$n_S = -\frac{\alpha(T)}{\beta(T)} \qquad (11.46)$$

Since the free energy must be a minimum, we have $\partial^2 F_S / \partial n_S^2 > 0$. So from Equation (11.45), β is positive, so from (11.42), α should be negative. For the phase transition of second order, we expect:

$$\alpha(T_c) = 0; \quad \beta(T) > 0, \qquad (11.47)$$

and

$$\alpha(T) = (T - T_c)\left(\frac{\partial \alpha}{\partial T}\right)_{T=T_c} \qquad (11.48)$$

This provides us with the necessary conditions above; for $T < T_c$, $\alpha(T) < 0$ and $\beta(T) > 0$. We can plot the form of the free energy, see Figure 11.10 (a), which should have a single solution for $T > T_c$, where $\psi = 0$, while just below T_c, the minimum energy will have non-zero $|\psi|$. We can now express the order parameter in terms of the phenomenological parameters $\alpha(T)$ and $\beta(T)$ as:

$$|\psi| = \begin{cases} 0 & T > T_c \\ \dfrac{(T - T_c)^{1/2}}{\beta^{1/2}(T)} \left(\dfrac{\partial \alpha}{\partial T}\right)^{1/2}_{T=T_c} & T < T_c \end{cases} \quad (11.49)$$

A curve of $|\psi|$ as a function of temperature is illustrated in Figure 11.10 (b).

From Equations (11.44) and (11.45), we obtain:

$$F_S = F_N - \frac{\alpha^2}{2\beta} \quad (11.50)$$

However, using Equation (11.22); $F_S - F_N = -\mu_0 H_c^2/2$, we have:

$$\frac{\alpha^2}{\beta} = \mu_0 H_c^2 \quad (11.51)$$

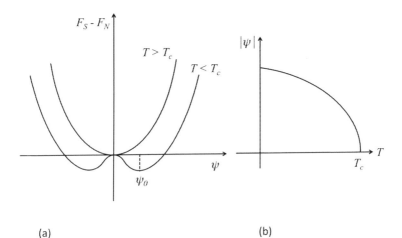

FIGURE 11.10: (a) Difference in free energy for the normal and superconducting states as a function of the order parameter. The free energy is reduced in the superconducting state for $T < T_c$, where ψ will have non-zero values. (b) Magnitude of the order parameter as a function of temperature in the GL model.

We can now express the critical field, near T_c, as:

$$H_c = \frac{(T_c - T)}{\sqrt{\mu_0 \beta}} \left(\frac{\partial \alpha}{\partial T}\right)_{T=T_c} \quad (11.52)$$

In the presence of a magnetic field, the effective wave-function must change. The GL theory really demonstrates its full potential once we include the magnetic field as it shows agreement with the Meissner - Ochsenfeld effect and the London equations. To include magnetic field effects it is important to make the following operator transformation:

$$\frac{\hbar}{i}\nabla \to \frac{\hbar}{i}\nabla - q\mathbf{A} \quad (11.53)$$

The relevant charge for superconductors is $-2e$, which wasn't clear until the BCS theory was developed. The link to GL theory was made by Gorkov, who saw the connection to the Cooper pairs. It turns out that a positive charge ($+2e$) can also be used, as we can visualize the Cooper pairs as holes. This has no effect on the GL theory. Using the above, we can now express the free energy as:

$$F_S = F_N + \int \left[\frac{\hbar^2}{2m_e}\left|\left(\frac{\hbar^2}{i}\nabla + 2e\mathbf{A}\right)\psi\right|^2 + \alpha|\psi|^2 + \frac{1}{2}\beta|\psi|^4\right] dV$$

$$(11.54)$$

This must now be minimized by taking the derivatives with respect to ψ and ψ^*. The resulting equation is rather complex and nonlinear, but the term of interest with respect to magnetic effects is that which involves the vector potential, $\mathbf{A}$:

$$-\frac{\hbar^2}{2m_e}\left(\nabla + \frac{2ei}{\hbar}\mathbf{A}\right)^2 \psi(\mathbf{r}) + (\alpha + \beta|\psi|^2)\psi(\mathbf{r}) = 0 \quad (11.55)$$

The current density in the superconducting state is given by the derivative of the free energy with respect to the vector potential:

$$\mathbf{j}_S = -\frac{\partial F_S}{\partial \mathbf{A}} \quad (11.56)$$

which allows us to obtain the supercurrent as:

$$\mathbf{j}_s = -\frac{ie\hbar}{2m_e^*}(\psi^*\nabla\psi - \psi\nabla\psi*) - \frac{(2e)^2}{m_e^*}|\psi|^2\mathbf{A} \qquad (11.57)$$

Equations (11.51) and (11.52) are the the fundamental equations of the GL theory and with them, Ginzburg and Landau were able to demonstrate nonlinear effects of fields, which are strong enough to affect n_s and its spatial variation. As such the GL model embodies the macroscopic quantum mechanical nature of the superconducting state.

11.6 ELEMENTS OF THE BCS THEORY OF SUPERCONDUCTIVITY

Any theory of superconductivity must account for a number of phenomena related to superconductivity, as for example outlined in Section 11.2. The Bardeen Cooper and Schrieffer (BCS) theory was published in 1957 and soon became recognized to be correct in its essential features and was able to explain a number of experimentally observed phenomena. It was the first truly microscopic theory and as such is preferred to the GL theory. The BCS theory can reproduce the results of the GL model.

Even before the full BCS model was published, there were a number of indicators of which way things were developing. In particular, the electron-phonon coupling was seen to have some important role to play. For example, the substitution of mercury with different isotopes indicated that the shift in the critical temperature was related to the frequency of lattice vibrations. While many properties of solids are independent of the elemental isotope, such as Fermi energy and the electronic heat capacity, some properties will vary. Lattice vibrations, as we saw in Chapter 5, depend very strongly on the atomic masses, varying as $M^{-\frac{1}{2}}$. This is known as the *isotope effect*, and hence the variation of T_c with isotope suggests a connection with lattice dynamics. Also, Fröhlich (1950) suggested that lattice vibrations would lead to an attractive interaction between electrons. The principal idea being that as an electron moves through a lattice of positive ions, it will displace them slightly in the direction of electron, see Figure 11.11. The ions being relatively heavy and slow moving with respect to the electron, relax back to their equilibrium positions only after the electron has passed. In this way we can imagine a wake left behind the electron, which acts like a "tube" of enhanced positive charge. This enhanced charge can attract a second electron and "drag" it along. Actually the second electron will also create its own wake and the two electrons can thus couple and move together through the lattice. The electrons need not be too close, thus avoiding Coulomb repulsion.

The suggestion of pairs of electrons being coupled in a bound state due to the mediation with phonons was made in 1956 by Leon Cooper. Cooper showed that the electrons can form a stable pair with their spins aligned in opposite directions even in the presence of other electrons, since the long range Coulomb repulsion between the electron pair is largely screened. Such a pair of electrons is named a *Cooper pair*. Cooper considered that the coherence length corresponds to the size of the pair and the energy gap, Δ_0, represents

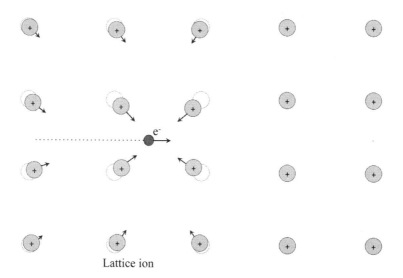

FIGURE 11.11: Schematic representation of the lattice deformation created by the Coulomb attraction between an electron and the positive ions of the crystal. As the electron moves through the lattice it leaves a wake of displaced ions behind it. This can create an attractive potential for other electrons and plays an important role in the electron - phonon coupling in the formation of the Cooper pair.

the pair binding energy. The energy gap refers to the separation of states: between the normal and superconducting electrons. On cooling through the critical temperature, there is a condensation of the numbers of electrons from the normal to the superconducting state. The existence of the energy gap has been measured experimentally using photon absorption for example, where the onset is found for:

$$h\nu = 2\Delta_0 \qquad (11.58)$$

The factor two arising from the fact that absorption requires the excitation of two electrons from the Cooper pair. Physically this corresponds to breaking the pair bond in the absorption of the photon and hence two free normal electrons are produced. The value of the gap varies with temperature, having a maximum value of Δ_0 at $T = 0$ and disappearing at T_c. The value of Δ_0 turns out to be proportional to $k_B T_c$.

In 1957 Bardeen, Cooper and Schrieffer published their now famous paper on the theory of superconductivity. At the heart of this theory is the electron-lattice interaction leading to the formation of the Cooper pairs in the superconducting state, which is mediated via the exchange of virtual phonons between the two electrons. The pairing requires that the spins of the electrons are coupled such that the electron with spin-up and momentum, **k** is paired with an electron with spin-down and momentum −**k**. The angular momentum of the pair is thus zero. The BCS theory demonstrated that superconductivity is a *cooperative phenomenon*, since there are interactions which couple electrons. Ferromagnetism is therefore also a cooperative phenomenon, since the exchange interaction couples the magnetic moments in the system.

In the following, we will only give a general qualitative overview of the theory; a quantitative description requires a level of mathematics beyond the scope of the book, where concepts such as Green's functions and many-body physics are necessary. More detailed accounts of the BCS theory can be found in a number of texts as well as the original paper itself.[1]

11.6.1 Electron - Phonon Coupling and Cooper Pairs

In this section, we will outline the main components to the BCS theory to indicate the principal reasoning behind the electron pair coupling. We start by considering a pair of electrons with an energy close to the Fermi energy, E_F, and lying close to the Fermi sphere. The electron will have a velocity therefore, close to the Fermi velocity, v_F. As we mentioned above, the electron will cause a deformation in the lattice, thus creating a (virtual) phonon. The free electrons can be considered as having a phonon cloud around them, and electrons can interact via this phonon cloud. The coupling can be envisaged therefore as an exchange of virtual phonons, as illustrated in Figure 11.12. The phonon interaction potential has the following form:

$$V_{ph}(\mathbf{q}, \omega) = \frac{|g_{\mathbf{q}\lambda}|^2}{\omega^2 - \omega_{\mathbf{q}\lambda}^2} \quad (11.59)$$

where $g_{\mathbf{q}\lambda}$ represents the matrix element for the electron with momentum, $\mathbf{k}$ being scattered to a momentum $\mathbf{k} + \mathbf{q}$. Since $g_{\mathbf{q}\lambda} \sim \sqrt{m_e/M}$ the coupling is relatively weak, where m_e and M are the masses of the electron and the ions of the crystal lattice.

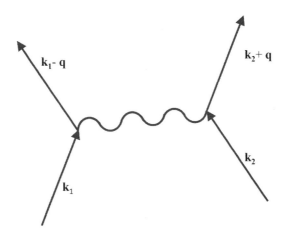

FIGURE 11.12: Two electrons with momenta, $\mathbf{k}_1$ and $\mathbf{k}_2$ interact via phonon coupling in the formation of the Cooper pair.

In the BCS model, the effective interaction was simplified to give:

$$V_{ph}^{eff}(\mathbf{q},\omega) = \frac{|g_{eff}|^2}{\omega^2 - \omega_D^2} \qquad (11.60)$$

where ω_D is the Debye frequency. This will give an attractive potential for electrons for $\omega < \omega_D$, while it is repulsive for $\omega > .\omega_D$. Recognition that the repulsive part plays no role was understood by BCS. Only electrons with energies within $\pm k_B T$ of the Fermi energy are of interest, which means that we consider the superconductive regime only for $\hbar\omega_D \gg k_B T$. This gives a simplified form of the BCS phonon potential:

$$V_{ph}^{eff}(\mathbf{q},\omega) = -|g_{eff}|^2; \qquad |\omega| < \omega_D \qquad (11.61)$$

The energy restriction means that the electron energies of interest are within a range $\pm\hbar\omega_D$ of the Fermi surface; i.e., $|E_\mathbf{k} - E_F| < \hbar\omega_D$. Therefore we expect $|g_{eff}|^2 \sim 1/M\omega_D^2$. But since $\omega_D \sim \sqrt{K/M}$, where K is an effective spring constant of the lattice, we thus expect $|g_{eff}|^2$ to be independent of M. The explanation of the isotope effect in the BCS theory arises since we consider only energies within $\hbar\omega_D$ of the Fermi surface, which varies as $M^{-\frac{1}{2}}$.

Consider a pair of electrons with states, E_1, $\mathbf{k}_1$ and E_2, $\mathbf{k}_2$, just above the Fermi energy. At low temperatures, $E_F \simeq E_F^0$ so that all states below E_F are occupied. If there is a weak phonon coupling, with other electrons considered non-interacting, and since the states below k_F are occupied, no states are available for $|\mathbf{k}| < k_F$. The phonon exchange between our two electrons means there is a continual change in wave-vector and using the conservation of momentum, can be expressed as:

$$\mathbf{k}_1 + \mathbf{k}_2 = \mathbf{k}_1' + \mathbf{k}_2' = \mathbf{q} \qquad (11.62)$$

where $\mathbf{q}$ is the phonon wave vector. Since we are limited to the range of energy (within $\hbar\omega_D$ above E_F^0), the only possible interactions will be as illustrated in Figure 11.13. The maximum energy reducing interaction will occur for the largest overlap between the two shaded

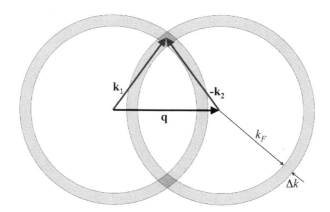

FIGURE 11.13: Two electrons with momenta, k_1 and k_2 can interact within the overlapping regions via the coupling with a phonon, **q**.

regions. This will obviously occur when $\mathbf{q} = 0$. This condition gives: $\mathbf{k}_1 = -\mathbf{k}_2 = \mathbf{k}$, i.e., when the momenta are equal and opposite.

The two-electron wave-function, $\psi(\mathbf{r}_1, \mathbf{r}_2)$, must satisfy the Schrödinger equation:

$$-\frac{\hbar^2}{2m_e}(\nabla_1^2 + \nabla_2^2)\psi(\mathbf{r}_1,\mathbf{r}_2) + V_{ph}(\mathbf{r}_1,\mathbf{r}_2)\psi(\mathbf{r}_1,\mathbf{r}_2) = E\psi(\mathbf{r}_1,\mathbf{r}_2)$$

$$= (\epsilon + 2E_F^0)\psi(\mathbf{r}_1,\mathbf{r}_2) \quad (11.63)$$

where ϵ is the energy of the Cooper pair with respect to the free state, with $V_{ph}(\mathbf{r}_1, \mathbf{r}_2) = 0$, where they would each have an energy $E_F^0 = \hbar^2 k_F^2 / 2m_e$. For the two-electron function written in the form of plane waves, the most general function can be written as:

$$\psi(\mathbf{r}_1,\mathbf{r}_2) = \psi(\mathbf{r}_1 - \mathbf{r}_2) = \frac{1}{\Omega}\sum_k g(\mathbf{k})e^{i\mathbf{k}\cdot(\mathbf{r}_1-\mathbf{r}_2)} \quad (11.64)$$

The quantity $g(\mathbf{k})$ represents the probability of finding the electron pair in the state $(\mathbf{k}, -\mathbf{k})$, which must be zero for $k_F > k > \sqrt{2m_e(E_F^0 + \hbar\omega_D)/\hbar^2}$. We now substitute the wave-function into the Schrödinger equation and multiply by $e^{-i\mathbf{k}'\cdot\mathbf{r}}$, where $\mathbf{r} = \mathbf{r}_1 - \mathbf{r}_2$, which yields:

$$\frac{\hbar^2}{2m_e}g(\mathbf{k}) + \frac{1}{\Omega}\sum_{k'} g(\mathbf{k}')V_{\mathbf{k},\mathbf{k}'} = (\epsilon + 2E_F^0)g(\mathbf{k}) \quad (11.65)$$

where $V_{k,k'}$ is the scattering matrix of the electron pair from state $(\mathbf{k}, -\mathbf{k})$ to $(\mathbf{k'}, -\mathbf{k'})$ and vice versa. The energy restriction to within $\hbar\omega_D$ still holds and for $V_{k,k'}$ to be attractive, it must be less than zero. Solution of Equation (11.65) leads to:

$$\epsilon = \frac{2\hbar\omega_D}{1 - e^{-2/V_0 g(E_F^0)}} \quad (11.66)$$

which, for weak coupling, $V_0 g(E_F^0) \ll 1$, the electron pair coupling energy is:

$$\epsilon \simeq -2\hbar\omega_D e^{-2/V_0 g(E_F^0)} \quad (11.67)$$

where V_0 in the interaction strength and $g(E_F^0)$ is the density of states at the Fermi level. This represents the two electron bound state energy

One important aspect is that the Pauli exclusion principle applies to both electrons. The two-electron wave-function was symmetric in coordinates $(\mathbf{r}_1, \mathbf{r}_2)$, under the exchange of electrons. However, a requirement of the Pauli exclusion principle is that the full wave-function be anti-symmetric under exchange. This can be achieved by including the spin components, where the spin wave-function must now be anti-symmetric. This requisite leads to the electrons in the Cooper pair having opposite spins: $(\mathbf{k}_\uparrow, \mathbf{k}_\downarrow)$.

11.6.2 The BCS Ground State

The BCS ground state refers to the configuration at $T = 0$ K, where all electrons must be occupied in pairs, such that $n = n_S$. The energy reduction from the normal to the ground state is represented above with energy, ϵ. To maintain the lowest energy state we require that in a scattering process, two electrons occupying states $(\mathbf{k}_\uparrow, \mathbf{k}_\downarrow)$ scatter as a pair to $(\mathbf{k'}_\uparrow, \mathbf{k'}_\downarrow)$, see Figure 11.14.

The total energy reduction is not a simple summation of of pair energies, since each pair energy depends on those already formed. The total energy minimum for the whole system must take into account all pair configurations including the one-electron kinetic

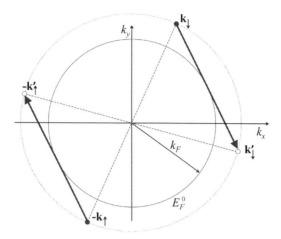

FIGURE 11.14: Two electrons with momenta, $\mathbf{k}_\uparrow$ and $-\mathbf{k}_\downarrow$ scatter as a pair to states $\mathbf{k}'_\uparrow$ and $-\mathbf{k}'_\downarrow$.

energy and the pair energy reduction due to collisions (i.e., phonon exchange). The kinetic energy takes the form:

$$E_K = 2 \sum_{\mathbf{k}} w_{\mathbf{k}} \xi_{\mathbf{k}} \qquad (11.68)$$

where

$$\xi_{\mathbf{k}} = \frac{\hbar^2 k^2}{2 m_e} - E_F^0 \qquad (11.69)$$

in which $\omega_{\mathbf{k}}$ is the probability that pair state $(\mathbf{k}_\uparrow, \mathbf{k}_\downarrow)$ is occupied. To evaluate the total energy reduction due to the pair collisions $(\mathbf{k}_\uparrow, \mathbf{k}_\downarrow) \leftrightarrows (\mathbf{k}'_\uparrow, \mathbf{k}'_\downarrow)$, we must use the Hamiltonian which takes into account that the annihilation of pair $(\mathbf{k}_\uparrow, \mathbf{k}_\downarrow)$ and the simultaneous creation of pair $(\mathbf{k}'_\uparrow, \mathbf{k}'_\downarrow)$, leads to an energy reduction of $V_{\mathbf{k},\mathbf{k}'}$. Given that any pair state can be occupied or unoccupied, the wave-function of the pair can be expressed as a combination of orthogonal states: $|1\rangle_{\mathbf{k}}$ and $|0\rangle_{\mathbf{k}}$, where the former represents the occupied state and the latter the unoccupied one. It is then possible to assemble the pair function as:

$$|\psi\rangle_{\mathbf{k}} = u_{\mathbf{k}} |1\rangle_{\mathbf{k}} + v_{\mathbf{k}} |0\rangle_{\mathbf{k}} \qquad (11.70)$$

In this alternative wave function to that represented in Equation (11.64), we have $v_k^2 = w_k$ and $u_k^2 = 1 - w_k$, being the probabilities of the state being occupied or unoccupied, respectively. The representation of the ground state of the system can be approximated as the product of such pair functions, such that the BCS wave function can be expressed as:

$$|\Psi_{BCS}\rangle_k = \prod_k (u_k |1\rangle_k + v_k |0\rangle_k) \quad (11.71)$$

The Hamiltonian for the scattering process takes the form:

$$\hat{\mathcal{H}} = -\frac{V_0}{\Omega} \sum_{kk'} \sigma_k^+ \sigma_{k'}^- \quad (11.72)$$

where σ_k^+ and $\sigma_{k'}^-$ are the Pauli spin operators[2]. It can thus be shown from the evaluation of $\langle \Psi_{BCS} | \hat{\mathcal{H}} | \Psi_{BCS} \rangle$, that the total energy of the Cooper pairs of the system is given by:

$$W_{BCS} = 2\sum_k v_k^2 \xi_k - \frac{V_0}{\Omega} \sum_{kk'} v_k u_k v_{k'} u_{k'} \quad (11.73)$$

It is useful to remember that $v_k = \sqrt{w_k} = \cos\theta_k$ and $u_k = \sqrt{1 - w_k} = \sin\theta_k$. The ground state is arrived at by minimizing this energy with respect to θ_k; $\partial W_{BCS} / \partial \theta_k = 0$. This allows the BCS superconducting ground state to be obtained as:

$$W_{BCS}^{(0)} = \sum_k \xi_k \left(1 - \frac{\xi_k}{E_k}\right) - \Omega \frac{\Delta_0}{V_0} \quad (11.74)$$

The energy difference between the ground state and the first excited state can be expressed as:

$$\Delta E_{BCS}^{(1)} = W_{BCS}^{(1)} - W_{BCS}^{(0)} = 2E_{k'} = 2\sqrt{\xi_{k'} + \Delta_0^2} \quad (11.75)$$

The quantity $\xi_{k'} = (\hbar^2 k'^2 / 2m_e) - E_F^0$ will be very small since it represents the kinetic energy of the Cooper pair being scattered out of their coupled state. Therefore the minimum excitation to break up the cooper pair can be expressed as $2\Delta_0$. This is the famous pair binding energy or energy gap that arises in the BCS theory. The density of states can evaluated from the a consideration of the excitation spectra, $E_{k'} = \sqrt{\xi_{k'}^2 + \Delta_0^2}$ and the fact that the number of charge carriers must be conserved: $g_S(E_k)dE_k = g_N(\xi_k)d\xi_k$. Since we are principally interested in energies in the region of Δ_0 around the Fermi level, we can write:

$$\frac{g_S(E_k)}{g_N(\xi_k)} = \frac{dE_k}{d\xi_k} = \begin{cases} 0 & E_k > \Delta_0 \\ \dfrac{E_k}{\sqrt{E_k^2 - \Delta_0^2}} & E_k < \Delta_0 \end{cases} \quad (11.76)$$

This produces a gap in the density of states in the superconducting state. The situation is illustrated in Figure 11.15 for absolute zero, $0 < T < T_c$ and for $T > T_c$.

11.6.3 Outcomes of the BCS Theory

The existence of an energy gap is a very important result and allows an experimental comparison to be made with theory. The gap energy reduces with temperature, from the maximum value at zero temperature and vanishes at T_c. This can be expressed in the form:

$$\Delta(T) = 1.74\Delta_0 \left(1 - \frac{T}{T_c}\right)^{1/2} \quad (11.77)$$

The zero temperature gap parameter, Δ_0 and the critical temperature T_c depend in the same way on the strength of the electron - phonon coupling and the density of states. For zero magnetic field it is possible to show that they are related by:

$$\frac{\Delta_0}{k_B T_c} = 1.764 \quad \text{or} \quad 2\Delta_0 = 3.528 k_B T_c \quad (11.78)$$

Now since it turns out that the gap parameter and the GL order parameter are related, the temperature dependence of Δ_0 is tha same as that illustrated in Figure 11.10 (b). A comparison with experiment was hailed as a major success of the BCS theory, see Figure 11.16.

In the case where a supercurrent, $\mathbf{j}_S$, flows, only a shift in the **k** value **K**/2 is permitted, where we account for the fact that each electron in the Cooper pair must experience the same shift. The energy gap in the superconductor is unchanged by the existence of

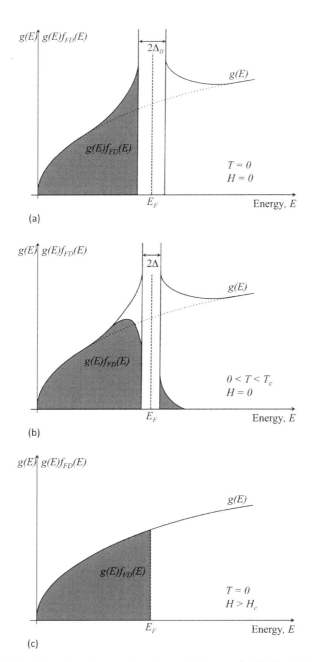

FIGURE 11.15: Density of states, $g(E)$, and occupied states, $g(E) f(E)$ for (a) $T = 0$ K, (b) $0 < T < T_c$ and (c) $T > T_c$. The energy gap is largest at 0 K and reduces with an increase of temperature. It vanishes at the critical temperature, T_c and the density of states is that for a normal metal.

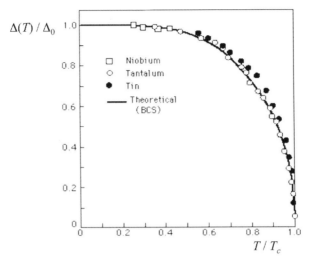

FIGURE 11.16: Temperature dependence of the energy gap, $\Delta(T)$. A comparison with experiment is shown for selected metals.

a current and is independent of **k**. An alteration of the **k**-state by an inelastic collision can only occur by excitation across the energy gap, 2Δ and would thus break the Cooper pair bond and reduce the number of pairs thus reducing the supercurrent. This must therefore be ruled out as a relaxation mechanism. This essentially means that the Cooper pair does not alter its energy, as long as its kinetic energy is below the 2Δ threshold. Since the supercurrent can be expressed as: $\mathbf{j}_s = -n_s e \mathbf{v} = -n_s e \hbar \mathbf{k}/m_e$, we can evaluate the critical current at which a large enough change of wave-vector can bring about the breaking of the Cooper pair bond. The energy can be estimated from taking $\mathbf{k} \to \mathbf{k} + \mathbf{K}/2$, from which we can write:

$$E = \frac{\hbar^2 (\mathbf{k} + \mathbf{K}/2)^2}{2m_e} = \frac{\hbar^2}{2m_e}\left(k^2 + \frac{K^2}{4} + \mathbf{k} \cdot \mathbf{K}\right) \quad (11.79)$$

Considering that $|\mathbf{K}| \ll k_F$ and that we are dealing with electrons near the Fermi energy, $k \simeq k_F$, we obtain the energy for pair dissociation as:

$$2\delta E = \frac{\hbar^2 k_F K}{m_e} \quad (11.80)$$

It is now a simple matter to obtain the critical current, for which $\delta E \geq \Delta_0$. Thus we obtain:

$$(j_S)_c = \frac{\Delta_0 n_S e}{\hbar k_F} \tag{11.81}$$

For example, in the case of Sn, the experimental value is $(j_S)_c = 2 \times 10^7 \text{Acm}^{-2}$.

By considering the magnetic field produced at the surface of a superconductor, it is possible to relate the critical current with a critical field. We can do this by considering a wire of radius, r carrying a supercurrent, j_S. We start from the Maxwell equation; $\nabla \times \mathbf{H} = \mathbf{j}$, which we can express as:

$$\int (\nabla \times \mathbf{H}) \cdot d\mathbf{S} = \oint \mathbf{H} \cdot d\mathbf{l} = \int \mathbf{j} \cdot d\mathbf{S} \tag{11.82}$$

This should be valid for a supercurrent with a magnetic field at the surface. Taking a closed path around the circumference of the wire and considering the current decay from the surface of the superconductor to be: $j_S = j_S^0 e^{-z/\lambda_L}$, from Equation (11.82) we obtain:

$$2\pi r H = 2\pi r \lambda_L j_S^0 \tag{11.83}$$

From the critical current density, Equation (11.81), we obtain the critical field:

$$H_c = (j_S)_c = \lambda_L \frac{\Delta_0 n_S e}{\hbar k_F} \tag{11.84}$$

where λ_L is the London penetration depth. Now, since the energy gap decays with temperature, Equation (11.77), then so should the critical field. This is yet another success of the BCS theory.

$$H_c = (j_S)_c = \lambda_L \frac{\Delta_0 n_S e}{\hbar k_F} \tag{11.85}$$

In quantum mechanics, we can write the generalized form of the flux (current) density as:

$$\mathbf{j} = \frac{\hbar}{2m_e} (\psi^* \nabla \psi - \psi \nabla \psi^*) \tag{11.86}$$

Taking into account the effect of a magnetic field, through Equation (11.53), and multiplying by the electric charge e, since we are dealing with an electric current, we obtain:

$$\mathbf{j} = -\frac{ie\hbar}{2m_e^*}(\psi^*\nabla\psi - \psi\nabla\psi^*) - \frac{q^2}{m_e^*}|\psi|^2 \mathbf{A} \quad (11.87)$$

c.f. Equation (11.57). Using the wave function of the form described above, Equation (11.64), we have: $\psi(\mathbf{r}) = \sqrt{n_S}\,e^{i\Theta(\mathbf{r})}$, where $\Theta(\mathbf{r})$ is a phase function. From this we obtain the supercurrent as:

$$\mathbf{j}_S = \frac{en_S}{2m_e^*}[\hbar\nabla\Theta(\mathbf{r}) - 2e\mathbf{A}] \quad (11.88)$$

we can now take the curl of this equation to obtain:

$$\nabla\times\mathbf{j}_S = -\frac{n_s}{m_e^*}e^2\mathbf{B} \quad (11.89)$$

This is the second London equation and is consistent with the Meissner - Ochsenfeld effect.

We now consider Equation (11.88) for a closed path around a superconductor ring, which is threaded by a magnetic field, **B**. Taking the path integral for Equation (11.8), we find:

$$\oint \mathbf{j}_S \cdot d\mathbf{l} = -\frac{e^2 n_s}{m_e^*}\oint \mathbf{A}\cdot d\mathbf{l} + \frac{en_s\hbar}{2m_e^*}\oint \nabla\Theta(\mathbf{r})\cdot d\mathbf{l} \quad (11.90)$$

Since the current around the loop inside the superconductor must be zero and using the identity: $\oint \mathbf{A}\cdot d\mathbf{l} = \int \mathbf{B}\cdot d\mathbf{S}$, we obtain:

$$\int \mathbf{B}\cdot d\mathbf{S} = \frac{\hbar}{2e}\oint \nabla\Theta(\mathbf{r})\cdot d\mathbf{l} = \frac{\hbar}{2e}2\pi N \quad (11.91)$$

We have identified here that the phase change around a loop must be zero or an integer multiple of 2π. The left-hand side of Equation (11.91) is just the magnetic flux, and this means that the flux trapped within the superconductor is quantized:

$$\Phi = \frac{h}{2e}N \quad (11.92)$$

This is referred to as *flux quantization*, where the basic unit of the quantization is $h/2e = 2.07 \times 10^{-15}$ Wb.

11.7 JOSEPHSON EFFECTS

The Josephson effects refer to the quantum tunneling of Cooper pairs from one superconducting region to another through a thin insulating barrier. Such a barrier usually takes the form of an oxide material, such as Al_2O_3, with a thickness of a few tens of Å. This type of structure is usually referred to as a *tunnel junction* and can be found in other tunnel applications, such as the magnetic tunnel junction, in which spin dependent tunneling between ferromagnetic layers can be used to make spintronic devices. See Section 13.12 (Chapter 13). The basic structure of the Josephson junction is illustrated in Figure 11.17.

The calculation of the current - voltage characteristics $(I - V)$ can be made via a consideration of the Schrödinger equation on

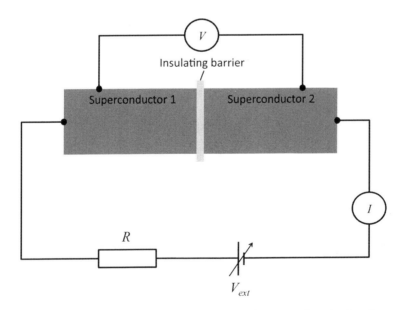

FIGURE 11.17: Electric circuit for measuring the tunnel current, I, through a Josephson junction with a variable external voltage source, V_{ext}.

either side of the insulating barrier, where the Hamiltonians $\mathcal{H}_1$ and $\mathcal{H}_2$ are valid. The Schrödinger equations take the form:

$$i\hbar \frac{\partial \Psi_1}{\partial t} = \mathcal{H}_1 \Psi_1 + T\Psi_2 \qquad (11.93)$$

and

$$i\hbar \frac{\partial \Psi_2}{\partial t} = \mathcal{H}_2 \Psi_2 + T\Psi_1 \qquad (11.94)$$

where $\Psi_{1,2}$ are the many-body wave-functions for the Cooper pair states in superconductors 1 and 2, respectively and T is a coupling constant for the tunnel junction and is related to the tunnel probability of the Cooper pairs between the two superconductors. This is a reflection of the fact that the wave-functions of the Cooper pairs in one superconductor do not entirely vanish in the other superconductor. Clearly for the case where $T = 0$ there will be no tunnel current and the superconductors will be uncoupled. The coupling strength with depend on the wave-functions of the Cooper pairs on either side of the barrier and importantly on the barrier itself, and in particular its thickness. Typically, an experiment will be made to measure the tunnel current of the Cooper pairs as a function of the applied voltage, V_{ext}. This will create a shift on the energy levels with respect to the Fermi level by an amount, qV_{ext}, where $q = -2e$ is the charge of the Cooper pair. Taking the energy scale with respect to the center of the barrier, we can express the above equations in the form:

$$i\hbar \frac{\partial \Psi_1}{\partial t} = \frac{qV_{ext}}{2} \Psi_1 + T\Psi_2 \qquad (11.95)$$

and

$$i\hbar \frac{\partial \Psi_2}{\partial t} = -\frac{qV_{ext}}{2} \Psi_2 + T\Psi_1 \qquad (11.96)$$

Taking into account that the wave functions must be normalized such that: $|\Psi|^2 = n_c = n_s/2$, where n_c is the density of Cooper pairs, which is evidently double that of the number of electrons in the superconductor, n_S. We now write the wave-function for the superconductor as:

$$\Psi_{1,2} = \sqrt{n_{c1,2}}\, e^{i\Theta_{1,2}} = \sqrt{n_{c1,2}}\,[\cos\Theta_{1,2} + i\sin\Theta_{1,2}] \quad (11.97)$$

We can now take the time derivative and substitute into the Schrödinger equation. Rearranging and separating the real and imaginary components, for Equation (11.95), we obtain:

$$\frac{\hbar}{2}\frac{\partial n_{c1}}{\partial t}\sin\Theta_1 - \left(\hbar\frac{\partial \Theta_1}{\partial t} + \frac{qV_{ext}}{2}\right) n_{c1} \cos\Theta_1 = T\sqrt{n_{c1}n_{c2}}\,\cos\Theta_2 \quad (11.98)$$

for the real part and:

$$-\frac{\hbar}{2}\frac{\partial n_{c1}}{\partial t}\cos\Theta_1 - \left(\hbar\Theta_1 + \frac{qV_{ext}}{2}\right) n_{c1} \sin\Theta_1 = T\sqrt{n_{c1}n_{c2}}\,\sin\Theta_2 \quad (11.99)$$

for the imaginary component. Similar expressions can be written for superconductor 2. We proceed by multiplying Equation (11.98) by $\cos\Theta_1$ and Equation (11.99) by $\sin\Theta_1$ and summing. We also multiply Equation (11.98) by $\sin\Theta_1$ and Equation (11.99) by $\cos\Theta_1$ and take the difference of the two. After such manipulations for both sets of equations for superconductors 1 and 2, we establish the following relations:

$$\Theta_2 - \Theta_1 = \frac{2_e V_{ext} t}{\hbar} + \delta_0 \quad (11.100)$$

and

$$\frac{\partial n_{c1}}{\partial t} = -\frac{\partial n_{c2}}{\partial t} = \frac{2T}{\hbar} n_c \sin(\Theta_2 - \Theta_1) \quad (11.101)$$

where we have simplified the expression by allowing the density of carriers to be equal in both superconductors on either side of the barrier ($n_{c1} = n_{c2} = n_c$), which does not affect the physics of the Josephson effect. We note that the factor δ_0 is a constant of integration, which was performed to obtain Equation (11.100) from the time derivatives of the phase factors $\Theta_{1,2}$, and has the effect of a change of phase of the signal. In the absence of an applied voltage, there should still be a tunnel current, the direction of which depends only on the sign of $\sin(\Theta_2 - \Theta_1)$, i.e., on the phase difference. The current density can be expressed from:

$$J_{CP} = q \frac{\partial n_c}{\partial t} = \frac{2qT}{\hbar} n_c \sin \delta_0 \qquad (11.102)$$

or

$$I_{CP} = I_0 \sin \delta_0 \qquad (11.103)$$

where we have taken into account that this is the current which flows for $V_{ext} = 0$. This is referred to as the *dc Josephson effect* and only requires that the circuit be closed such that the charges can be removed from the superconductors. Given that there is no resistance in the superconductors, no voltage will be registered on the voltmeter; $V = 0$. If we apply an external potential, the current will increase rapidly to a maximum and becomes unstable and a voltage will appear across the junction, which depends on the externally applied tension and the resistance, R in the external circuit. The current will no longer be due to the Cooper pair tunnel effect, but from single electron tunneling due to the break up of the paired electrons. The voltage required to break up the Cooper pair is given by: $2\Delta(T)/e$. The current - voltage characteristics are shown schematically in Figure 11.18.

Referring to Equation (11.101), we can express the tunnel current of Cooper pairs, which is proportional to the time variation of the pair density, as:

$$I_{CP} \propto q \frac{\partial n_c}{\partial t} = \frac{2qT}{\hbar} n_c \sin(\omega_{CP} t + \delta_0) \qquad (11.104)$$

where we have $\omega_{CP} = 2eV_{ext}/\hbar$, which gives the ac angular frequency of the tunnel current of the Josephson junction. This is known as the *ac Josephson effect*. A 1mV applied potential will produce a signal with a frequency of 3×10^{12} Hz. We can write the ac Josephson current as:

$$I_{CP} = I_0 \sin(\omega_{CP} + \delta_0) \qquad (11.105)$$

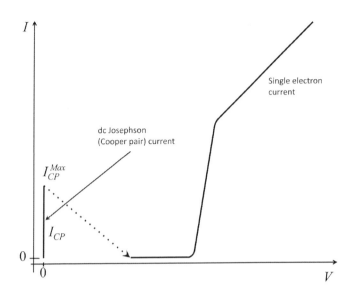

FIGURE 11.18: Idealized current - voltage characteristics for a Josephson junction, where a Cooper pair tunnel current, I_{CP} exists in the absence of an applied voltage. This rapidly reaches a maximum value, I_{CP}^{Max}, which is reached with a small applied potential equal to $2\Delta(T)/e$ and is therefore temperature dependent.

We can understand the generation of the ac signal from a dc input as the energy of the electron pair being converted to photon energy.

If one of the superconductors is replaced by a normal metal, the quasi-particle tunneling can still occur, depending on the position of the chemical potential on either side of the insulating barrier. This effect was actually observed before the Josephson effect was postulated, and was experimentally observed in 1960 by I. Giaever, and is known as *Giaever tunneling*.

It is possible to demonstrate the coherence of the superconducting state by constructing twin junction device, see Figure 11.19. This is analogous to an interferometer in optics, in which the incident beam (current in the case of our superconductor) is split into two equal paths and then made to overlap in space, at some later position. The interference pattern itself can only be produced if the signals in the two branches are coherent and the electron pairs have a wave nature. This is another example of a macroscopic quantum effect. The device shown in Figure 11.19 is called a *superconducting quantum interference device* or SQUID, for short.

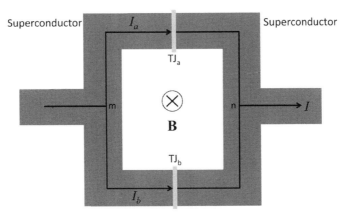

FIGURE 11.19: Illustration of a double Josephson junction device. This is known as a superconducting quantum interference device (SQUID). The supercurrent in the superconductor to the left is split into currents I_a and I_b, which pass through tunnel junctions TJ_a and TJ_b, respectively. Each will therefore have a current with a phase difference of δ_a and δ_b. The total current $I = I_a + I_b$.

Given that each Josephson junction has a response given by Equation (11.103), the interference effect will be seen through the addition of the two signals:

$$I = I_0(\sin \delta_a + \sin \delta_b) = 2I_0 \cos\left(\frac{\delta_a - \delta_b}{2}\right) \sin\left(\frac{\delta_a + \delta_b}{2}\right) \quad (11.106)$$

When the loop containing the two Josephson junctions has no magnetic field the phase shifts at the junctions, of δ_a and δ_b, will be fixed. Once a magnetic field, **B** (vector potential, **A**) penetrates this region, then a further phase change will be produced. The additional phase change between two points, say x and y, in the superconductor due to the vector potential can be expressed as:

$$\delta_{x-y} = \int_x^y \nabla \Theta \cdot d\mathbf{l} = \frac{2e}{\hbar} \int_x^y \mathbf{A} \cdot d\mathbf{l} \quad (11.107)$$

We can now write the phase shifts for the two paths via the tunnel junctions, from point m to point n, as:

$$\delta_{m-n} = \delta_a + \frac{2e}{\hbar} \int_a \mathbf{A} \cdot d\mathbf{l} \quad (11.108)$$

$$\delta_{m-n} = \delta_b + \frac{2e}{\hbar} \int_b \mathbf{A} \cdot d\mathbf{l} \tag{11.109}$$

Since these equations must be equivalent we obtain:

$$\delta_b - \delta_a = \frac{2e}{\hbar} \oint \mathbf{A} \cdot d\mathbf{l} = \frac{2e}{\hbar} \int \mathbf{B} \cdot d\mathbf{S} \tag{11.110}$$

The closed loop integral arises from the fact that the line integrals in Equations (11.108) and (11.109) are in opposite directions. This shows that the total phase is controlled, and can be adjusted, by the magnetic flux passing through the loop between the junctions. Introducing the arbitrary initial phase difference, we now have:

$$\delta_a = \delta_0 - \frac{e}{\hbar} \int \mathbf{B} \cdot d\mathbf{S} \tag{11.111}$$

and

$$\delta_b = \delta_0 + \frac{e}{\hbar} \int \mathbf{B} \cdot d\mathbf{S} \tag{11.112}$$

Substituting in Equation (11.106) yields:

$$I = 2I_0 \sin \delta_0 \cos \left(\frac{e}{\hbar} \int \mathbf{B} \cdot d\mathbf{S} \right) \tag{11.113}$$

Finally, we can insert the magnetic flux obtained from Equations (11.91) and (11.92), which gives:

$$I = 2I_0 \sin \delta_0 \cos(N\pi) \tag{11.114}$$

This means that the current output is oscillatory with the magnetic field (flux), with maxima occurring when integral numbers of flux units pass through the loop.

The SQUID as a device has a number of clear applications, the most obvious being its use in magnetometry. The SQUID magnetometer is actually one of the principal applications of superconducting materials. It is the most sensitive of magnetometers and is used as a research tool to study low dimensional magnetic systems.

One of the most sensitive magnetic measurements available is the micro-SQUID, which has been used to study the magnetic switching fields and anisotropies in a single magnetic nanoparticle.[3] The SQUID is so sensitive that is can also be used to detect the minute magnetic fields generated in the human brain and has other medical and nondestructive detection applications.[4]

11.8 HIGH-TEMPERATURE SUPERCONDUCTORS

We briefly introduced the topic of high-temperature superconductivity (HTS) in the introduction. Here we noted that the topic really took off in the mid to late 1980s, with the research work of Bednorz and Müller and the discovery of the high transition temperature of the oxide $YBa_2Cu_3O_{7-\delta}$ (sometimes referred to as YBCO) and related compounds. This material is referred to as a 123 compound, which refers to the relative content of the various metals in the chemical formula and has a critical temperature of $T_c = 92$ K. This is significant since it has a temperature above the boiling point of liquid nitrogen (77K), which means that it is much simpler to study and use than low temperature materials since the cryogenic handling is much simplified and there is no requirement for the use of liquid helium, which is significantly more expensive. Before 1986 all superconductivity experiments required low temperature cryogenics, with the highest transition temperatures being in the range of 15–23 K.

The existence of HTS was and is still a significant problem in solid state physics from a theoretical point of view. It would appear that the BCS theory is not applicable in the high temperature regime of these materials at least not in the form that it was originally envisaged by BCS. Despite this, some form of charge carrier pairing must be a component and any electron phonon coupling must be particularly strong.

The YBCO compound has a rather complex crystalline structure, as illustrated in Figure 11.20 (a). The structure consists of planes of CuO_2 alternating within the structures for oxygen atoms in pyramid and rectangular planar coordination, as shown in the figure. These latter form oxygen chains along the b-axis of the lattice. Oxygen

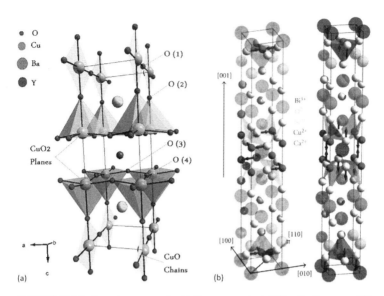

FIGURE 11.20: Crystal structures for HTS compounds: (a) YBa$_2$Cu$_3$O$_{7-\delta}$ type 123 (b) Bi$_2$SrCa$_2$Cu$_2$O$_{10}$, type 2122 and Bi$_2$Sr$_2$Ca$_2$Cu$_2$O$_{10}$, type 2223.

vacancies (indicated as δ for off-stoichiometric compounds) play an important role in the superconductivity of the oxide. The CuO$_2$ planes are a common feature of all superconductors with $T_c > 50$ K discovered up to around 1990 and also play a crucial role in the electrical properties of the superconducting phase. Such materials, with CuO$_2$ planes in their crystalline structure are commonly referred to as *cuprates* The YBCO type compounds exhibit a significant anisotropy in its electronic properties between the c-axis and the ab-plane.

The HTS materials generally exhibit rather high upper critical fields, typically reaching hundreds of Tesla. YBCO, for example has a value of H_{c2} of about 340 T. The value also depends on the orientations of the field with respect to the crystalline axes, being between 5 to 7 times lower for fields applied along the c-axis. A measurement of the critical field also allows an estimation of the coherence length, which is about 3–5 Å in the c-direction and 20–30 Å in the ab-plane.

Related to the YBCO like compounds are the Bi and Tl containing materials, with typical stoichiometries Bi$_2$Sr$_2$Ca$_2$Cu$_3$O$_{10}$ or

$Tl_2Ba_2Ca_2Cu_3O_{10}$. Such materials are again referred to in terms of the relative content of the various metals in the chemical formula, such as 2223 or 2212. These have structures of the form illustrated in Figure 11.20(b). As with the YBCO compounds, the crystalline structure consists of alternating layers with CuO_2 planes, but have a much larger unit cell. We note that the unit cell has specific regions in which the CuO_2 planes are found.

While there is still no overall accepted theory of the superconductivity in HTS materials, there are a great many published works. As we mentioned above, the Cooper pair mechanism is still expected to play a central role in the superconductivity of these systems, though not necessarily as electrons but in the form of paired holes. The hole concept is exactly the same as that outlined for semiconductors, where a missing electron can be treated like a positive charge carrier, and hence we can speak of p-type conductivity or *p-type superconductivity*, and is determined from the sign of the Hall constant, as in the case of semiconductors. Since the critical temperatures are that much higher in HTS, we can expect correspondingly high values for the band gap energies, which in this case would fall in the region of 20–30 meV. This means that the characteristic values of $\Delta_0/k_B T_c$ are in the region of 3–4 as compared to a value of 1.764 in the traditional BCS theory. The pairing mechanism between the holes has not yet been identified and the energies indicated here are too high for the normal phonon pairing model. It may be possible that some form of phonon enhancement is responsible due to structural properties, though this mechanism has yet to be identified. It is also possible that coupling between the difference CuO_2 planes plays some role in the coupling.

One rather crucial aspect of the high-temperature superconducting materials is the dependence of the critical temperature on the oxygen deficiency in the stoichiometry of the compound. As with most oxide systems, the oxygen atom readily removes two electrons from the metal atoms. Therefore, any oxygen deficiency will affect the charge neutrality of the unit cell. It is noted in the YBCO materials, $YBa_2Cu_3O_{7-\delta}$, where δ indicates the O deficiency, that above $\delta = 0.7$, the ceramics lose their superconducting properties and become antiferromagnetically ordered insulators. Even before this transition, the increase of δ is accompanied by a general decrease of the transition temperature, T_c. At some level, it would appear that the electron affinity of the oxygen atoms is related to the production of the hole states in these ceramics and thus the off stoichiometric oxygen deficiencies will remove the necessary holes from the solid. It is thought that the CuO_2 planes in both the YBCO, 2122 and 2223 superconductors, are in some way responsible for the superconductivity in the HTS class of materials. The planar structures give rise to some form of quasi-two-dimensional transport of charge carriers in their paired hole states, as seen with the anisotropic conductive properties of these materials. It is further postulated that the alkali or rare-earth metals and oxygen atoms between these layers act as charge carrier reservoirs which can remove electrons from the CuO_2 planes, thus creating the Cooper pairs in the conduction planes of the structure.

The YBCO compound has a large T_c and low Fermi velocity, meaning that the coherence length, Equation (11.42), will be relatively small and smaller than the unit cell size. The coherence length is related to the extension of the Cooper pair wave function, meaning that the pair size should be much smaller than those in the low temperature metal superconducting state. On the other hand, the low carrier density will mean a large penetration depth. This means that the HTS compounds should be type II superconductors with elevated values of the upper critical field, H_{c2}.

Despite the many difficulties and complexities of the HTS systems, there is much interest and technological advances that would stem from such an understanding also provide much motivation for the continued hunt to understand this class of superconductor.

As a final remark, we may point to the superconductivity observed in the fullerene (C_{60}) crystals doped with potassium and rubidium. These can form a regular fcc type lattice with semiconducting properties and are found to have semiconducting band gaps of 1.5–1.9 eV. Transition temperatures in such systems have been found in the region of 20–40 K, with their superconductivity being explained within the BCS model. With recent developments in the related graphene type materials, maybe progress will soon be made in this direction too.

11.9 SUMMARY

Superconductivity is a remarkable natural phenomenon which has captured the imagination of many scientists. The phenomenon itself was first observed as a total disappearance of the electrical resistivity of the material. It was soon found that the absence of resistance has a number of related consequences seen as the Meissner - Ochsenfeld effect, i.e., the elimination of magnetic flux from the interior of the superconductor and perfect diamagnetism. In fact, the magnetic properties of superconductors is an integral component of the superconductive phenomenon. These arise from the surface currents generated by the magnetic field which form within a thin layer at the surface of the material and create fields which oppose those that are externally applied.

It was soon found that there are two classes of superconductor. A type I material, which eliminates all magnetic fields from its interior up to a critical field, H_c, at which point the materials undergoes a phase transition to the normal resistive state. The type II material has two critical magnetic fields. Below the first it acts like a type I superconductor, while between the first and second critical fields magnetic flux can penetrate the material in filaments, which under certain circumstances self organize into regular arrays. These vortex states are so named due to the vortex currents that circulate the filaments, and are also referred to as Abrikosov or Shubnikov phases. Within the filaments the electrons move in the normal state while in the regions between the filaments the material is in the superconducting state. This is also referred to as the mixed state.

While much experimentation progressed in the subsequent years after its discovery, a theoretical description of the superconductive state was much more elusive. It was only in the 1930s that any real progress was made, with the London theory and the Ginzburg - Landau model. The former is essentially based on the application of the Maxwell equations and the latter on thermodynamics and the theory of phase transitions, where an order parameter is used to describe the state of the system. While providing much insight into the phenomenological description of the superconducting state, it was not sufficient to describe the microscopic state of the system. Significant progress really had to wait until well after the end of the second world war, and by the mid-1950s an understanding of a microscopic description was beginning to form. Fröhlich in 1950 had suggested some form of phonon interaction should be important, based on the isotope effect. However, it was not until 1956 that Cooper postulated that a pairing of electrons was a necessary condition to describe the superconducting state. The following year saw the major breakthrough in superconductivity, and after its discovery, possibly singly *the* most important. This was the publication of the so-called BCS theory, after its authors, Bardeen, Cooper and Schrieffer, who provided a robust quantum mechanical microscopic theory of the Cooper pair and its role in superconductivity.

The pairing of electrons with opposite spins was found to be the ground state of the electron gas, where coupling was provided by an electron - phonon mechanism. This gives rise to a band gap separating the Cooper pair (superconducting) states from the single unpaired electron states in the solid. The band gap is a form of Cooper pair binding energy and is related to the critical temperature of the superconducting state. This was a major triumph of the BCS model. The BCS model was also able to cope with the Meissner - Ochsenfeld effect and the Ginzburg - Landau model. It furthermore was able to predict the flux quantization inside the superconductor.

Based on the quantum mechanical description of the superconducting state, progress was made in the early 1960s in the theory and experimental fields relating to tunneling effects between superconductors and metals through an insulating barrier (Giaever tunneling) and between two superconductors (Josephson effects). In the latter an ac and a dc effect were identified. These developments permitted some of the first real applications of superconductors in the SQUID device. This is constructed from two branched of a superconductor, each containing a Josephson junction. It is found that the interference that is observed allows a measure of the coherence of the Cooper pair state and is extremely sensitive to magnetic fields penetrating the loop between the Josephson junctions. This has been used to make the worlds most sensitive detector of magnetic fields and is regularly used in the SQUID magnetometer.

The study of materials exhibiting superconductivity has developed ever since the discovery of superconductivity in 1911 by Kamerlingh - Onnes. Most materials that showed superconductivity had very low transition temperatures, and usually only up to around the 20 K region. Compounds and alloys exhibit somewhat higher transition temperatures. It was however not until 1986 that significant progress was made in the development of cuprate ceramics. In particular, the $YBa_2Cu_3O_{7-\delta}$ compound showed a transition temperature of around 95 K, and is significant in being above the temperature of liquid nitrogen (77 K), which provides for a much cheaper and more easily manageable cryogen. Cuprates have CuO_2 planes in their crystalline structure which are thought to form the conducting channels. While some form of pairing mechanism is thought to hold, it is not between electrons but holes. Hole states are produced via oxygen uptake of electrons, though the mechanism for coupling does not work in the same way envisaged in the BCS theory. The explicit details of this mechanism are the goals of much research in high-temperature superconductivity.

REFERENCES AND FURTHER READING

Basic Texts

- J. R. Hook and H. E. Hall, *Solid State Physics*, J. Wiley and Sons, Chichester, England (1999)
- H. P. Myers, *Introductory Solid State Physics*, Taylor and Francis, London (1998)

Advanced Texts

- J. F. Annett, *Superconductivity, Superfluids and Condensates*, Oxford University Press, Oxford (2004)
- H. Ibach and H. Lüth, *Solid State Physics: An Introduction to the Principles of Materials Science*, Springer, Berlin (1996)
- A. K. Saxena, *High-Temperature Superconductors*, Springer, Heidelberg (2010)

- C. P. Poole, H. A. Farach and R. J. Creswick, *Superconductivity*, Academic Press, California (1995)
- M. Tinkham, *Introduction to Superconductivity*, McGraw-Hill, New York, (1996)

EXERCISES

Q1. Use the Gibbs free energy, as expressed in Equation (11.17), to obtain the condition:

$$\frac{dH_c(T)}{dT} = \frac{S_N - S_S}{M_S - M_N} \quad (11.115)$$

where the function G is continuous across the phase boundary between the normal (N) - superconducting (S) phase boundary.

Q2. Demonstrate that the decay of the magnetic field at the surface of a superconductor conforms to the relation:

$$\mathbf{B}(x) = \mathbf{B}(0)e^{-x/\lambda} \quad (11.116)$$

Q3. Consider a superconducting slab of thickness, A, which is subject to an applied magnetic field of B. Find the functional form of the magnetic field profile inside the slab.

Q4. Derive Equation (11.52) from the Ginzburg - Landau theory.

Q5. Consider the Gibbs Free energy at the boundary of the normal and superconducting states, where $G_S(T,H) = G_N(T,H)$. With the use of Equation (11.17) demonstrate the relation:

$$-S_S \delta T - \mu_0 V M_S \delta H = -S_N \delta T - \mu_0 V M_N \delta H \quad (11.117)$$

Q6. From the above result, further show that the latent heat per unit volume for the phase transition can be expressed as:

$$L = -\mu_0 T H_c \frac{dH_c(T)}{dT} \quad (11.118)$$

Q7. Use the BCS wave-function, Equation (11.71), to evaluate the total energy of the Cooper pair system using the Hamiltonian expressed in Equation (11.72).

Q8. Use the form of the wave-function given in Equation (11.64) to obtain the current density (Equation (11.86)).

Q9. Demonstrate the flux quantization condition as expressed in Equation (11.92).

Q10. Derive Equations (11.100) and (11.101) for the Josephson junction.

NOTES

[1] See for example: M. Tinkham, *Introduction to Superconductivity*, 2e, McGraw-Hill, New York (1996); BCS, Phys. Rev. **108**, 1175 (1957).

[2] Given by: $\sigma_k^+ = \frac{1}{2}(\sigma_k^{(1)} + i\sigma_k^{(2)})$ and $\sigma_k^- = \frac{1}{2}(\sigma_k^{(1)} + i\sigma_k^{(2)})$, where $\sigma_k^{(1)} = \begin{pmatrix} 0 & 1 \\ 1 & 0 \end{pmatrix}_k$, $\sigma_k^{(2)} = \begin{pmatrix} 0 & -i \\ i & 0 \end{pmatrix}_k$. Now since $|1\rangle_k = \begin{pmatrix} 1 \\ 0 \end{pmatrix}_k$ and $|0\rangle_k = \begin{pmatrix} 0 \\ 1 \end{pmatrix}_k$, we obtain: $\sigma_k^+|1\rangle_k = 0$; $\sigma_k^+|0\rangle_k = |1\rangle_k$ and $\sigma_k^-|1\rangle_k = 0$; $\sigma_k^-|0\rangle_k = 0$.

[3] See for example, M. Jamet *et al.*, Phys. Rev. B, **69**, 024401 (2004).

[4] See J. Ouellette, The Industrial Physicist, **4**(**2**), 20, (1998), for an overview of some applications of the SQUID.

CHAPTER 12

DIELECTRIC MATERIALS

"Every man, in his own opinion, forms an exception to the ordinary rules of morality."

—William Hazlitt

"If you cannot get rid of the family skeleton, you may as well make it dance."

—George Bernard Shaw

"I became insane, with long intervals of horrible sanity."

—Edgar Allan Poe

12.1 INTRODUCTION

The interaction of electromagnetic radiation with matter provides an important way of classifying solids and has a number of consequences on how we look at materials. We can consider this interaction from a classical macroscopic point of view, where we apply the Maxwell equations, and describe the process via material constants and thus distinguish their responses and hence categorise the solid. Alternatively, we can think of the interaction from a microscopic description, where we might think of the absorption of a photon as creating a quantum excitation in the solid, such as a phonon or an exciton. Of course the nature of the interaction will depend on the wavelength (energy) of the incident photon and the specifics of the material in terms of its lattice properties and free electrons etc. More complex forms of interaction can also occur, which depend on the polarization state of the light and if it is a pulsed form, on the duration of that pulse. Such considerations will not be elaborated here in this introductory text and we will not consider some of the more complex nonlinear responses that some materials have.

In general, dielectric materials are insulating substances in which there are few or no free charge carriers and typically some form of ceramic or polymer. This means that an applied electric field can penetrate a significant distance into the their crystals. While we have discussed the importance of the electronic properties of materials in previous chapters, we also need insulating materials to prevent the passage of electrical current in regions where we do not want it to flow. The insulating coatings on wires and cables are an important application of dielectric, for instance. The use of dielectrics is also a crucial component in the construction of capacitors, where the static dielectric properties of the material between the electric plates are an essential part of the capacitors ability to store charge.

The fact that dielectric materials have no free carriers can also be viewed as it having a large enough band gap to prevent the excitation, via thermal ionization, of electron - hole pairs. The optical properties of solids is intimately related to the existence or not of free charge carriers in the material. In the case of a material with many free charge carriers such as electrons, the interaction of the electric field with the electrons causes the mirror like appearance that most metallic substance have. This is due to the excitation and re-emission of radiation below the skin depth of the material. Actually the response in strongly frequency dependent. In the case of an insulating material, the penetration of light can go much further into the solid and many dielectric materials are indeed transparent. The behavior can be evaluated from the frequency-dependent dielectric constant, $\epsilon(\omega)$, or in an equivalent manner, the refractive index of the material; $n = \sqrt{\epsilon(\omega)}$. The presence of an electric field can distort the lattice, shifting the center of charge from the centers of negative and positive charge of the atom, thus creating an electric dipole moment. The electrical and optical properties can then be determined by the ability of the dielectric materials to form dipoles in the presence of the electric field.

In ionic crystals, where there are long range electrostatic forces between the ions, the crystal lattice can be deformed by the application of electric fields. As with the case of magnetic materials, it is sometimes easier to treat such situations as if there was a specific field within the material which produces this intrinsic effect. It is often much simpler to then deal with these effective fields as a sum of vectors along with any applied electric field component.

12.2 SOME BASIC PROPERTIES OF DIELECTRIC MATERIALS

12.2.1 Electrical Conductivity

Despite these materials having very small values of conductivity, its value is by no means zero. Both free electrons and the migration of ions can contribute to the electrical conductivity in ceramics. As was shown in the Chapter 9, we can estimate the number of free charge carriers in an intrinsic semiconductor from the law of mass action:

$$np = N_C(T)P_V(T)e^{-\Delta/k_B T} \qquad (12.1)$$

So for a band gap, Δ, of a few eV, the number of charge carriers present at room temperature should be less than one electron or hole per cubic meter. However, electrically active impurities that act as donors or acceptors can dominate and significantly contribute to the charge carrier concentrations. For example, even for a materials with only one impurity per million atoms (1 ppm), which acts as a donor will, when fully ionized produce a carrier concentration of around $10^{22} m^{-3}$. The conductivity of such a sample can be calculated as outlined in Chapter 9.

12.2.2 Ionic Conduction

Ions can migrate through an ionic compound through the mechanism of vacancy diffusion, which can be driven by an electric field rather than by a concentration gradient. This should be evident from the continuity equations that were established in Chapter 9. The current density for such a process can be expressed as:

$$\mathbf{j}_i = n_i Z_i e \mu_i \mathcal{E} \tag{12.2}$$

where Z_i is the charge on the ion and μ_i the ionic mobility. We can use the Einstein relation to relate the diffusion constant with the mobility; $D_i = k_B T \mu_i / e$, from which we can write:

$$\mathbf{j}_i = \frac{n_i Z_i e^2 D_i}{k_B T} \mathcal{E} = \sigma_i \mathcal{E} \tag{12.3}$$

Therefore for an ionic concentration of $n_i = 10^{28} \mathrm{m}^{-3}$ and $D_i = 10^{-12} \mathrm{m}^2 \mathrm{s}^{-1}$, the ionic conductivity will be given by $\sigma_i = n_i Z_i e^2 D_i / k_B T \simeq 10^{-5} \mathrm{Sm}^{-1}$.

The ionic conductivity is an important parameter in a common battery and provides the conduction pathways to combine with electrons flowing through an external load on the battery. In many cases, the battery has a liquid electrolyte, such as KOH, which provides the ionic conductivity. However, there is much interest in the use of β-alumina solid electrolyte (BASE) which can be used in high power Na-S batteries. The Na$^+$ ions can be complexed with Al$_2$O$_3$ to form β-alumina, which acts as a substrate with channels through which the Na$^+$ ions can flow and can furnich conductivities of about 1 Sm^{-1}.

12.2.3 Dielectric Breakdown

One of the primary uses for dielectric materials is insulating applications. Therefore it is important to understand under what conditions they maintain their integrity for such a role. As such, we need to determine the *dielectric strength* or *breakdown potential* of the material. Since the conductivity is not zero, for strong applied electric fields some current will flow. If there are weak points in the dielectric, more leakage current will flow at these points and cause local (Joule) heating, which in turn can increase the number of free carriers excited into the conduction band. This will increase the current and at some point, for a sufficient electric field will lead to an avalanche increase in the carrier density and thermal runaway, which can cause a catastrophic breakdown.

Defects, such as inclusions and cavities in the dielectric can lead to an increase in the local field causing an arcing across a cavity, thus ions and electrons can be accelerated by these fields further eroding the walls of the cavity, thus increasing the size of the cavity and again leading to the eventual breakdown of the dielectric medium. While small, the ionic conductivity can transport impurities from the surface of the solid into its bulk. Such a process can be assisted by humidity and an acid environment. Impurities can accumulate and form conductive pathways in the material, which will lead to thermal runaway effects and electrolytic breakdown.

Even in the absence of defect assisted mechanisms, the principal requirement for the dielectric breakdown is a sufficiently large electric field to reach a critical level, such that electrons can be extracted from the valence band (or impurity states) and then to be accelerated by the electric field. Once they have sufficient energy they can then excite further electrons from the valence or impurity states and bring about the avalanche breakdown of the dielectric. The specific value of the breakdown potential is material dependent, with polystyrene showing one of the highest values for dielectric strength, being about 140 MVm^{-1}, while ceramics typically have values in the region of 10–20 MVm^{-1}.

Certain electrical components and devices are susceptible to electrostatic damage and are shipped in conductive packaging and must be installed carefully using an antistatic wristband. This is due to the thin dielectric layers in FET channels, which due to their dimensions are particularly at risk.

12.3 ELECTROSTATICS AND THE MAXWELL EQUATIONS

The optical and electrical properties of dielectrics are largely derived from their ability to form electric dipole moments in the presence of an electric field. We can consider that such a dipole as consisting of a positive and negative charge, q, separated by a distance, x, giving the electric dipole moment:

$$\mathbf{p} = q\mathbf{x} \tag{12.4}$$

We define the polarization of the material in the same way as the magnetization is defined in a magnetic material, which is just the density of moments:

$$\mathbf{P} = \frac{\sum_n \mathbf{p}_n}{V} \tag{12.5}$$

or simply the dipole moment per unit volume. Macroscopically we can introduce the electric displacement as:

$$\mathbf{D} = \epsilon \boldsymbol{\mathcal{E}} = \epsilon_0 \boldsymbol{\mathcal{E}} + \mathbf{P} \tag{12.6}$$

the constant ϵ_0 is called the permittivity of vacuum (or free space) having a value of $8.854 \times 10^{-12} \mathrm{Fm}^{-1}$. In the case of a crystalline material, the dielectric properties are generally anisotropic and ϵ is a rank two tensor. This means that the $\mathbf{P}$ and $\boldsymbol{\mathcal{E}}$ can be in different directions. In the description we give below, we only consider the case of an isotropic medium for which ϵ is scalar. Frequently it is convenient to define a relative dielectric constant, which we can express as: $\epsilon_r = \epsilon/\epsilon_0$. In this case we can write:

$$\epsilon_r \boldsymbol{\mathcal{E}} = \boldsymbol{\mathcal{E}} + \frac{\mathbf{P}}{\epsilon_0} \tag{12.7}$$

or

$$\mathbf{P} = \epsilon_0 (\epsilon_r - 1) \boldsymbol{\mathcal{E}} = \epsilon_0 \chi_e \boldsymbol{\mathcal{E}} \tag{12.8}$$

where we have introduced the static electric susceptibility $\chi_e = (\epsilon_r - 1)$. The electric susceptibility relates the degree of polarization with the applied electric field. We can combine the above with Equation (12.6) to obtain:

$$\begin{aligned} \mathbf{D} &= \epsilon_0 \boldsymbol{\mathcal{E}} + \epsilon_0 \chi_e \boldsymbol{\mathcal{E}} = \epsilon_0 (1 + \chi_e) \boldsymbol{\mathcal{E}} \\ &= \epsilon_0 \epsilon_r \boldsymbol{\mathcal{E}} \end{aligned} \tag{12.9}$$

where the constant of proportionality, ϵ_r, is also referred to as the relative permittivity or relative dielectric constant.

We can write the analogous relationships with relation to the magnetic properties, where we have a relation between the magnetic induction, **B**, the applied magnetic field, **H**, and the magnetization, **M**, which is expressed as:

$$\begin{aligned}\mathbf{B} &= \mu_0(\mathbf{H}+\mathbf{M}) = \mu_0(1+\chi_m)\mathbf{H} \\ &= \mu_0\mu_m\mathbf{H} = \mu\mathbf{H}\end{aligned} \qquad (12.10)$$

See Chapter 10 for more details. Equations (12.9) and (12.10) are consistent with the four Maxwell equations:

$$\nabla \cdot \mathbf{D} = \rho \quad \text{or} \quad \nabla \cdot \boldsymbol{\varepsilon} = \frac{\rho}{\epsilon_0} \qquad (12.11)$$

$$\nabla \times \boldsymbol{\varepsilon} = -\frac{\partial \mathbf{B}}{\partial t} \qquad (12.12)$$

$$\nabla \cdot \mathbf{B} = 0 \qquad (12.13)$$

$$\nabla \times \mathbf{H} = \mathbf{J} + \frac{\partial \mathbf{D}}{\partial t} \quad \text{or} \quad \nabla \times \mathbf{B} = \mu_0 \mathbf{J} + \mu_0\epsilon_0 \frac{\partial \boldsymbol{\varepsilon}}{\partial t} \qquad (12.14)$$

where ρ is the charge density and **J** is the current density.

It is instructive to illustrate how the static electric properties of dielectric materials are used and an excellent example is in the capacitor, which is used as a charge storage device. The capacitor consists, in its basic form, of a metallic sandwich with an insulating filling. When the metallic plates are connected to an external source, electrons will accumulate on one of these and induce an equal but opposite charge (Q) on the other plate, as illustrated in Figure 12.1 (a). Once the potential between the plates has become equal to that of the source, charges will cease to flow, with the capacitor being fully charged for that particular potential, V.

The capacitance is defined as the charge stored on the capacitor, divided by the applied voltage, which is written as:

$$C = \frac{Q}{V} \qquad (12.15)$$

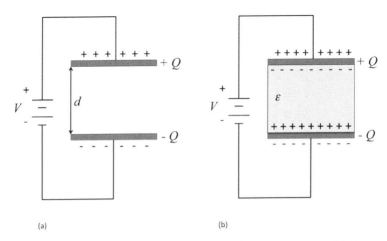

FIGURE 12.1: Parallel plate capacitor connected to an external potential, V, which charges the plates with a charge of ±Q for (a) an air filled gap and (b) a dielectric filled gap between the metallic plates.

where the unit is given in Farads (F), which by the above definition is equivalent to Coulombs per volt (CV^{-1}). The charge stored is equal to the product of the area of the plates, A, the electric field, ε, across the plates and the dielectric constant, ϵ of the material between the plates:

$$Q = \epsilon A \varepsilon \qquad (12.16)$$

From a dimensional point of view, we note that the permittivity has units of CV^{-1}m^{-1}, and since the electric displacement $D = \epsilon \varepsilon$, it has units of charge displacement per unit area. Since the electric field corresponds to the potential applied divided by the separation of the plates ($\varepsilon = V/d$), we can express the capacitance as:

$$C = \frac{\epsilon A}{d} = \frac{\epsilon_0 \epsilon_r A}{d} = \frac{\epsilon_0 (1 - \chi_e) A}{d} \qquad (12.17)$$

In the case where there is no material between the parallel plates, the permittivity will be that of free space, ϵ_0. However, when a dielectric material is located between the plates of the capacitor, the electric field (which from plus to minus in our sign convention) will align the electric dipoles formed in the dielectric producing a negatively charged surface in the dielectric adjacent to the positive plate of the capacitor. In a similar way there will be a positive surface charge in the dielectric adjacent to the negative plate, see Figure 12.1 (b). The surface charges cause an increase in the charges on the plates, thus increasing the total charge stored by ΔQ. We can now write the total charge as:

$$Q = Q_0 + \Delta Q = DA = \epsilon_0 \mathcal{E} + P \qquad (12.18)$$

where P is the polarization of the dielectric. Thus we note that the additional charge on the capacitor was created by the polarization of the dielectric and the capacitance takes the form:

$$C = \frac{\epsilon_0 \epsilon_r A}{d} \qquad (12.19)$$

and is directly related to the relative dielectric constant of the medium between the parallel plates.

12.4 THE LOCAL FIELD APPROXIMATION

We can use the macroscopic relations of electrostatics to relate the polarization with the electric field. We can do this by approximating the local displacement of the ions due to the applied electric field by an effective local field, $\mathcal{E}_{loc}(\mathbf{r})$. This local field is based on similar concepts used in magnetics, where we consider the effect of the local environment on the magnetization state of an ion. In the present case we substitute the polarization for the magnetization. To do this we can consider the regions near and far from a particular ion, as suggested by Lorentz. The ion in question is generally considered to be enclosed within an imaginary sphere, where the contributions of the dipoles within the sphere surface are counted individually, while everything outside is treated in a continuum approximation of the effective field. For an atom in the dielectric under the influence of an external charge, we can express the local field as:

$$\mathcal{E}_{loc}(\mathbf{r}) = \mathcal{E}_0(\mathbf{r}) + \mathcal{E}_{dep}(\mathbf{r}) + \mathcal{E}_{surf}(\mathbf{r}) + \mathcal{E}_{dip}(\mathbf{r}) \qquad (12.20)$$

where the first term corresponds to the field due to external charges, the second term accounts for depolarizing effect (in the same we have demagnetizing fields in ferromagnetic samples) due to induced surface charges on the outer surface of the sample. The contribution ε_{surf} is the field produced at the center of an imaginary cavity in the sample interior by the polarization induced surface charges on the surface of this cavity and ε_{dip} is the field at the center of this cavity due to discrete dipoles distributed on atomic sites inside this cavity, excluding that at the center position itself.

We envisage the sample geometry as illustrated in Figure 12.2, where we can sum the first two terms in Equation (12.20) with ε_{ext}, where we consider the effect of depolarization due to the outer surfaces. The cavity shown is considered to be a sphere with surface charges induced by the outer surface. For materials with cubic symmetry the dipolar contributions inside the sphere will vanish due to symmetry considerations. We can now evaluate the charge density on the surface of the spherical cavity as:

$$\varepsilon_{surf}(\mathbf{r}) = 2\pi \int_0^\pi r^2 \sin\phi \cos\phi \frac{P\cos\phi}{4\pi\epsilon_0 r^2} d\phi = \frac{P}{3\epsilon_0} \quad (12.21)$$

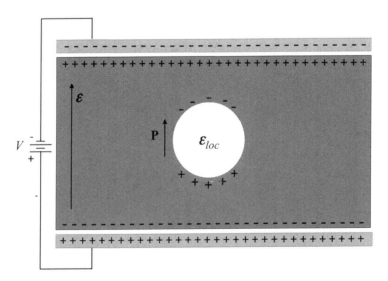

FIGURE 12.2: Application of an electric field, ε, can lead to the displacement of the centers of positive and negative charge, leading to the polarization, **P**, of the dielectric medium.

The local field can now be expressed in the form:

$$\boldsymbol{\mathcal{E}}_{loc}(\mathbf{r}) = \boldsymbol{\mathcal{E}}_{ext}(\mathbf{r}) + \frac{P}{3\epsilon_0} \tag{12.22}$$

The local field can be used to express the polarizability, α_i, of the ith atom, such that:

$$\mathbf{p}_i = \alpha_i \boldsymbol{\mathcal{E}}_{loc}(\mathbf{r}) \tag{12.23}$$

The total polarization will then be expressed as:

$$\mathbf{P} = \boldsymbol{\mathcal{E}}_{loc}(\mathbf{r}) \sum_i \alpha_i = \boldsymbol{\mathcal{E}}_{loc}(\mathbf{r}) N\alpha \tag{12.24}$$

where we assume all atoms to have the same polarizability, α and N is the number density of atoms. We now substitute this into Equation (12.22) and rearrange to obtain the local field as:

$$\boldsymbol{\mathcal{E}}_{loc}(\mathbf{r}) = \frac{\boldsymbol{\mathcal{E}}_{ext}(\mathbf{r})}{1 - N\alpha/3\epsilon_0} \tag{12.25}$$

This is now back substituted into Equation (12.24), to give the polarization:

$$\mathbf{P} = \frac{N\alpha \boldsymbol{\mathcal{E}}_{ext}(\mathbf{r})}{1 - N\alpha/3\epsilon_0} \tag{12.26}$$

This gives the electric susceptibility as:

$$\chi_e = \frac{\mathbf{P}}{\epsilon_0 \boldsymbol{\mathcal{E}}_{ext}(\mathbf{r})} = \frac{N\alpha}{\epsilon_0(1 - N\alpha/3\epsilon_0)} \tag{12.27}$$

We can write the relative dielectric constant as $\epsilon_r = 1 + \chi_e$, from which we can write:

$$\epsilon_r = 1 + \chi_e = 1 + \frac{N\alpha}{\epsilon_0(1 - N\alpha/3\epsilon_0)} = \frac{1 + 2N\alpha/3\epsilon_0}{1 - N\alpha/3\epsilon_0} \tag{12.28}$$

This expression can be rewritten in the following way:

$$\frac{N\alpha}{3\epsilon_0} = \frac{\epsilon_r - 1}{\epsilon_r + 2} \tag{12.29}$$

The above equation is commonly known as a the *Clausius - Mossotti relation*, which relates the atomic polarization with the macroscopic dielectric constant. This is an important expression since we can use it to predict the optical properties of insulators based on the Maxwell equations.

12.5 THE DIELECTRIC FUNCTION

Thus far we have only considered that the relative permittivity is a material (dielectric) constant. However, this should be more correctly considered as a frequency dependent material parameter, referred to as the *dielectric function*, $\epsilon(\omega)$. As with the application of a magnetic field, all materials exhibit some form of response to the existence of an electric field. In general, this interaction is understood in terms of the distortion of the charge distribution in the atoms and leads to a displacement of the centers of positive and negative charges, see Figure 12.3.

The relative dielectric constant from electronic polarization is quite small, but plays an important role in the optical properties of solids. In general, the electric and magnetic fields are time dependent functions, as described in Equations (12.12) and (12.14). However, it is often more convenient to present these in the form

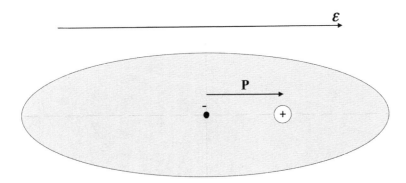

FIGURE 12.3: Application of an electric field, ε, can lead to the displacement of the centers of positive and negative charge, leading to the polarization, **P**, of the dielectric medium.

of frequency dependent variables using a Fourier transform, as given by:

$$\boldsymbol{\varepsilon}(t) = \int_{-\infty}^{\infty} \boldsymbol{\varepsilon}(\omega) e^{-i\omega t} \, d\omega \qquad (12.30)$$

and

$$\mathbf{D}(t) = \int_{-\infty}^{\infty} \mathbf{D}(\omega) e^{-i\omega t} \, d\omega \qquad (12.31)$$

Since $\boldsymbol{\varepsilon}(t)$ and $\mathbf{D}(t)$ are real functions, we have: $\boldsymbol{\varepsilon}(\omega) = \boldsymbol{\varepsilon}^*(-\omega)$ and $\mathbf{D}(\omega) = \mathbf{D}^*(-\omega)$. We can relate these Fourier coefficients with the dielectric function as follows:

$$\mathbf{D}(\omega) = \epsilon_0 \epsilon(\omega) \boldsymbol{\varepsilon}(\omega) \qquad (12.32)$$

Using Equation (12.14), for the dielectric medium in an oscillating electric field, we can write:

$$\nabla \times \mathbf{H} = \sigma \boldsymbol{\varepsilon}(\omega) - i\omega \epsilon_0 \epsilon(\omega) \boldsymbol{\varepsilon}(\omega) = [\sigma - i\omega \epsilon_0 \epsilon(\omega)] \boldsymbol{\varepsilon}(\omega) \qquad (12.33)$$

We can now express a frequency-dependent conductivity as:

$$\tilde{\sigma}(\omega) = \sigma - i\omega \epsilon_0 \epsilon(\omega) \qquad (12.34)$$

Equation (12.33) can be expressed alternatively in the form:

$$\nabla \times \mathbf{H} = -i\omega \epsilon_0 \tilde{\epsilon}(\omega) \boldsymbol{\varepsilon}(\omega) = \frac{\partial \mathbf{D}}{\partial t} \qquad (12.35)$$

where we have used a generalized dielectric function:

$$\tilde{\epsilon}(\omega) = \epsilon(\omega) + \frac{i\sigma}{\epsilon_0 \omega} \qquad (12.36)$$

Conductive phenomena are taken into account from the σ term. The use of oscillatory fields can blur the differences between the roles of free and bound electrons and we can describe the response in terms of the frequency dependent conductivity, Equation (12.34) or in terms of the frequency dependent dielectric function, Equation (12.36). In static fields this differences is very marked and the distinction between the conductivity and the dielectric response is more evident.

The dielectric susceptibility and dielectric functions are complex and related; $\chi(\omega) = \epsilon(\omega) - 1$. It is possible to show that the real and imaginary parts of the dielectric function are related, where $\epsilon(\omega) = \epsilon_1(\omega) + i\epsilon_2(\omega)$ and we have:

$$\epsilon_1(\omega) = 1 + \frac{P}{\pi}\int \frac{\epsilon_2(\omega')}{\omega' - \omega} d\omega' \qquad (12.37)$$

and

$$\epsilon_2(\omega) = -\frac{P}{\pi}\int \frac{\epsilon_1(\omega') - 1}{\omega' - \omega} d\omega' \qquad (12.38)$$

Equations (12.37) and (12.38) are known as the *Kramers - Kronig relations*. The parameter P denotes the *principal value* of the integral. The Kramers - Kronig relations are important since, for example, they can be used to calculate one part of the dielectric function from a measurement of the other without the need for a separate measurement.

12.5.1 Electronic Polarization

All materials undergo a certain degree of polarization due to the distortion created in their electronic orbits when subject to an electric field. When this applied electric field is oscillatory we can treat the problem as a form of damped harmonic oscillation, where a restoring force is present, which acts to return the system to its equilibrium state. We can characterize this restoring force with the constant ω_0, which is the characteristic resonance frequency of the system. The equation of motion can be expressed as:

$$\ddot{x} + \gamma\dot{x} + \omega_0^2 x = -\frac{e}{m_e}\mathcal{E}(\omega) \qquad (12.39)$$

where γ is the damping constant, which characterizes the lifetime of a normal mode of vibration. The system is driven by the time varying electric field $\varepsilon(\omega)$. The solution to this form of equation for a field which varies as: $\varepsilon(\omega) = \varepsilon_0(\omega)e^{i\omega t}$, can be expressed as:

$$x(t) = \tilde{x}(\omega)e^{i\omega t} \qquad (12.40)$$

where the complex amplitude is given by:

$$\tilde{x}(\omega) = -\frac{e\varepsilon(\omega)}{m_e(\omega_0^2 - \omega^2 - i\gamma\omega)} \qquad (12.41)$$

This can easily be verified by substituting Equation (12.40) in (12.39). The dipole moment induced by the oscillating electric field takes the form: $p = -e\tilde{x}(\omega)$ and the electronic polarizability, defined as $\chi_e \epsilon_0$, can be written in the form:

$$\alpha(\omega) \equiv \frac{p}{\varepsilon(\omega)} = -\frac{e\tilde{x}(\omega)}{\varepsilon(\omega)} = \frac{e^2}{m_e(\omega_0^2 - \omega^2 - i\gamma\omega)} \qquad (12.42)$$

The resonant frequency of the material, ω_0 typically lies in the range $10^{15} - 10^{16}$ rads^{-1}, which lies in the ultraviolet region of the electromagnetic spectrum. When the damping is weak and small with respect to ω_0, the expression for low frequencies, $\omega < \omega_0$, can be approximated as:

$$\alpha(\omega) = \frac{e^2}{m_e(\omega_0^2 - \omega^2)} \qquad (12.43)$$

This expression is for a single electron, therefore extending to a multi-electron atom we need to account for the ionic species and sum them to obtain the dielectric function:

$$\epsilon_r(\omega) = 1 + \frac{P(\omega)}{\epsilon_0 \varepsilon(\omega)} = 1 + \frac{1}{\epsilon_0}\sum_i n_i \alpha_i(\omega) \qquad (12.44)$$

Using the expression for the polarizability including damping term, we obtain, for a single atomic species:

$$\epsilon_r(\omega) = 1 + \frac{Ne^2}{\epsilon_0 m_e(\omega_0^2 - \omega^2 - i\gamma\omega)} \qquad (12.45)$$

It is customary to separate the dielectric function in terms of the real and imaginary components, where we use; $\epsilon(\omega) = \epsilon_1(\omega) + i\epsilon_2(\omega)$ and we can thus obtain the real and imaginary parts of the dielectric function as:

$$\epsilon_1(\omega) = 1 + \frac{Ne^2}{\epsilon_0 m_e} \frac{\omega_0^2 - \omega^2}{(\omega_0^2 - \omega^2)^2 + \gamma^2 \omega^2} \quad (12.46)$$

$$\epsilon_2(\omega) = \frac{Ne^2}{\epsilon_0 m_e} \frac{\gamma \omega}{(\omega_0^2 - \omega^2)^2 + \gamma^2 \omega^2} \quad (12.47)$$

These are illustrated in Figure 12.4, where $\epsilon_2(\omega)$ has the form of a damped resonance curve, the linewidth (full width half maximum - FWHM) being given by γ.

12.5.2 Ionic Polarization

Ionic materials consists of oppositely charged ion cores which are displaced by the application of an external electric field. This produces an ionic polarization in the solid and can have important effects in the infrared absorption. Even in covalent materials, such as Ge and Si, large susceptibilities are found experimentally due to the directional nature of the covalent (sp^3) bond, which means that the center of the electronic charge is somewhere between the ion cores that make up the crystal lattice.

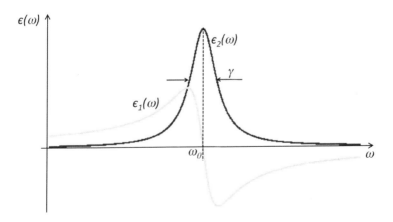

FIGURE 12.4: General functional forms of $\epsilon_1(\omega)$ and $\epsilon_2(\omega)$ in the region of a resonance at $\omega = \omega_0$.

When evaluating the dielectric response of an ionic crystal we should bear in mind that there are three possible modes of vibration that can be stimulated by the passing of an electromagnetic wave: we can can produce two transversal modes and one longitudinal mode. The specific excitation depends on the relative orientation (axis of polarization) of the incident electromagnetic radiation with respect to the axes of the crystal lattice. The equations of motion for the case of a linear chain of ions, with masses m and M (for the two types of ion) can be written as:

$$m\ddot{u}_{2n} - K(u_{2n+1} + u_{2n-1} - 2u_{2n}) = -e\varepsilon_0 e^{-i\omega t} \quad (12.48)$$

$$M\ddot{u}_{2n} - K(u_{2n+1} + u_{2n-1} - 2u_{2n}) = e\varepsilon_0 e - i\omega t \quad (12.49)$$

where we have introduced a driving force of the form: $-e\varepsilon e^{-i\omega t}$. Actually we have already calculated a very similar problem to this in Chapter 5 for the normal modes of vibration in a diatomic crystal. The difference here is that there is an additional driving term in the equation of motion, cf. Section 5.3. The solution for the displacement, u, of the ions takes a similar form to that given in Equations (5.24) and (5.25), and we can express the solutions as follows:

$$\omega^2 mA - 2K[A - B\cos(ka)] = e\varepsilon_0 e^{-ika} \quad (12.50)$$

$$\omega^2 MB + 2K[A\cos(ka) - B] = -e\varepsilon_0 e^{-ika} \quad (12.51)$$

Where A and B are the complex amplitudes of the ionic vibrations. We can simplify the above expressions for solutions near $k = 0$, from which we obtain:

$$\omega^2 mA + 2K(B - A) = e\varepsilon_0 \quad (12.52)$$

$$\omega^2 MB - 2K(B - A) = -e\varepsilon_0 \quad (12.53)$$

The quantity $(B - A)$ is of interest since it represents the oscillating change of position of the two charges and hence that of the dipole moment, which has a moment expressed as $e(B - A)$. It is often useful to evaluate the reduced mass which we obtain as: $\mu = mM/(M+m)$. Subtracting Equation (12.52) from (12.53) yields:

$$\left(\frac{2K}{\mu} - \omega^2\right)(B-A) = \frac{e\varepsilon_0}{\mu} \qquad (12.54)$$

For the free oscillation solution, which we can obtain by setting $\varepsilon_0 = 0$, and corresponds to the optical transverse mode, which at $k = 0$ gives:

$$\omega_T^2 = \frac{2K}{\mu} \qquad (12.55)$$

Substituting back into Equation (12.54) we obtain:

$$(\omega_T^2 - \omega^2)(B-A) = \frac{e\varepsilon_0}{\mu} \qquad (12.56)$$

The ionic polarization can now be expressed using the above relation, where we omit damping, as:

$$P_{ion} = N_{ion}(B-A)e = \frac{N_{ion}e^2\varepsilon_0}{\mu(\omega_T^2 - \omega^2)} \qquad (12.57)$$

In a similar way we can express the ionic polarizability as:

$$\alpha_{ion} = \frac{p_{ion}}{\varepsilon_0} = \frac{e^2}{\mu(\omega_T^2 - \omega^2)} \qquad (12.58)$$

The dielectric function can now be expressed as:

$$\epsilon_r(\omega) = 1 + \chi_e = 1 + \frac{P_{ion}}{\epsilon_0 \varepsilon_0} = \epsilon(\infty) + \frac{e^2}{\epsilon_0 \mu(\omega_T^2 - \omega^2)} \qquad (12.59)$$

It is possible to extract the low frequency ($\omega \ll \omega_T$) solution, where the static response takes the form:

$$\epsilon(0) = \epsilon(\infty) + \frac{N_{ion}e^2}{\epsilon_0 \mu \omega_T^2} \qquad (12.60)$$

which is real and positive. We now re-write Equation (12.59) as:

$$\epsilon_r(\omega) = \epsilon(\infty) + \frac{[\epsilon(0) - \epsilon(\infty)]\omega_T^2}{(\omega_T^2 - \omega^2)} \qquad (12.61)$$

The functional form of this equation is illustrated schematically in Figure 12.5, where the function diverges as the frequency approaches ω_T from above and below. The longitudinal optical frequency, ω_L, is observed as the crossing point of the frequency axis, where we can write $\epsilon(\omega_L) = 0$, and we obtain:

$$\epsilon(\infty) + \frac{[\epsilon(0) - \epsilon(\infty)]\omega_T^2}{(\omega_T^2 - \omega_L^2)} = 0 \qquad (12.62)$$

such that

$$\omega_L^2 = \omega_T^2 \left\{ 1 - \frac{[\epsilon(0) - \epsilon(\infty)]}{\epsilon(\infty)} \right\} = \omega_T^2 \frac{\epsilon(0)}{\epsilon(\infty)} \qquad (12.63)$$

Equation (12.63) is known as the *Lyddane-Sachs-Teller relation*, or LST, which relates the transverse and longitudinal optical frequencies to the static dielectric constant and refractive index.[1] This is consistent with Equation (12.6), where for the polarization being parallel to the incident wave vector we have $\boldsymbol{\varepsilon} = -\mathbf{P}/\epsilon_0$ and $\epsilon = 0$ for longitudinal modes. On the other hand, for transversal optical modes we have the polarization perpendicular to the

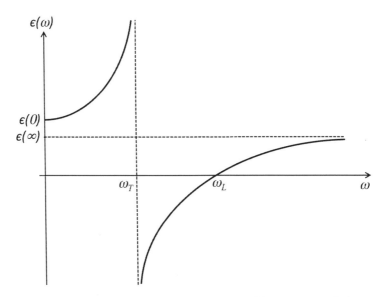

FIGURE 12.5: Dielectric function in the vicinity of the transverse optical mode for an ionic crystal.

incident wave vector and $\varepsilon = 0$, for which $\epsilon = \infty$, corresponding to the divergence at $\omega = \omega_T$.

From the Maxwell equations it is possible to establish the general form of the electromagnetic wave equation, which we can express as:

$$-\nabla^2 \varepsilon = \frac{\omega^2}{c^2} \epsilon(\omega) \varepsilon \qquad (12.64)$$

Now given the general form of the oscillatory electric field, we obtain:

$$\epsilon(\omega) = \frac{k^2 c^2}{\omega^2} \qquad (12.65)$$

where k is the wave vector of the electromagnetic wave. We can now depict the dispersion relation for the ionic crystal with the aid of the LST relation. The solutions represent coupled electromagnetic and mechanical wave, which are called *polaritons*. The dispersion relation is shown in Figure 12.6.

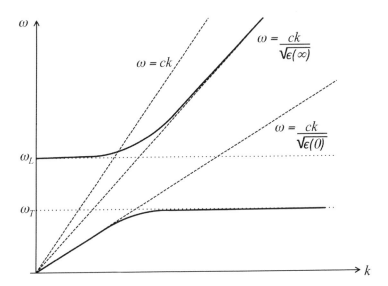

FIGURE 12.6: Dispersion relations for a phonon - polariton in an ionic crystal.

We see from this that the dispersion relation has two branches, one which lies below ω_T and the other above ω_L. In the case of the lower branch, we have $\omega = \omega_T$, which describes the electric field accompanying the transverse optical mode. For low wave vectors, however, the frequencies fall off and vanish as $kc/\sqrt{\epsilon_0}$, and corresponds to the case of the propagation of an electromagnetic wave in a medium of dielectric constant ϵ_0. In the upper branch, for the higher frequencies we have a similar situation, in which we have the propagation of an electromagnetic wave in a medium of dielectric constant ϵ_∞. However, for the lower k values, the frequencies level off to a constant value of ω_L. Between the frequencies ω_T and ω_L there are no solutions to the dispersion relation.

The polariton dispersion relation can be verified experimentally using the Raman spectroscopy techniques, in which the energy and momentum change of a photon is observed from inelastic scattering processes. Reflectivity measurements also offer another experimental method of observing phenomena related to the dispersion curve. The reflectivity can be expressed in terms of the dielectric constant as:

$$r = \left(\frac{\sqrt{\epsilon}-1}{\sqrt{\epsilon}+1} \right)^2 \qquad (12.66)$$

This shows that as $\epsilon(\omega) \to \infty$, the reflectivity will tend to unity. This means that in the region between ω_T and ω_L, the dielectric should act as a perfect mirror. This frequency region, where the reflectivity is 100%, is referred to as the *reststrahlen* band, deriving from the German for "residual rays".

12.5.3 The Total Dielectric Function

The total dielectric function covers the full electromagnetic spectrum and the different contributions show specific features over this range. We include the electronic (UV), ionic (IR), and dipolar (microwave) contributions to the frequency dependent dielectric constant which has a variation as illustrated in Figure 12.7.

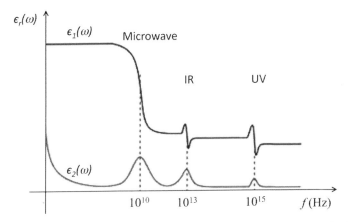

FIGURE 12.7: Schematic diagram of the total dielectric function showing the dipolar orientation, ionic and electric contributions. These features are occur at microwave, infrared (IR), and ultraviolet (UV) frequencies, respectively. The imaginary component of the dielectric function $(\epsilon_2(\omega))$ is also referred to as the dielectric loss factor.

12.6 FERROELECTRICS

Ferroelectric materials bear much in common with ferromagnetic solids. The ferroelectric is a material which exhibits a spontaneous dipole moment without the application of an applied electric field. This is analogous to the spontaneous magnetization observed in ferromagnetic materials in the absence of an external magnetic field. Ferroelectrics exhibit hysteretic behavior of the polarization as a function of the electric field applied to the sample and also form domain structures. Also in common with ferromagnetic materials, ferroelectrics also have a temperature dependence of the polarization, where the spontaneous polarization disappears at a critical (or Curie) temperature, T_c. At temperatures above T_c, the material becomes paraelectric with the general behavior of the dielectric susceptibility following a Curie - Weiss law:

$$\chi_e = \frac{C}{T - T_c} \tag{12.67}$$

where C is a constant.

The origin of a permanent electric dipole moment can be seen from crystal symmetry considerations of an ionic crystal. The conditions for ferroelectricity can be given as follows: (i) There must be a single rotation axis along the dipole moment; (ii) there can be no mirror planes perpendicular to the rotation axis and (iii) there must be no center of inversion in the crystal. Of the 32 point groups in crystallography, only 10 satisfy these conditions. Materials in which the above is satisfied and a permanent dipole exists are called *pyroelectrics*. This is because their dipole moment changes with temperature, and is related to thermal expansion of the crystal lattice. Pyroelectrics are not necessarily ferroelectrics and may have coercive fields greater than their breakdown field, or their Curie temperatures may be above their melting temperature. All ferroelectrics must however be, by definition, pyroelectrics. Some properties of ferroelectric compounds are listed in Table 12.1.

Of the ferroelectric materials discovered, the main classes are the phosphate and titanate groups. The latter group is probably the simplest to understand in terms of the origin of its ferroelectricity. This class of materials generally crystallize in the cubic perovskite structure, see Figure 12.8. The deformation of the $BaTiO_3$ structure occurs at a temperature below 393 K, where the centers of the oxygen (O^{2-}) ions and the titanium (Ta^{4+}) ions become displaced creating a permanent dipole moment.

Typically the cubic pervskite ferroelectrics form with the dipole moment locked into one six possible orientations. If these are randomly oriented, there will be no net polarization. The material can be poled by heating near to the Curie temperature and applying a strong electric field to align the dipoles in

TABLE 12.1: Properties of selected ferroelectric compounds.

Material	Chemical Formula	T_c (K)	P (μCcm^{-2})	at T (K)
KDP	KH_2PO_4	123	4.75	96
Deuterated KDP	KD_2PO_4	213	4.85	180
Barium Titanate	$BaTiO_3$	393	26.0	300
Lead Titanate	$PbTiO_3$	763	> 50	300
Strontium Titanate	$SrTiO_3$	32	3.0	4.2
Potassium Niobate	$KNbO_3$	710	30.0	523

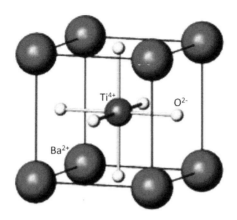

FIGURE 12.8: Cubic perovskite structure of $BaTiO_3$.

a particular orientation and cooling under an applied electric field. After such treatment, reversal of the polarization requires a substantial electric field. At the Curie temperature, the lattice expands so that the cubic configuration becomes the stable phase and the material loses its spontaneous polarization and becomes paraelectric.

The second-order ferroelectric transition can be modelled using the Curie - Weiss theory. Assuming a local field that is proportional to the polarization of the form: $\varepsilon_{Loc} = \varepsilon + \gamma P/\epsilon_0$, where γ is a constant of proportionality, for $T > T_c$, we find $P = Np^2 \varepsilon_{Loc}/k_B T$ and the susceptibility can be expressed as:

$$\chi_e = \frac{P}{\epsilon_0 \varepsilon} = \frac{Np^2 \varepsilon_{Loc}/\epsilon_0 k_B T}{1 - \gamma Np^2 \varepsilon_{Loc}/\epsilon_0 k_B T} = \frac{T_c/\gamma T}{1 - T_c/T} = \frac{T_c/\gamma}{T - T_c} \quad (12.68)$$

where $T_c = \gamma Np^2/\epsilon_0 k_B$. The above equation has the form of the Curie - Weiss law, Equation (12.67), from which we can determine the Curie constant as $C = Np^2/\epsilon_0 k_B$. We follow a similar approach to that used in the analysis of the magnetic moment to express the polarization as:

$$P = Np \tanh\left(\frac{p\varepsilon_{Loc}}{\epsilon_0 k_B T}\right) = Np \tanh\left(\frac{p\gamma P}{\epsilon_0 k_B T}\right) \quad (12.69)$$

This condition can be used to solve for P. Given that the tanh function cannot exceed unity, as $T \to T_c$ we have $\gamma P p / \epsilon_0 k_B T_c = P / Np$. From this we can establish that:

$$\gamma = \frac{\epsilon_0 k_B T_c}{Np^2} \tag{12.70}$$

such that:

$$\frac{P}{Np} = \frac{P}{P_{sat}} = \tanh\left(\frac{P}{P_{sat}} \frac{T_c}{T}\right) \tag{12.71}$$

This allows the temperature dependence of the polarization to be expressed as:

$$\frac{P}{P_{sat}} = \frac{T}{T_c} \operatorname{arctanh}\left(\frac{P}{P_{sat}}\right) \tag{12.72}$$

This variation is illustrated in Figure 12.9.

As a further analogy with magnetic materials, there also exist antiferroelectrics in which there can exist two directions for a dipole state, which are antiparallel and thus cancel their net dipole moment.

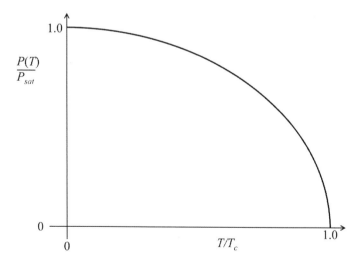

FIGURE 12.9: Temperature dependence of the polarization according to the Curie - Weiss law for spontaneous polarization.

Examples of materials which exhibit an antiferroelectric dipole structure are the oxides: NH_4PO_4, $PdZrPO_8$ and $NaNbO_3$.

12.7 PIEZOELECTRICS

When certain crystalline solids are mechanically deformed or elastically strained, they can become electrically polarized due to the displacement of the centers of positive and negative charge. Conversely, when an electric field is applied to these materials in specific orientations, the crystal becomes strained in a process called *electrostriction*. Such materials are called *piezoelectrics*. These properties can only occur in crystals which lack a center of inversion. Given these attributes, these materials are capable of converting mechanical energy into electrical energy and vice versa. Piezoelectric materials are to be found in many applications in which micropositioning is of importance and has numerous technological applications, such as pressure and strain gauges, as well as thermostats and the spark generators found in cookers and lighters. In addition to this they are to be found in applications for precision actuators, such as in scanning tunneling microscopes and translation stages.

Of the materials that exhibit piezoelectric behavior, we find barium titanate, Rochelle salt and lead zirconium titanate (PZT), which have large piezoelectric coefficients. The response of a piezoelectric material is, to a first approximation, linear between the electric and mechanical displacements and arises from the linear electrostriction related to Hooke's law.

12.8 MULTIFERROIC MATERIALS

An altogether more exotic and indeed rarer class of material is the so-called *multiferroic* and possess both two or more of the ferroic order parameters. That is, has ferroelectric, ferromagnetic and/or ferroelastic ordering. Magnetoelectric coupling refers to the induced magnetization from an applied electric field or the electric polarization arising from an external magnetic field. While the

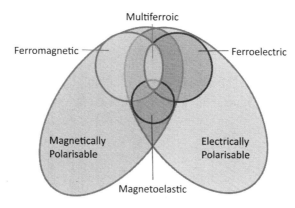

FIGURE 12.10: Relationships between multiferroic and magnetoelastic materials.

existence of electrically and magnetically polarizable materials is relatively rare, the occurrence of both types of polarizability in the same material is exceedingly rare. The complexity of the different types of order is illustrated in Figure 12.10.

Multiferroics are grouped into two types. In the type-I multiferroic, the ferroelectricity and magnetism have different sources and appear to be independent of one another, though some weak coupling does exist. In general the ferroelectricity persists to higher temperatures than the ferromagnetism. The most studied, and hence most popular, of the multiferroic materials is the perovskite crystal $BiFeO_3$. The multiferroic properties are quite weak, though can be enhanced by making them in thin film form. This is a good example of a type-I multiferroic. On the other hand, the type-II multiferroic has a much stronger coupling in which the magnetism causes the ferroelectricity. In such a case, the ferroelectricity disappears only in the magnetically ordered state of the crystal. The polarization strength in the type-II multiferroic are found to be relatively weak. One example of this class of multiferroic is the oxide, $TbMnO_3$.

12.9 OPTICAL PROPERTIES OF SOLIDS

Having introduced the dielectric function, we are now in a position to consider the optical properties of insulating materials. In this section, we will give a brief overview in this section of some of the more important considerations based on the previous sections and also indicate how the optical properties of metallic systems differ from those of non-conductive media.

12.9.1 The Wave Equation

It is a simple matter to derive the wave equation for electromagnetic waves from the Maxwell equations. We can start by taking the curl of Equation (12.12):

$$\nabla \times (\nabla \times \boldsymbol{\varepsilon}) = -\nabla \times \frac{\partial \mathbf{B}}{\partial t} \tag{12.73}$$

from which we obtain:

$$\nabla(\nabla \cdot \boldsymbol{\varepsilon}) - \nabla^2 \boldsymbol{\varepsilon} = -\mu_0 \nabla \times \frac{\partial \mathbf{H}}{\partial t} \tag{12.74}$$

We now take the time derivative of Equation (12.14) to obtain:

$$\nabla \times \frac{\partial \mathbf{H}}{\partial t} = \frac{\partial}{\partial t}\left(\frac{\partial \mathbf{D}}{\partial t}\right) = \frac{\partial^2 \mathbf{D}}{\partial t^2} = \epsilon_0 \frac{\partial^2 \boldsymbol{\varepsilon}}{\partial t^2} \tag{12.75}$$

Given $\nabla(\nabla \cdot \boldsymbol{\varepsilon}) = 0$, from Equations (12.74) and (12.75) we find:

$$\nabla^2 \boldsymbol{\varepsilon} = \mu_0 \epsilon_0 \frac{\partial^2 \boldsymbol{\varepsilon}}{\partial t^2} \tag{12.76}$$

This is the three-dimensional wave equation and has a counterpart which can be expressed in terms of the magnetic field **H** in a similar manner. We recognize that the propagation of the electromagnetic wave in vacuum occurs at a velocity given by:

$$c = \frac{1}{\sqrt{\mu_0 \epsilon_0}} \tag{12.77}$$

where we note that c is the velocity of light in free space. However, for the propagation of the electromagnetic wave in a non-conductive medium, the velocity will be modified to:

$$v = \frac{1}{\sqrt{\mu \epsilon}} = \frac{1}{\sqrt{\mu_0 \epsilon_0 \epsilon_r}} \tag{12.78}$$

Dielectric Materials • 475

where for a non-magnetic material we write: $\mu = \mu_0$. We can now express the refractive index of the medium, which is defined as the ratio of the velocity of the electromagnetic wave in free space and the medium, as:

$$n = \frac{c}{v} = \sqrt{\frac{\mu_0 \epsilon_0 \epsilon_r}{\mu_0 \epsilon_0}} = \sqrt{\epsilon_r} \qquad (12.79)$$

12.9.2 Transmission and Reflection Coefficients

An important aspect of the electromagnetic wave is the orthogonality of the electric and magnetic components, which can be quite easily shown and is incorporated in the formulation of the Maxwell equations. We can simplify the problem by choosing the direction of propagation of the electromagnetic wave, which we will do so by using the x-direction. This then allows us to express the various components with respect to this direction and then evaluate the specific components of interest. From Maxwell Equations (12.12) and (12.14) we can express the variation of the y-component of the electric field and the corresponding z-component of the magnetic field varying in the x-direction, which gives:

$$\frac{\partial \varepsilon_y}{\partial x} = -\frac{\partial B_z}{\partial t} = -\mu \frac{\partial H_z}{\partial t} \qquad (12.80)$$

and

$$\frac{\partial H_z}{\partial x} = \frac{\partial D_y}{\partial t} = -\epsilon \frac{\partial \varepsilon_y}{\partial t} \qquad (12.81)$$

Taking the derivatives of the solutions to the wave equation, $\varepsilon_y = \varepsilon_0 e^{i(kx+\omega t)}$ and $H_z = H_0 e^{i(kx+\omega t)}$ we obtain:

$$ik\varepsilon_y = -i\omega\mu H_z \quad \text{or} \quad ik\varepsilon_0 = -i\omega\mu H_0 \qquad (12.82)$$

$$-ikH_z = i\omega\epsilon\varepsilon_y \quad \text{or} \quad -ikH_0 = i\omega\epsilon\varepsilon_0 \qquad (12.83)$$

Multiplying Equation (12.82) by $\epsilon\varepsilon_0$ and Equation (12.83) by μH_0 and adding the results gives:

$$\epsilon\varepsilon_0^2 = \mu H_0^2 \qquad (12.84)$$

or alternatively we can write:

$$\frac{|\boldsymbol{\varepsilon}|}{|\mathbf{H}|} = \sqrt{\frac{\mu}{\epsilon}} \tag{12.85}$$

The energy flow for the electromagnetic wave is represented by the Poynting vector, defined as: $\mathbf{S} = \boldsymbol{\varepsilon} \times \mathbf{H}$. Using the above expression we can thus write:

$$|\mathbf{S}| = \varepsilon^2 \sqrt{\frac{\epsilon}{\mu}} = H^2 \sqrt{\frac{\mu}{\epsilon}} \tag{12.86}$$

Considering an electromagnetic wave incident (normally) from vacuum to a flat surface of a non-conducting medium, with material constants, ϵ and μ. Given the conservation laws applying to ε and H, the incident (i), reflected (r) and transmitted (t) components must adhere to the relations:

$$\varepsilon_i = \varepsilon_r + \varepsilon_t \quad \text{and} \quad H_i = H_r + H_t \tag{12.87}$$

It is now a simple matter to write the following expression using the above equations:

$$(\varepsilon_i + \varepsilon_r)\sqrt{\frac{\epsilon_0}{\mu_0}} = \varepsilon_t \sqrt{\frac{\epsilon}{\mu}} = (\varepsilon_i - \varepsilon_r)\sqrt{\frac{\epsilon}{\mu}} \tag{12.88}$$

Taking the case of a non-magnetic medium ($\mu = \mu_0$) we obtain:

$$\frac{\varepsilon_r}{\varepsilon_i} = \frac{\sqrt{\epsilon/\mu} - \sqrt{\epsilon/\mu_0}}{\sqrt{\epsilon/\mu} + \sqrt{\epsilon/\mu_0}} = \frac{\sqrt{\epsilon} - \sqrt{\epsilon}}{\sqrt{\epsilon} + \sqrt{\epsilon}} = \frac{n-1}{n+1} \tag{12.89}$$

The intensity for any component can be expressed as $|\varepsilon|^2$, which means that the reflection coefficient can be expressed as:

$$R = \left|\frac{\varepsilon_r}{\varepsilon_i}\right|^2 = \left|\frac{n-1}{n+1}\right|^2 = \frac{(n-1)^2}{(n-1)^2} \tag{12.90}$$

The generalized expression for the reflection coefficient between two media, with refractive indices n_1 and n_2, takes the form:

$$R = \left|\frac{n_1 - n_2}{n_1 + n_2}\right|^2 \tag{12.91}$$

Since the intensity of the reflected and transmitted beams must equal to that of the incident beam, we can express the relation for the transmission coefficient as:

$$T = 1 - R = 1 - \frac{(n-1)^2}{(n+1)^2} = \frac{4n_1 n_2}{(n_1+n_2)^2} \qquad (12.92)$$

Introducing the complex refractive index, defined as: $\tilde{n} = N + iK$, where N is the real part and K the imaginary part. These are generally referred to as the optical constants, though they do depend on frequency. Using this, we can write the complex form of the reflection coefficient as:

$$\tilde{R} = \left(\frac{\tilde{n}_1 - \tilde{n}_2}{\tilde{n}_1 + \tilde{n}_2}\right)\left(\frac{\tilde{n}_1^* - \tilde{n}_2^*}{\tilde{n}_1^* + \tilde{n}_2^*}\right) = \frac{N_1^2 + N_2^2 + K_1^2 + K_2^2 - 2N_1 N_2 - 2K_1 K_2}{N_1^2 + N_2^2 + K_1^2 + K_2^2 - 2N_1 N_2 - 2K_1 K_2} \qquad (12.93)$$

the symbols marked with an asterisk (*) denote the complex conjugate. For the case where one of the media is air, this expression becomes:

$$\tilde{R} = \frac{(N-1)^2 + K^2}{(N+1)^2 + K^2} \qquad (12.94)$$

12.9.3 Absorption of Electromagnetic Waves

The refractive index and the wave vector for the propagating electromagnetic wave are related by:

$$k = \frac{\omega}{v} = \omega\sqrt{\epsilon\mu_0} = \frac{\omega n}{c} \qquad (12.95)$$

The corresponding complex relations take the form:

$$\tilde{k} = \frac{\omega\sqrt{\epsilon\mu_0}}{c\sqrt{\epsilon_0\mu_0}} = \frac{\omega\tilde{n}}{c} = \frac{\omega(N+iK)}{c} \qquad (12.96)$$

Using our previous functional form of the solution to the wave equation we write:

$$\mathcal{E}_y = \mathcal{E}_0 e^{i(\tilde{k}x + \omega t)} = \mathcal{E}_0 e^{i[\omega(N+iK)x/c + \omega t]} = \mathcal{E}_0 e^{i(\omega Nx/c + \omega t)} e^{-\omega Kx/c} \qquad (12.97)$$

We now see that the final exponential term acts as a damping component and we see that this arises from the imaginary part of the complex refractive index. Therefore the propagating wave will lose energy as a function of distance into the medium and for this reason the constant K is referred to as the *extinction coefficient*. Since the intensity is proportional to $\varepsilon\varepsilon^*$, we note that the intensity of the electromagnetic wave will decay according to *Beer's law*:

$$I(x) = I_0 e^{-2\omega K x/c} = I_0 e^{-\alpha x} \qquad (12.98)$$

where the absorption coefficient takes the form:

$$\alpha = \frac{2\omega K}{c} \qquad (12.99)$$

12.9.4 Optical Properties of Dielectrics

The insulating properties of dielectrics have an important influence on their optical behavior. They tend to be transparent in the visible region of the electromagnetic spectrum in their single crystalline and amorphous state. The latter being a glassy solid. In the polycrystalline phase, the solids tend to be opaque due to multiple scattering processes at grain boundaries. However, if the wavelength of the radiation is much greater than the average grain size, the material becomes transparent.

We saw earlier that the refractive index of a material is related to the dielectric constant via the expression:

$$\tilde{n}(\omega) = N(\omega) + iK(\omega) = \sqrt{\epsilon_r(\omega)} \qquad (12.100)$$

This expression can be used to determine the optical properties of dielectrics. We can apply the dielectric function of Equation (12.28) and the polarizability, Equation (12.43). This is valid for the visible and near ultra-violet region of the electromagnetic spectrum. As we move into the infra-red region, the absorption can be evaluated through the dielectric function given in Equation (12.61).

We previously considered the optical modes near $k = 0$. We can extend this to the region of small k to explore the dispersion relation where photons and phonons can couple. Equation (12.61) can be rewritten as:

$$\epsilon_r(\omega) = \frac{\epsilon(0)\omega_T^2 - \epsilon(\infty)\omega^2}{\omega_T^2 - \omega^2} \qquad (12.101)$$

Substituting from the LST relation, Equation (12.63), we obtain:

$$\epsilon_r(\omega) = \epsilon(\infty)\left(\frac{\omega_L^2 - \omega^2}{\omega_T^2 - \omega^2}\right) \qquad (12.102)$$

We now obtain the dispersion relation for the photon - phonon coupling, where using Equation (12.65), we find the quadratic relation in ω^2:

$$\omega^4 - \left(\omega_L^2 + \frac{k^2 c^2}{\epsilon(\infty)}\right)\omega^2 + \frac{k^2 c^2}{\epsilon(\infty)}\omega_T^2 = 0 \qquad (12.103)$$

The form of the dispersion relation is illustrated in Figure 12.6. The two solutions for ω^2 give the transverse and longitudinal phonon bands.

In addition to this, there are a number of other optical absorption phenomena which also affect the optical properties of dielectric media. We have already mentioned the effects of grains and grain boundaries on the transmission of light in these materials. Scattering losses to the transmission of radiation can also occur via inclusions or other defects in the solid. For example, small metallic particles can give color to glasses and Rayleigh scattering can also be a factor. In Chapter 4, we saw that point defects in ionic crystals give rise to color centers (see Section 4.2.4). Other impurities, such as Cr in what would otherwise be a colorless transparent crystal of Al_2O_3 give the red color to rubies, while Ni produces a blue hue to the sapphire crystal.

If the photon energy is equal to or greater than the band gap of the insulator (or semiconductor), the photon can be absorbed, transferring its energy to produce an electron - hole pair. While it is customary to consider that the material is transparent to radiation of frequencies below the band gap energy, it is possible for photons with energy below the band gap energy to raise electrons from the valence band and instead of "entering" the conduction band, the electron is attracted to the hole state forming a bound pair called an *exciton*. These can be described by hydrogen like states with energies just below the cut-off point for absorption, usually appearing as a peak. The exciton is a quasi-particle which has no charge and can move through the crystal, eventually decaying to the ground state with the emission of a photon.

Cubic crystals have three equivalent principal axes which interact with light in the same way. By this we mean that light passing through the crystal will be refracted in an identical manner without the alteration of its polarization state irrespective of its orientation. Such a crystal is said to be optically isotropic. In crystal systems with lower symmetry, this will not be the case and the medium is optically anisotropic. Certain axes, such as the c-axis in hexagonal crystals, have a unique symmetry and are referred to as the *optical axis*. Light which propagates along such axes will act in the same way as it does in an isotropic material and is called the *ordinary ray*. However, light which enters the crystal at some angle with respect to the optical axis will be split into two rays; the ordinary ray, with refractive index n_O, and the extraordinary ray, which has refractive index n_E. The difference between these two refractive indices defines the *birefringence* of the crystal. The ordinary and extraordinary rays propagate with the electric fields vibrating in different planes and travel at different speeds through the crystal. The light emerging from such a crystal can have different polarization states. These properties are exploited, for example, in the Nicol prism to select specific polarizations.

In Equation (12.8) we gave the polarization as a linear function of the electric field. However, this is a simplified approximation and can be more generally expressed as a nonlinear function of the form:

$$\mathbf{P} = \epsilon_0 [\chi_e^{(1)} \boldsymbol{\varepsilon} + \chi_e^{(2)} \boldsymbol{\varepsilon}^2 + \chi_e^{(3)} \boldsymbol{\varepsilon}^3 + \ldots] \quad (12.104)$$

where $\chi_e^{(n)}(n>1)$ are the nonlinear susceptibilities. These higher order terms can give rise to some interesting phenomena, such as frequency doubling ($\chi_e^{(2)}$) or second harmonic generation. Other higher order effects include *three-* and *four-wave mixing*, which rely on the $\chi_e^{(2)}$ and $\chi_e^{(3)}$ terms, respectively. Such effects allow the up and down conversion of photons and are frequently used in laser devices.

Dielectric Materials • **481**

While the majority of scattering events between light and solids is elastic (or Rayleigh scattering), inelastic processes can occur, whereby there is a transfer of energy from the photon to the crystal or vice versa. This type of scattering is usually called *Raman scattering*, where the transfer of energy from the photon to the crystal can give rise to the excitation of vibrational modes or phonons. This means that the energy (frequency) of the photon will be reduced in a process called *Stokes scattering*. The inverse process can occur in a crystal in which there is a reasonable population of phonons (at higher temperatures).

The transfer of energy from phonons to the photon increases the photon energy (frequency) in the *anti-Stokes scattering* process. In 1928, C. V. Raman discovered this effect, for which he was awarded the 1930 Nobel Prize. Raman spectroscopy, which evolved from this discovery, is an important diagnostic tool and can be used to evaluate the phonon spectrum and remotely calculate the temperature of a solid from this.

The existence of free electrons in a material can radically change the optical properties of a solid. The electrons in a metal will respond to a time varying electric field, but can only follow the oscillations up to a certain cut-off point defined by the relaxation time. This allows us to define the plasma frequency of the metal. We have discussed the optical properties of metals in Chapter 6, Section 6.10.

12.10 SUMMARY

Since dielectric materials are electrically insulating, their response to an applied electric field will not result in the conduction of charge carriers. This does not mean they not interesting. As we have seen in this chapter, there are many important consequences of their physical properties that provide material solutions to specific technological problems. Of particular importance are the applications for charge storage in capacitors and their optical properties, which derive from the consideration of the frequency response to a time varying electric field.

The production of electric dipole moments under the application of an electric field gives rise to polarization phenomena and can be categorized as electron, ionic and dipolar. These contributions have very different frequency responses and in turn will affect the dielectric function (constant) in different regions of the electromagnetic spectrum for example. At the high end of the spectrum the electronic polarization arises from the electron orbitals being displaced from the nuclear charge by an oscillating electric field of frequencies in the ultraviolet range, where the dielectric constant is designated as $\epsilon(\infty)$. The ionic polarization, which occurs in ionic and covalently bonded solids, depends on the charge separation between different atoms in the crystal. The characteristic response is usually observed in the infrared region of the electromagnetic spectrum, where the dielectric constant is denoted by $\epsilon(0)$.

In systems where the dipoles are free to move under the influence of an applied electric field, a paraelectric response can be encountered. The alignment will occur against thermal fluctuations, which tend to randomize the dipoles. When the dipoles are fixed in the solid, a spontaneous polarization can occur. Such a solid is called a ferroelectric and is analogous in physical description to a ferromagnetic material. The analogy is fairly robust between the electric polarization in the formed and the magnetization in the latter. Indeed both can be described with a Curie - Weiss law for their electric and magnetic susceptibilities, respectively and have ordering temperatures called the Curie temperature. In pyroelectrics, the dipole moment is temperature dependent. In systems without a center of symmetry, a distortion of the lattice can produce a potential difference due to a non-compensated charge separation, while an applied electric field can conversely cause a deformation of the crystal lattice. These are piezoelectrics and have a number of important applications in sensor and actuator devices.

Materials which exhibit both ferroelectric and ferromagnetic properties (and/or ferroelasticity) are called multiferroics and exist in two forms, referred to as type I and type II. In the former, the origin of dipole and magnetic moments are largely independent and the ordering temperatures are different. In the latter, type II materials, the magnetism of the solid produces a ferroelectric ordering and both have the same ordering temperature.

The optical properties of dielectric materials are largely determined by their dielectric constants which vary as a function of the frequency of electromagnetic radiation. However, a number of more exotic behaviors can be observed in these materials and depend on the crystalline order of the solid. Effects such as nonlinear susceptibility and birefringence can produce some interesting responses for the interaction of light with matter. The transmission and reflection coefficients of the solids can be described with the aid of Maxwell's equations and the refractive index.

REFERENCES AND FURTHER READING

Basic texts

- J. S. Blakemore, *Solid State Physics*, Cambridge University Press, Cambridge (1985)
- H. P. Myers, *Introductory Solid State Physics*, Taylor and Francis, London (1998)

Advanced texts

- H. Ibach and H. Lüth, *Solid State Physics: An Introduction to Principles of Materials Science*, Springer-Verlag, Berlin (1996)
- N. W. Ashcroft and N. D. Mermin, *Solid State Physics*, Saunders College, Philadelphia (1976)
- R. J. Naumann, *Introduction to the Physics and Chemistry of Materials*, CRC Press, Baton Rouge (2009)
- M. E. Lines and A. M. Glass, *Principles and Applications of Ferroelectrics and Related Materials*, Oxford University Press, Oxford (2001)

EXERCISES

Q1. Compare the electrical capacitance of a parallel plate capacitor, with plate area $5\,\text{mm}^2$ and separation $60\,\mu\text{m}$, in vacuum and with a dielectric between the plates with a relative dielectric constant of 24. Explain this result.

Q2. Demonstrate the form of Equation (12.21).

Q3. Derive fully Equation (12.36).

Q4. Show the steps involved in the formulation of the Kramers - Kronig relations.

Q5. Explain the origin of the terms in the equation of motion as expressed in (12.39). Further show that Equation (12.41) is a valid solution of the equation of motion.

Q6. In Section 12.5.2, we discussed the modes of vibration for a diatomic crystal, whose ions have masses M and m. Show that the reduced mass for this system can be expressed as:

$$\mu = \frac{Mm}{M+m} \tag{12.105}$$

Q7. Derive the Lyddane - Sachs - Teller (LST) relation.

Q8. Use the Maxwell equations to derive the wave equation in terms of the magnetic field component of an electromagnetic wave. (c.f. Equation (12.76).)

Q9. Evaluate the reflectivity of an electromagnetic wave perpendicularly incident from the air side of an interface with silicon. (Note: you will need to look up the dielectric constant of Si.)

Q10. Use the complex form of the refractive index to obtain Equation (12.94). What is the physical significance of the imaginary component of the refractive index?

Q11. Evaluate the limits of the reststrahlen band for GaAs, where $\epsilon(0) = 12.4$, $\epsilon(\infty) = 10.9$ and $\nu_L = 8.5\,\text{THz}$.

NOTE

[1] Note that the refractive index is related to the dielectric constant via the relation: $n = \sqrt{\epsilon(\omega)}$. We will discuss the optical properties of solids in Section 12.9, where we will derive some of the more important relations used here.

CHAPTER 13

NANOTECHNOLOGIES AND NANOPHYSICS

"You have enemies? Good. That means you've stood up for something, sometime in your life."
—Winston Churchill

"The function of education is to teach one to think intensively and to think critically. Intelligence plus character - that is the goal of true education."
—Martin Luther King, Jr.

"I'm getting worried about the cats, they've been looking at me funny. I think they have been reading Skinner again..."
—Hank Kahney

13.1 INTRODUCTION

Nanotechnologies refer to a broad range of scientific disciplines which study the behavior of objects of nanometric dimensions. The exact size of the object, which is the subject of nanotechnologies, is rather subjective and depends on the physical property that we are considering and how the size and shape of the object behaves on characteristic length scales. As such it is better to use a flexible definition, at least to start with. In broad terms, we can generally consider objects with spatial dimensions from 1 to a few 100s of nm to form the domain of nanotechnologies.

In terms of the physical properties of materials, the reduction of the size of an object leads to the deviation of the bulk properties and depends principally on surface and confinement effects. In this book, we have been considering the relationship between the physical properties of a material with respect to its crystalline order and chemical composition. In this regard, the science of nanotechnologies can be seen as an extension of solid state physics and rightly belongs in a book of this nature. In the early chapters, we have seen that the physical properties of solids depends intimately in the crystalline structure and atomic nature of the material. That is, the organization or arrangement of the atoms. Also the elements within the structure are of vital importance since they can provide electrons to the system, and thus controls interatomic forces, and all the known physical attributes of the solids, including mechanical, electrical, optical, and magnetic properties.

So we should now ask, how does the size of a solid object affect its physical properties? This is one of the central questions in the subject of nanotechnologies and one which will be addressed in this chapter. As a first basic level reply, we can consider two fundamental aspects of a nano-object: surface effects and confinement effects. In the former we need to consider how the surface can modify the physical properties of a material, while in the latter it is the restriction of the space available to electrons which give rise to quantum phenomena that can alter the way a material can behave. This all sounds rather complex and abstract, but we will clarify these issues to give a more satisfactory explanation in the following sections.

Nanotechnologies have attracted much media attention in recent years, capturing the imagination of the wider public domain. It is rare that topics related to solid state physics reach such a broad spectrum of the public in such a prolonged and profound manner. Part of the reason is the extraordinary new opportunities that are now available to society and extends well beyond the realms of solids state physics and physics itself. The unprecedented position of science today is now following very different paths to those traditionally available. We are seeing new horizons opening up with interdisciplinary research from the cross fertilization between many branches of scientific study. Today we can see biologists working with computer scientists, physicists researching with medics, and so on. The manipulation and control of tiny objects is indeed allowing us to find new ways to study nature and gives many options for future development. Novel nano-solutions can be seen popping up in all areas of science. New topics are emerging as major areas of research. New courses are developing into new career opportunities for future scientists. As an example we can consider the subject of *Nanomedicine*, a branch of science in which the physics of nanomaterials is just as important as the consideration of the biochemistry of these objects. Nanomedicine and nanobiology are subjects in their own right. This is no mere fancy speculation, billions of pounds, dollars, euros, and yen are being invested in this technology. Governments do not like spending public funds without some definite hope of long term gains. Many large companies and corporations are also pumping in large sums of cash into research in nanotechnologies.

So where did all this begin? It didn't just happen overnight. Read any book or article on nanotechnology and chances are the name of Richard Feynman will come up. Feynman was a visionary physicist who in 1959 delivered a talk to the American Physical Society entitled "There is Plenty of Room at the Bottom", in which he envisioned what are the consequences of measuring and manipulating objects on the nanoscale. What we call nanoscience has been around for a number of decades. Much of the technologies emerged from the electronics industry and the preparation of thin films for electronic and optoelectronic devices, such as transistors and photovoltaic cells. Indeed an important advance was the development of the integrated circuit (IC) and its subsequent miniaturization; this was one of the driving forces behind nanotechnologies. The IC was the brainchild of Jack Kilby, an employee of Texas Instruments, who came up with the idea of integrating the components of a circuit on to a monolithic block of a semiconductor. The patent was filed in February of 1959, though work was begun in the previous year. Kilby was finally awarded a Nobel Prize for his invention in 2000. The idea of incorporating several components onto the same block of semiconductor was also developed by Robert Noyce some six months after Kilby. Noyce went on to co-found the Intel Corporation in 1968 and was popularly known as the "Mayor of Silicon Valley".

The semiconductor industry was at the center of the microelectronics revolution in the 1960s and 1970s, and the development of miniaturization technologies. Once the IC had been established, it soon became the dominant technology, with the size reduction following the so-called *Moore's law*, in which size reduction was found to follow a logarithmic law, which holds to this day. Another way of measuring this development is by stating that the number of transistors on integrated circuits doubles approximately every two years. The Moore's law as a concept, was named after another co-founder of Intel, Gordon E. Moore, who wanted to know how the semiconductor industry would develop over the coming years. The general idea of Moore's law is also found in other technologies, such as computer processing speed, memory capacity, sensors, and even the number and size of pixels in digital cameras. All of this is related to the development of the technologies for the production of devices. Again the semiconductor industry lead the way.

The technologies developed in the 1960s with vacuum technology and thin film development were of vital importance as was the evolution of photolithographic techniques. This latter is of enormous importance in the production of integrated circuits as it defines a reliable technique for the repeated production of device components. These methods have been subsequently applied to the development and fabrication of other types of devices. Thin film technologies are of critical importance to research in solid state physics, where they are routinely used for the development of most forms of materials, including semiconductors, magnetic materials, and ferroelectrics. The branch of physics called surface science would probably not exist without it, going hand-in-hand with thin film preparation techniques. Devices such as the electron microscope, the scanning tunneling microscope (STM) and electron spectroscopies all rely on vacuum technologies. At the time of writing (2013), the STM was used for making the world's smallest movie, where atoms were manipulated to make a sequence of images.

Quantum mechanics is the domain of physics at the atomic level. As we consider objects of increasing size quantum behavior will give way to classical physics. Quantum mechanics rapidly becomes extremely difficult once we start dealing with objects with many atoms and we need to consider how we can describe a system as a function of its size. This is no simple task. We usually decide how to treat the physics of an object based on which is easier to model. The limits of scales are generally considered as the microscopic and macroscopic, where quantum and classical physics dominate. In the intermediate region we have the *mesoscopic* scale of objects and while having many atoms, are still subject to the laws of quantum mechanics and really marks the separation from the macroscopic phenomena that belong to the classical domain.

Clearly the subject of this chapter is very extensive and we will just provide a flavour of some of the principal physics involved and some of the main workings for device applications. In the following sections we will outline the principles of surface physics and why it is of importance in nanophysics and nanotechnologies. Subsequent sections will consider the quantum confinement effects mentioned earlier. This will be discussed with respect to how electrons confined in low dimensional systems, or nano-objects and how this can radically alter the electronic transport properties of materials. We will see how this is closely related to the optical properties of solids, where well-defined nanomaterials can be used to tune optical transitions for specific applications. We shall also consider how the magnetic properties of materials are subject to modification and manipulation by making low dimensional structures and coupling them together to produce new properties and devices. Unfortunately, the enormous number of developments in nanotechnologies is far too great for us to give anything but a brief overview of the subject and since we are considering the physical properties of solids we cannot go beyond this area in to other emerging technologies. We will limit our discussion to the physical principles which are at the root of nanotechnologies.

13.2 THE PHYSICS OF SURFACES

Let us commence by considering what a we mean by a physical surface. A surface can be envisaged as a terminating plane of a crystal. Being a surface, this atomic plane will lack the symmetry of the bulk as a whole plane of atoms (above the surface) is, in effect, missing. This will therefore mean that the normally occupied bonds to this plane are absent and will severely affect the atomic environment. Such missing bonds are referred to as *dangling bonds*. To accommodate this situation, the surface atoms will frequently re-arrange their relative positions, often giving rise to changes in symmetry and interatomic separation. The reorganization of the atomic symmetry is referred to as *surface reconstruction* and is often defined in symmetry terms with respect to the bulk symmetry of the same bulk atomic plane as that "exposed" on the surface. The region that we define as the surface is that region, which can extend several atomic planes from the exposed plane, where there is a modification of the normal bulk crystalline structure. This zone of the crystal is called the *selvedge*. The alteration of surface atomic planes is schematically illustrated in Figure 13.1.

With respect to the measured physical properties, the surface layer can be expected to behave rather differently to the normal bulk atoms, where for ultrathin films of a few monolayers. Indeed, in this case the number of surface atoms can be a very significant proportion of the atoms in the layer. In terms of measuring these properties, we require techniques which are surface sensitive, otherwise these

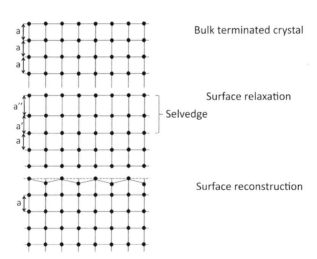

FIGURE 13.1: Modification of atomic planes at the surface of a crystal. The upper figure shows a bulk terminated crystal which is just terminated in a particular lattice plane with no modification of structure. The middle figure considers the relaxation of atomic planes in the direction perpendicular to the surface. The lower figure shows the periodic displacement of atoms at the surface plane, as occurring in surface reconstructions.

properties will be effectively invisible to us. There are some specific surface analytical tools which were developed and are used for this purpose, and we will mention some of these a little later. Alternatively, by making ultrathin structures, we essentially only have surface atoms and if we can ensure that our measurement is either element specific or limited to the range of the surface, we can gain surface sensitivity in this way. For example a thin magnetic film deposited on a nonmagnetic substrate can be measured using the ferromagnetic resonance technique, which while being a bulk technique, will only deliver a magnetic response from the thin film region of the sample. An ideal surface can be considered to be an atomically flat crystalline plane with no defects. However, as with bulk crystals, such perfect situations are rare and there are many of the types of defects that we discussed in Chapter 4. In addition to these we can have stepped and rough surfaces as well as foreign adsorbates. The strength of the interaction between the surface and an adsorbate will depend on whether there is an exchange of electrons. When this is the case, the strong bond formed is termed *chemisorption*, for weaker bonds, such as with van der Waals forces, the bonding is termed *physisorption*. Here no electrons are exchanged between neighboring atoms, but their states are modified and give rise to interatomic interactions.

The study of surfaces requires ultra-high vacuum (UHV) conditions. Without this the sample surface would rapidly become coated in atoms from the atmosphere and mask any surface effects that might be of interest. UHV provides the necessary conditions to maintain a clean surface for the periods of time required to perform measurements and for the deposition of thin films. Thin film technologies are of enormous interest for electronics, optical, and magnetic devices and therefore the considerations of surface properties are an important aspect of these developments. To illustrate the point, we can consider the atoms or molecules of a gas whose behavior depend principally on pressure and temperature. The density of particles in a certain volume can be expressed as:

$$n = \frac{p}{k_B T} \quad (13.1)$$

At room temperature we can write: $n = 2.5 \times 10^{20} p$ in units of m^{-3} with pressure in Pa, or $n = 2.5 \times 10^{22} p$ in units of m^{-3}, with pressure in mbar [1]. The mean free path for collisions between molecules takes the form:

$$\lambda = \frac{1}{n\sqrt{2}\pi r^2} = \frac{1}{\sqrt{2}\pi r^2} \frac{k_B T}{p} \tag{13.2}$$

In the case of N_2 molecules at room temperature, we find $\lambda = 6.6 \times 10^{-3}/p$ in units of m with pressure in Pa, or $\lambda = 6.6 \times 10^{-3}/p$ in units of cm with pressure in mbar. We can also evaluate the rate of impact of the atoms/molecules of the gas with the surface using the following expression:

$$J = \frac{p}{\sqrt{2\pi m k_B T}} = \frac{p N_A}{\sqrt{2\pi M k_B T}} \tag{13.3}$$

This can be evaluated for the case of N_2 molecules at room temperature, from which we can write: $J = 2.9 \times 10^{22} p$ in units of $m^{-2}s^{-1}$ with pressure in Pa, or $J = 2.9 \times 10^{20} p$ in units of $cm^{-2}s^{-1}$ with pressure in mbar. We can illustrate these principal parameters as a function of pressure, as shown in Table 13.1.

The time taken to completely cover the surface with a monolayer of residual gas, assuming a sticking coefficient of 1 (i.e., all atoms that impinge on the surface remain there) is given by: $\tau \simeq 4 \times 10^{-6}/p$, in units of s. We see from Table 13.1 that it is only under UHV conditions that we have the required conditions for performing any reasonable analysis of a physical surface.

Vacuum systems have finite dimensions and the flow rates are limited by these physical dimensions. We can consider such situations through the *Knudsen number*, which we define as:

TABLE 13.1: Vacuum parameters as a function of pressure. (Data partially adapted from Chambers, Fitch and Halliday.)

p (mbar)	n (m^{-3})	λ	J (cm^{-2}s^{-1})	τ (s)
$10^3 = 1$ atm	2.5×10^{25}	6.6×10^{-6} cm	2.9×10^{23}	4×10^{-9}
1	2.5×10^{22}	6.6×10^{-3} cm	2.9×10^{20}	4×10^{-6}
10^{-3}	2.5×10^{19}	6.6 cm	2.9×10^{17}	4×10^{-3}
10^{-6} (HV)	2.5×10^{16}	66 m	2.9×10^{14}	4
10^{-10} (UHV)	2.5×10^{12}	660 km	2.9×10^{10}	4×10^4 (~ 11 hours)

$$K_n = \frac{\lambda}{D} \qquad (13.4)$$

where D is the diameter of the tubing in the vacuum system. This allows us to separate the pumping regimes in a vacuum system. When the Knudsen number $K_n < 1 (\lambda < D)$, there will be many collisions between the atoms and molecules of the gas and it behaves as a fluid. This means that the gas can be sucked away, due to the intermolecular collisions, using a pumping systems, such as a rotary pump. This regime is called *viscous flow*. Once the pressure decreases below a certain limit this will no longer occur and we have $K_n > 1 (\lambda > D)$. When this occurs the interactions between the gas atoms and molecules starts to become negligible and the interactions with the walls of the chamber become dominant. For $K_n \gg 1 (\lambda \gg D)$, there is no way the gas molecules can be pumped away using a traditional pump. This regime is called *molecular flow*. The only way to reduce the pressure in this situation is by removing the atoms and molecules using capture methods, such as with a turbo molecular pump or cryopump. There are in fact many types of vacuum pump whose use depends on the conditions required.

13.2.1 Surface Structure

Since the surface of a crystal is a two-dimensional periodic array, we only require two lattice vectors to define its structure. In doing this we will define the Bravais lattice for two dimensions. The definitions we described in Chapter 2 for crystallography are equally valid for surfaces, though are simplified since only two dimensions are required. The 14 Bravais lattices that occur in three dimensions are reduced to five in two dimensions. These are illustrated in Figure 13.2.

When a surface reconstructs, the surface symmetry is altered with respect to the bulk terminated plane. There are different notations to define the surface reconstruction, which are based on a comparison of the unreconstructed and reconstructed surface. The most commonly used is the *Wood's notation*, which we will outline here. We start by noting that the terminated plane has a specific crystallographic plane (hkl) and a chemical description of the element, S. We must now consider the new, reconstructed symmetry with respect to this surface. We can define translation vectors:

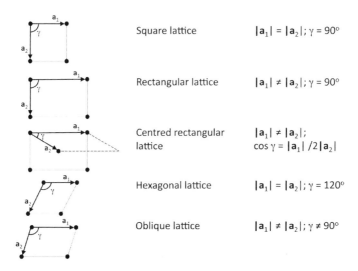

FIGURE 13.2: Bravais lattices for two dimensional crystals/surfaces.

$$|\mathbf{a}_{s1}| = m|\mathbf{a}_1| \quad \text{and} \quad |\mathbf{a}_{s1}| = n|\mathbf{a}_2| \quad (13.5)$$

where $\mathbf{a}_{sn}$ is the surface translation vector. We can also take into account any rotation that may be produced by indicating an angle ϕ. We can now refer to the superstructure surface (reconstruction) as:

$$S(hkl)K(m \times n) - R\phi \quad (13.6)$$

where K can be either "p" (primitive) or "c" (centered), according to the Bravais lattice type. If there is no symbol given, then it is implicitly a primitive type reconstruction. Some simple examples are illustrated in Figure 13.3.

There are several techniques available to observe the surface structure of crystals. The most commonly used are electron diffraction techniques such as LEED (low-energy electron diffraction) and RHEED (reflection high-energy electron diffraction), which were briefly outlined in Chapter 3, see Section 3.6.2. The scanning tunneling microscope (STM) can also be used to visualize the structure of surfaces. The STM is a rather unusual microscope in that we do not observe the surface directly via the incidence radiation. Rather, the image is constructed as a map of the electrical current between

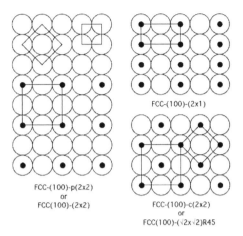

FIGURE 13.3: Surface reconstructions for an fcc (100) plane.

a very fine tip and the surface of the sample as a function of the position in the two directions of the surface. The technique is limited in that only conducting surfaces can be measured. As the name suggests, the current is a tunnel current and is therefore a quantum effect. The metallic tip is placed in close proximity to the surface, this allows the wave functions for electrons in the tip to overlap with those at the surface. The electrons are transferred by the quantum tunneling effect. The distance between the tip and the surface is typically in the region of $d = 5-10$ Å. Applying a bias voltage between tip and surface a tunnel current can flow across a gap, which takes a general form of the Fowler - Nordheim equation:

$$j = D(V)\frac{V}{d}e^{-A\phi_B^{1/2}d} \qquad (13.7)$$

here $D(V)$ represents the joint (tip - surface) density of states between tip and surface, A is a constant and ϕ_B is the effective barrier height. This equation shows that the tunnel current is a very sensitive function of the barrier width, d, and gives the STM techniques its exceptional depth resolution. A change of the barrier by 1 Å can change the current by an order of magnitude. The lateral spatial resolution also derives from the fact that around 90% of the current originates from the "last" atom on the tip and the surface atom closest to it. Resolutions of around 1 Å are attainable.

The STM acquisition can be performed either by maintaining the current constant and altering the height of the tip above the surface, with feedback to adjust the height controlled via a piezoelectric driver. This essentially means we adjust the tip height to maintain a constant gap between tip and surface. The measurement will then be that of height position, obtained from the feedback voltage from the piezoelectric driver, as a function of the xy position. Alternatively, the STM can be run in constant height mode and we map the tunnel current as a function of the xy position. In Figure 13.4, we show some STM images, illustrating the exceptional spatial (atomic) resolution. In the first image we can see the Si (111)-(7 × 7) surface reconstruction. The second image shows a ring of Fe adatoms on a Cu (111) surface. This illustrates that atoms can be manipulated with the tip, first they can be attracted to it, moved to a specific position and then released[2]. An important artifact is seen in the center of the ring, where we clearly see what looks to be a ripple pattern, much like that on water when a drop has fallen in. This demonstrates that the STM as a technique is sensitive to the local density of states at the surface, and should be a warning as to how we should interpret the STM images. Indeed, what we observe is not really an image of the surface atoms, but the electron density at the surface. The image in Figure 13.4 (b) should be thus interpreted as the standing electron wave due to the confinement of electrons in the ring structure.

In fact, this property is exploited in the scanning tunneling spectroscopy technique in which a measurement of the tunnel current is made as a function of the applied bias potential between tip and

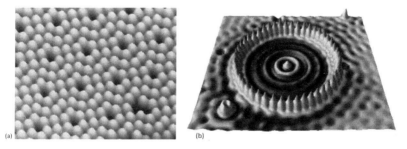

FIGURE 13.4: Scanning tunneling microscopy images of (a) the Si (111) - (7 × 7) reconstruction and (b) a ring of Fe atoms on a Cu (111) surface.[Image (a) reprinted with the kind permission of Professor Michael Trenary, University of Illinois at Chicago.]

surface. This contains information about the surface density of states as well as that of the tip itself. Forward and reverse bias potentials usually give rise to very different spectra due to the distinct nature of the initial and final states for the tunneling process between the tip and a surface.

The STM technique lead to the development of other so-called *scanning probe microscopies*, which are based broadly in the interaction of a tip with a sample surface. The most common examples are atomic force microscopy (AFM) and magnetic force microscopy (MFM). In the former, the electrostatic forces between a tip and the surface produce a deflection of a cantilever upon which the tip is mounted. A position sensor is used to detect this movement as a function of the tip position on the surface. The AFM technique is routinely used to provide surface roughness profiles in thin films. The latter technique, MFM, is essentially the same, where the tip is now a magnetic material and is used to probe the magnetic state of a magnetic film or surface and can be readily used to obtain magnetic domain images. The STM was designed and built in the early 1980s by Gerd Binnig and Heinrich Rohrer, who were awarded the Nobel Prize in Physics in 1986.[3]

13.2.2 Surface Composition and Excitation States

Once we have established the surface structure, we would like to know what is at the surface. The chemical composition of the surface can tell us a lot about the sample. Firstly, whether there is any contamination and the chemical composition of the surface itself. To measure this we require methods which are either element specific or allow us to distinguish between the various atomic species. Furthermore, we must be careful to only probe the surface and not the interior of the sample. To do this we can use one of several forms of *electron spectroscopy*.

The basic principles of electron spectroscopies rely on the excitation of electrons from the sample surface via the incidence of typically electron or x-ray beams. The excitation can produce the emission of characteristic energy electrons and x-rays from which it is possible to identify the atomic species from where it originated. This is possible since the electron states for each atom are slightly different and hence have very specific energies for the emitted radiations.

The bombardment of a surface with electrons or photons produces secondary electrons with energies from a few eV up to around 2 keV. The surface sensitivity of electron spectroscopy techniques arises from the fact that electrons of energies in this range are strongly scattered. This can be seen from the inelastic mean free path of electrons as a function of kinetic energy, which forms a universal graph and is largely independent of elemental species, see Figure 13.5. From this figure we see that the greatest surface sensitivity arises for energies in the range of 20 to 200 eV, where the inelastic mean free path is less than 10 Å or 1 nm. In this range of energies there is a strong interaction between electrons and atoms, and they therefore cannot penetrate deeply into the crystal beyond the surface region. At lower energies, they have insufficient energy to interact with the atoms in the solid, while at larger energies they have a lower scattering cross-section due to higher velocities and can penetrate further into the crystal.

There are many variations of electron spectroscopies, which partially depend on the incident (exciting) radiation and partially on the detection system used. The main techniques used for studying surfaces are Auger electron spectroscopy (AES), electron energy loss spectroscopy (EELS) and photoelectron spectroscopy. All of these methods analyze the electrons emitted from the sample after

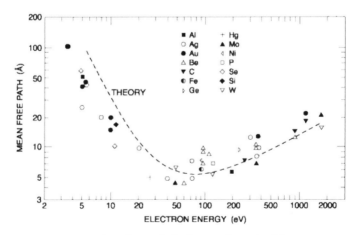

FIGURE 13.5: Universal curve: inelastic mean free path of electrons as a function of their energy taken from a variety of materials.

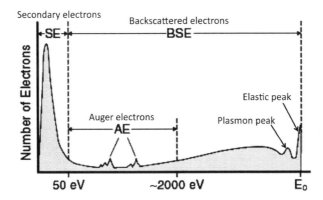

FIGURE 13.6: Illustration of the secondary electron spectrum, showing the various types of backscattered electron peaks from a solid.

excitation. The general form of the secondary electron spectrum is illustrated in Figure 13.6. The principal features are: (i) Sharp elastic peak at the primary electron beam energy, E_0, and corresponds to electrons from the primary beam being elastically backscattered to the detector; (ii) Plasmon peaks, which have a specific energy loss from the primary energy due to the excitation of plasmons and more precisely surface plasmons; (iii) Fixed energy peaks, which arise from the excitation and emission of Auger electrons; (iv) secondary electron peak, which is a broad peak due to electrons which under go multiple scattering processes within the solid and emerge with energies of less than 50 eV and originate from within 10 Å of the sample surface.

Probably the most common and useful of the electron spectroscopies is that of AES and is principally used as a chemical identification technique. The technique relies on the fact that characteristic electrons are emitted with energies specific to the elements, which is due to the fact that all atoms have slightly different energy levels for their electrons. This is due to the different nuclear charges and the screening effects for the various electron shells in the atom. The Auger process is illustrated in Figure 13.7, where an incident beam (electron or x-ray) excites and removes a core electron (for example, from the K shell). The hole is rapidly filled with an electron from a less tightly bound level, for example the L_2 level. Further de-excitation occurs via the emission of an electron from another, less tightly

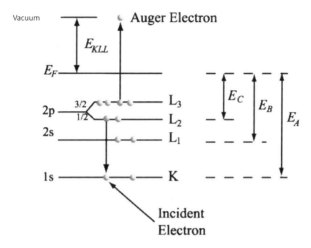

FIGURE 13.7: Illustration of the emission of an Auger electron from an atom. The process illustrated is that of the KL_2L_3 Auger emission.

bound state, e.g., the L_3 level. The emitted electron is referred to as an *Auger electron* and has a very specific energy depending on the three levels involved in the process. In the example shown, these are the KL_2L_3 and the energy is denoted as $E_{KL_2L_3}$ or simply, E_{KLL}. The evaluation of this energy is not as simple as it might appear. A first approximation can be expressed as:

$$E_{KL_2L_3} = E_K - E_{L_2} - E_{L_3} - \Phi \qquad (13.8)$$

where $\Phi = E_{vac} - E_F$ is the work function of the material. The problem is that when the electron is emitted the atom is in an ionized state and the energy levels suffer a shift due to the alteration of screening effects.

Auger electron spectroscopy is an extremely powerful technique and can give information well beyond just chemical identification. Since the Auger emission process depends on the energy levels of the atoms in the solid, any alteration in these will be measured as a shift in the energy of the Auger peak. This can be used to identify ionization states and surface oxidation. Furthermore, surface composition maps can be studied to obtain a chemical profile of a sample using the scanning Auger microscope.

Electron energy loss spectroscopy (EELS) comes in a variety of forms and can be used to study core level excitations (100 - 10^4 eV), plasmons and electronic interband transitions (1 - 100 eV) and excitation of vibrational surface states (10^{-3} - 1 eV). All of these processes involve a specific energy loss from the electrons of the primary beam. This discrete energy quantum is transferred to the sample surface, with the energy peak occurring at a set energy below that of the primary peak. It is possible to distinguish between an Auger peak and an EELS peaks by adjusting the primary energy of the incident electron beam. Any peak that remains constant is an Auger feature while the shift of the peak by the same amount as the primary energy will be an EELS process. The energy of the loss process is simply:

$$\Delta E = E_0 - E_s \tag{13.9}$$

where E_s represents the remaining kinetic energy of the backscattered electron. In the case where a core electron is excited to an empty state in the conduction band, see Figure 13.8, we can express the scattered electron energy as:

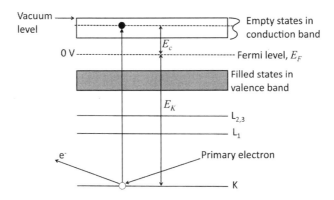

FIGURE 13.8: Electron energy scheme for the excitation of a core (K) level electron in a semiconductor by a primary electron. The excitation will reduce the energy of the primary electron by an amount equal to the energy between initial and final states, $\Delta E = E_K + E_c$. The process can be measured using the EELS techniques and can effectively measure the density of empty states in the conduction band of the semiconductor.

$$E_s = E_0 - \Delta E = E_0 - E_K - E_c \qquad (13.10)$$

The EELS technique can also be used to measure the plasma excitations of free electron gases in metallic systems, where the plasmon energy is given by $\hbar \omega_p = \hbar \sqrt{ne^2/\epsilon_0 m_e^*}$, see Section 6.10. The technique can also be used to measure surface plasmon resonances, with frequency, $\omega_{sp} = \omega_p/\sqrt{2}$. The measurement of low energy excitations, such as surface or adatom vibrations, requires high resolution electron analyzers and is termed high resolution electron energy loss spectroscopy (HREELS).

Photoelectron spectroscopies use photons to excite electrons in solids and can be used to study the densities of occupied states near the crystal surface. The technique is broadly based on the photoelectric effect, where an electron, initially in a bound state, E_i, absorbs a photon of energy, $\hbar\omega(= h\nu)$, and is ejected from the surface of the solid with an energy:

$$E_{kin} = \hbar\omega - E_i - \Phi \qquad (13.11)$$

In order to detect the ejected electron the following conditions must be satisfied: (i) The photon energy must be sufficient to allow the electron to escape the solid; $\hbar\omega \geq E_i + \Phi$, (ii) The electron velocity must be directed towards the outer surface and (iii) The electron must not lose energy in collisions with other electrons on its way to the surface.

Depending of the photon energy used in the excitation process, photoelectron spectroscopy can be considered as: x-ray photoelectron spectroscopy (XPS, also known as ESCA - electron spectroscopy for chemical analysis) or ultraviolet photoelectron spectroscopy (UPS). In XPS, the incident photon energy typically falls in the range 100 eV - 10 keV. A consequence of these high energies is that deep core levels can be excited and processes similar to those in AES can occur and hence this spectroscopy is frequently used as a chemical identification tool. The UPS technique uses much lower electron energies; around 10 - 50 eV and is typically used for the study of the density of occupied states in the valence band. An important variant of UPS is the use of angular resolved measurements, termed ARUPS (angular resolved ultraviolet photoelectron spectroscopy). This is a very powerful tool since it measures the electron yield as a function of orientation and provides a direct measurement of the surface dispersion relation and hence the band structure of the solid.

13.3 LOW DIMENSIONAL SYSTEMS

The term *low dimensional structure* or LDS, refers to any system in which one or more of its spatial dimensions are reduced and in principle can lead to quantization effects. We will discuss this in further detail in the next section of this chapter. Low dimensional systems include thin films, multilayers, nanowires as well as quantum dots and nanoparticles.

As we noted above, the symmetry of a crystal has an important role in defining the physical properties of a solid. Therefore, since a surface has an atomic symmetry which is different from its bulk, we can expect the physical properties of the surface to differ from those of the bulk. Now we can begin to see how the size of an object can influence its physical properties. This can be seen by considering the proportion of atoms at a surface to the number of atoms in the object itself. It so happens that these numbers only become important for objects of a nanometric size. In a normal object, visible to the naked eye, the number of atoms at the surface is only a very small percentage of those in the whole object. It is therefore possible to neglect surface effects for large, or what we term bulk, objects. However, if we consider a thin film, many atoms are on the surface and surface properties can dominate the object. This will also be true for nanosized entities. We can demonstrate this with a simple example. We will consider a hypothetical solid which has a simple cubic structure. We will scale the linear dimension of a cube shaped object and calculate the numbers of atoms on the surface and those in the bulk. We consider, for this exercise, that bulk atoms have the full coordination of six nearest neighbors, while surface atoms have lower coordinations. Of course we could also consider edge and corner atoms, each having a specific coordination number. For the purposes of evaluation we consider a lattice parameter of

2.5 Å, which is reasonable for metals. For our linear dimensions we have an object with length, width and height x, being a cube, which has a discrete number of atoms; i.e., $x = na$, where a is the lattice parameter and there will be $n+1$ atoms in the linear dimension. The total number of atoms in the cube of material is

$$n_V = \frac{V}{a^3} = (n+1)^3 \qquad (13.12)$$

where V is the volume of the object and a^3 is the volume of a unit cell. Since we are considering a simple cubic structure there is only one atom per unit cell. The number of atoms in the bulk, with full coordination must be:

$$n_B = (n-1)^3 \qquad (13.13)$$

The total number of atoms at the surface of the object will simply be the difference between the above two equations:

$$n_S = n_V - n_B = (n+1)^3 - (n-1)^3 = 6n^2 + 2 \qquad (13.14)$$

We can now construct a table of sizes for our cubic object and evaluate the number of atoms on its surface and in its bulk. From this we can see what are the proportions of each type of atom and hence conclude at what point the material can be expected to have its physical properties dominated by the surface. The variation of the proportion of surface and bulk (core) atoms is illustrated in Table 13.2 and Figure 13.9. From the figure we see that the proportion of surface atoms reaches around the 50% mark at a linear dimension of a few nm. However, even for particles as large as 10 nm, there is still a significant proportion of surface atoms, which can already influence the physical properties of the nanoparticle. Other crystalline structures will differ in absolute numbers, but should follow a similar trend with respect to the surface and bulk proportions of atoms and hence their subsequent physical properties.

With respect to the fabrication of nanostructures, there are an enormous range of techniques available, from the very sophisticated to the very elementary. In broad terms these follow one of two

TABLE 13.2: Cubic shaped nanoparticle with simple cubic structure. Length, x; Volume, x^3; length, n in number of lattice spacings; number of atoms in object, n_V; number of bulk or core atoms, n_B; number of surface atoms, n_S; proportion of bulk atoms, n_B/n_V and proportion of surface atoms, n_S/n_V.

$x = na$ (nm)	$V = x^3$ (nm³)	n	n_V $(n+1)^3$	n_B $(n-1)^3$	n_S $n_V - n_B$	n_B/n_V	n_S/n_V
0.5	0.125	2	27	1	26	0.037	0.963
0.75	0.422	3	64	8	56	0.125	0.875
1	1	4	125	27	98	0.216	0.784
1.25	1.953	5	216	64	152	0.296	0.704
2.25	11.391	9	1000	512	488	0.512	0.488
3.75	52.73	15	4096	2744	1352	0.670	0.330
5	125	20	9261	6859	2402	0.740	0.260
7.5	421.87	30	29791	24389	5402	0.819	0.181
10	1000	40	68921	59319	9602	0.860	0.140
12.5	1953	50	132651	117649	15002	0.890	0.110
18.75	6592	75	438976	405224	33752	0.920	0.080
25	15625	100	1030301	970299	60002	0.940	0.060

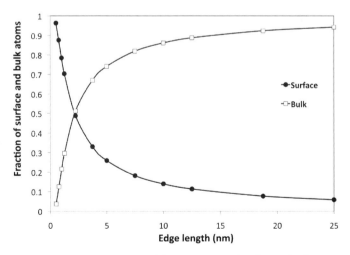

FIGURE 13.9: Variation of the proportions of surface and bulk atoms as a function of the linear dimensions of a nanoparticle. This is the example given in Table 13.2 for a cubic particle with a simple cubic structure and a lattice parameter of 2.5 Å.

approaches. The so called *top-down* approach can be considered as starting with a large block of the material of study and removing the excess to leave behind the final structure of interest. This can be achieved by a number of techniques, such as photolithography, electron beam lithography and focused ion beam milling. These are all fairly complex processing techniques. Due to the complexities of such methods, we cannot go into the details here. The other approach is called *bottom-up* and contemplates the fabrication of nanostructures via the joining together of constituent components. The use of the STM to manipulate atoms to form well defined structures is a good example of a bottom-up approach. Indeed, there is much research on going into the use of biological structures, such as DNA, for their use as building blocks for more complex structures. Again the consideration of the details of these technologies goes beyond the scope of a single chapter on nanotechnologies. There are many books dedicated to these technologies, see for example the book by M. Madou[4].

In the discussion of the physical properties of low dimensional structures, we will frequently refer to *characteristic length scales*. Such a concept arises from the fact that specific properties depend on physical parameters which require a certain length over which they act. This is best demonstrated with an example. Size quantization effects in electronic properties occurs at length scales related to the Fermi wavelength:

$$\lambda_F \frac{2\pi}{k_F} = \frac{h}{\sqrt{2m_e^* E_F}} \quad (13.15)$$

where E_F is the Fermi energy for a metal. The Fermi wavelength decreases with the density of electrons in the metal, n_d, where the d refers to the dimensionality of the system. The explicit forms are given in Table 13.3.

TABLE 13.3: Fermi wavelength for different dimensionalities.

d	λ_F
3	$2^{2/3}(\pi/3m_3)^{1/3}$
2	$\sqrt{2\pi/n_3}$
1	$4/n_1$

When an object has dimensions comparable to the characteristic length, we can expect a divergence of its behavior from that in bulk materials. We will discuss further examples of characteristic lengths in the following sections, with regards to the electronic, optical, and magnetic properties of nanostructures.

It is worth pointing out that the size dependence of various physical quantities can be studied via *scaling laws*. These define how the parameter of interest varies with length scales. For example, we can consider the scaling law for electrical resistance, which follows the equation: $R = \rho l/A$, where l is the length of a sample and A is its cross-sectional area. Thus we see, from a dimensional analysis, that R scales as L^{-1}. We can similarly evaluate the scaling law of any physical parameter or property to assess how it will behave under size reduction.

13.4 ELECTRONIC AND OPTICAL PROPERTIES OF NANOSTRUCTURES

The electronic properties of solids can alter radically upon miniaturization down to the nanoscale. A good indication of this change can be considered from the length scale associated with normal resistive scattering, i.e., the scattering mean free path, $\lambda_m = v_F \tau_m$. Here v_F represents the Fermi velocity, given by: $v_F = \sqrt{2E_F/m_e^*}$ and τ_m is the relaxation time of the charge carriers. Given this definition, we say that the charge carriers travel, on average, a distance λ between collisions, or in other words the electrons will travel a distance of up to λ_m without being elastically scattered. Considering some typical values for metals, we have for Cu: $v_F \simeq 1.6 \times 10^6$ ms^{-1} and $\tau_m = 2.7 \times 10^{-14}$ s, which gives a mean free path for scattering of around 40 nm. Therefore in structure of these dimensions made of Cu, an electron can traverse the object without any scattering. Such a process is referred to as *ballistic transport*. So the normal resistive type behavior has no meaning in this sense for nanostructures with dimensions below the mean free path for scattering. We mentioned the Fermi wavelength above as a characteristic length scale, and when the dimensions of a nanostructure reaches these dimensions we can observe quantization effects.

This is possible since standing waves can be produced due to the boundary conditions at the surfaces of the object. At low temperatures, the phase shifts between electrons can occur from scattering events and a dephasing will occur. Such considerations only make sense in situations where there is a coherence between electrons. It is possible to calculate the dephasing time due to scattering and hence a *phase coherence length*, λ_ϕ. At low temperature this can reach length scales that are much larger than the mean free path of electrons and can be of the order of μm. Another important consideration is the charge that can be stored on a nano-object and charging effects can become important when the electrostatic energy, $U = e^2/C < k_B T$, where C is the capacitance of the object. We will discuss many of these concepts in the coming sections.

We have indicated here a number of parameters that are important for defining the transport properties in nanostructured devices and can dominate the behavior of these systems. In Table 13.4, we summarize these parameters.

The size reduction of an object can be defined in 1, 2, or 3 dimensions. For 1D size reduction only one direction of the object is reduced, and the form will be that of a thin film. In terms of confinement effects this would mean that our electrons or charge carriers are free to move in two directions and limited in the third direction

TABLE 13.4: Important length scales for transport properties in low dimensional structures.

Macroscopic object	Mesoscopic object
$L \gg \lambda$	$L \leq \lambda$
Diffusive transport	Ballistic transport
$L \gg \lambda_\phi$	$L \leq \lambda_\phi$
Incoherent	Phase coherence
$L \gg \lambda_F$	$L \leq \lambda_F$
No size quantization	Size quantization
$e^2/C < k_B T$	$e^2/C < k_B T$
No single electron charging effects	Single electron charging effects

due to size reduction and quantization effects. Such a situation in semiconductors is referred to as a *quantum well* and the electrons are referred to as a *two-dimensional electron gas* or 2DEG for short. Size reduction in two dimensions leads to a wire shape also called a *quantum wire*. In this case the charge carriers have only freedom to move along the wire length, i.e., we have a 1D object, and quatization effects occur in the two lateral directions. For three dimensional size reduction, the object will have limited dimensions in all three directions. Such an object is called a *quantum dot*, or a 0D device, where quantization effects dominate in all directions. We have already introduced the concept of the quantum well, see Section 9.6, with respect to device applications.

13.4.1 Size Reduction and Energy Quantization

We have indicated above that the reduction of the size of an object will severely alter the way in which electrons can exist in the space available, i.e., within the boundaries of the object. This effect becomes critical when the de Broglie wavelength associated with the electron is of the same order of magnitude as the space in which it can move. In such a situation, the edges of the object will act as boundaries and the boundary conditions will define the allowed wavelengths (or the wave-functions) and hence energies that the electrons can have. Actually we have already met a situation which has a similar effect. Electrons in atoms are confined by the electric potential of the nucleus. The discrete electron energy levels in atoms are defined by the effective potential that the electrons in the atoms are subject to due to the nuclear charge and from the screening effects of the other electrons that constitute the atom. This will produce a localizaton of the electrons and hence their energy states become discretised. The simplest case of confinement that we can consider is that of an infinite potential outside the nanostructure (quantum well), such that the wavefunction must vanish at the boundary. This means that the stationary states (standing waves) in 1D will correspond to those with wavelengths given by: $\lambda_n = 2L/n$, where L is the spatial thickness of the sample available for the motion of the electron and n an integer, which defines the mode (and energy) of the electron state. The energy corresponding to the various discrete states can be evaluated from the general form of the energy:

$$E_n = \frac{\hbar^2 k_n^2}{2m_e^*} \tag{13.16}$$

where the wave vector k_n is given by $k_n = 2\pi/\lambda_n$, such that we obtain:

$$E_n = \frac{\hbar^2 \pi^2 n^2}{2m_e^* L^2} \tag{13.17}$$

This expression only corresponds to the energy contribution in the direction of size reduction where quantization effects occur. The full energy term is given by an equation of the form of (9.100) for a quantum well:

$$E_{QW} = E_n + E(k_y, k_z) = \frac{\hbar^2}{2m_e^*}\left(\frac{\pi^2 n_x^2}{L_x^2} + k_y^2 + k_z^2\right) \tag{13.18}$$

where we have considered that the reduced dimension is in the x-direction. Extending to the quantum wire, with two reduced dimensions, the energy takes the form:

$$E_{QWire} = E(n_x, n_y) + E(k_z) = \frac{\hbar^2}{2m_e^*}\left(\frac{\pi^2 n_x^2}{L_x^2} + \frac{\pi^2 n_y^2}{L_y^2} + k_z^2\right) \tag{13.19}$$

Finally, the corresponding energy for 3D size reduction, i.e., for a quantum dot, takes the form:

$$E_{QD} = \frac{\hbar^2}{2m_e^*}\left(\frac{\pi^2 n_x^2}{L_x^2} + \frac{\pi^2 n_y^2}{L_y^2} + \frac{\pi^2 n_z^2}{L_z^2}\right) \tag{13.20}$$

In Figure 13.10, we show the basic forms of these energy dispersion curves for the quantum well, Equation (13.18), see Figure 13.10(a), and for the quantum wire, Equation (13.19), see Figure 13.10(b). The energies for the quantum dot are fully quantized and can be defined by the set of three quantum numbers (n_x, n_y, n_z). (Note that this excludes the electron spin. However, this is not a problem since the electrons are spin degenerate in terms of their energy.)

The size reduction in one or more directions will drastically alter the density of states in the solid. For the case of a three-dimensional solid we found (see Chapter 6, Section 6.7):

Nanotechnologies and Nanophysics • 511

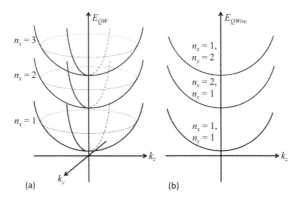

FIGURE 13.10: Energy dispersion curves for (a) a quantum well structure, and (b) a quantum wire. We note that each discrete state will give rise to a sub-band labelled with quantum numbers referring to the discrete energy state in the directions of size reduction.

$$g_{3D}(E)dE = \frac{1}{2\pi^2}\left(\frac{2m_e^*}{\hbar^2}\right)^{3/2} E^{1/2} dE \qquad (13.21)$$

In the case of a two-dimensional material (2DEG) a similar approach can be used to that for three dimensions. This yields a 2D density of states of the form:

$$g_{2D}(E)dE = \frac{m_e^*}{\pi\hbar^2} dE \qquad (13.22)$$

The 2D density of states does not depend on the energy. Taking into account that there will be quantization effects in one dimension, i.e., in the quantum well, the density of states will take a step-like function, where the steps occur at the energies corresponding to the discrete energy steps in the direction of reduced dimensions. The exact step positions will therefore depend on the thickness of the quantum well, as given in Equation (13.18). The density of states takes on a staircase like function given by:

$$g_{2D}(E)dE = \frac{m_e^*}{\pi\hbar^2} \sum_i \Theta(E - E_i) dE \qquad (13.23)$$

where $\Theta(E - E_i)$, is the *Heaviside function*. Extending now to the case of the one dimensional system (i.e., reduced size in two dimensions), the density of states takes the form:

$$g_{1D}(E)dE = \frac{1}{\pi}\left(\frac{2m_e^*}{\hbar^2}\right)^{1/2}\frac{1}{E^{1/2}}dE \qquad (13.24)$$

This variation has an energy dependence that goes as the inverse root of the energy. To consider more than one energy level, due to the discrete nature of the energy from quantization effects we can re-write the density of states in the form:

$$g_{1D}(E)dE = \frac{1}{\pi}\left(\frac{2m_e^*}{\hbar^2}\right)^{1/2}\sum_i\frac{n_i\Theta(E-E_i)}{(E-E_i)^{1/2}}dE \qquad (13.25)$$

For the case of a quantum dot (0D), only discrete energy levels will exist, as given in Equation (13.20), such that the density of states will consist of Dirac delta functions at the energies defined by the quantization in 3D. In Figure 13.11, we show a schematic illustration of the form of the densities of states for the various types of nanostructure.

The effect of quantum confinement can also be arrived at from the Heisenberg uncertainty principle. If we, for simplicity sake, consider confinement in 1D in the x-direction, we can say that the particle (electron) is confined within the space of length, $L_x = \Delta x$. In

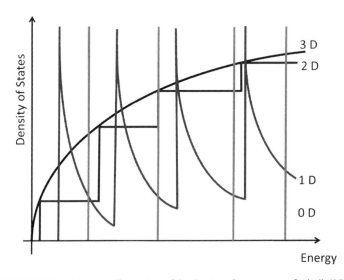

FIGURE 13.11: Schematic illustration of the density of states curves for bulk (3D), quantum wells (2D), quantum wires (1D), and quantum dots (0D).

terms of the uncertainty principle, the corresponding uncertainty in the linear momentum of the particle will be:

$$\Delta p_x = \frac{\hbar}{\Delta x} \qquad (13.26)$$

The additional energy produced by the confinement in the x-direction can be expressed by:

$$E_{\Delta x} = \frac{(\Delta p_x)^2}{2m_e^*} = \frac{\hbar^2}{2m_e^* \Delta x^2} \qquad (13.27)$$

This will correspond to a zero point energy and will be significant if it has a value greater than or equal to the kinetic energy of the electron due to thermal motion in the direction of quantization. This means that to observe some form of quantum effect the following condition must be satisfied:

$$\Delta x \leq \frac{\hbar}{\sqrt{m_e^* k_B T}} \qquad (13.28)$$

From this we can see that for an effective mass of say $0.1\, m_e$ at room temperature, significant quantum effects can be expected for confinement distances of around 5 nm. Clearly, for lower temperature, quantum effects will be more pronounced and Δx will increase.

13.4.2 Quantum Point Contacts

Quantum wires (QWR) represent a quasi-one-dimensional system for transport properties. To define the quantum wire we need to define its dimensions more carefully. The QWR must have a width which is of the order of the Fermi wavelength; $W \sim \lambda_F$. The wire will have diffusive transport if its length is greater than the mean free path; $L \gg \lambda_m$, where the electrons will undergo many scattering events before it can emerge at the other end of the wire under the action of an electric field, see Figure 13.12 (a). When the QWR length is below the mean free path, $L < \lambda_m$, the electron will suffer one or no collisions, except with the walls of the wire itself. This is the case of the ballistic regime, see Figure 13.12 (b). For QWRs which are very short, such that $W \simeq L \ll \lambda_m$, the QWR becomes a *quantum point contact* (QPC), see Figure 13.12 (c). In this case there is very little chance

514 • Solid State Physics

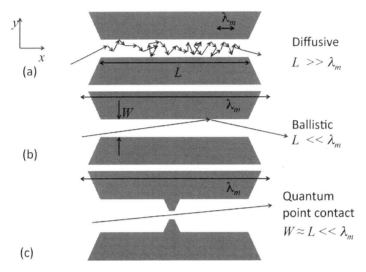

FIGURE 13.12: Schematic illustration of quantum wires in (a) the diffusive regime; (b) the ballistic regime and (c) a quantum point contact.

that the electron will be scattered in its passage thought the device. (This situation is analogous to the case of viscous and molecular flow regimes in vacuum technology, see Section 13. 2.)

We can consider the passage of current through a QWR or QPC by conceptualising the device as a left and right reservoir for charge carriers being connected by a waveguide. In this approach, the left and right reservoirs are nothing more than the electrical contacts to the device (which are supplied with electrons from an external power source) and the waveguide is the QPC itself, see Figure 13.13.

The problem seen in this way resembles a tunnel barrier problem, where a particle is incident from left or right on a barrier, in this case modelled as a QPC. The transmission and reflection coefficients can be represented in terms of the wave functions of the particles to the left and right of the QPC, or in a more general way the device under test (DUT). This approach can be used in a very general manner to obtain the scattering or S - parameters from the DUT, as is used in waveguide technology. Since we are essentially dealing with a 1D problem, we can express the wave functions as a product:

$$\psi(x,y,z) = \psi_{\perp}(y,z)\psi_{\parallel}(x) \qquad (13.29)$$

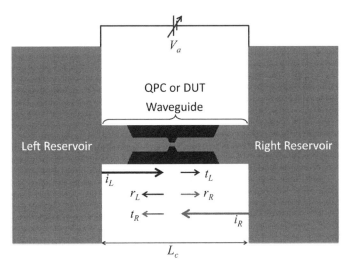

FIGURE 13.13: QPC (or DUT - device under test) as waveguide system connected to left and right reservoirs. Also indicated are the electrons incident from left and right ($i_{L,R}$), transmitted from left and right ($t_{L,R}$) and reflected from left and right ($r_{L,R}$).

where $\psi_\perp(y,z)$ is the transverse component and $\psi_\parallel(x)$ the translational component in the direction of electron motion of the wave function. Left moving and right moving electrons can be expressed as:

$$\psi_{\parallel,L}(x) = e^{ik_L(x-x_L)} + r_L e^{-ik_L(x-x_L)}; \quad x \sim x_L \quad (13.30)$$
$$= t_R e^{ik_R(x-x_R)}; \quad x \sim x_R \quad (13.31)$$

and

$$\psi_{\parallel,R}(x) = t_L e^{-ik_L(x-x_L)}; \quad x \sim x_L \quad (13.32)$$
$$= e^{-ik_R(x-x_L)} + r_R e^{-ik_R(x-x_R)}; \quad x \sim x_R \quad (13.33)$$

where $x_{L,R}$ denote the positions of the interfaces to the left and right of the QPC with the respective reservoirs. This shows that the wave functions have incident, reflected ($r_{L,R}$) and transmitted ($t_{L,R}$) components (amplitudes) in both directions, see Figure 13.13. Continuity of the wave function and its first derivative allow the following relations to be established:

$$k_L(1-|r_L|^2) = k_R |t_R|^2 \quad (13.34)$$

$$k_R(1-|r_R|^2) = k_L |t_L|^2 \tag{13.35}$$

Also from $|r_L|^2 = |r_R|^2$ and $t_R^* t_L = t_R t_L^*$ we obtain:

$$k_R^2 |t_R|^2 = k_L^2 |t_L|^2 \tag{13.36}$$

The expression for the current density:

$$j = \frac{-i\hbar}{2m_e^*}(\psi^* \nabla \psi - \psi \nabla \psi^*) \tag{13.37}$$

allows the evaluation of incoming and outgoing currents, which are expressed in terms of the transmission and reflection amplitudes and the electron velocity. For example if we have a bias applied to the QPC device, V_a, we can evaluate the input current as: $j_{in} = v_L$, and the outgoing current as: $j_{out} = v_R |t_R|^2$, where $v_{L,R} = \hbar k_{L,R}/m_e^*$. The ratio of these currents will define the left-to-right transmission coefficient:

$$T_{L-R}(E) = \frac{j_{in}}{j_{out}} = \frac{k_R}{k_L}|t_R|^2 \tag{13.38}$$

Similarly we can write:

$$T_{R-L}(E) = \frac{k_L}{k_R}|t_L|^2 \tag{13.39}$$

where we have $T_{L-R}(E) = T_{R-L}(E) = T(E_\parallel)$ and $E_\parallel = \hbar^2 k_x^2/2m_e^*$. Therefore the transmission coefficients are the same for both directions for incoming electrons. In terms of the reflection coefficient we have:

$$R(E) = \frac{j_R}{j_{in}} = |r_L|^2 = |r_R|^2 \tag{13.40}$$

and we must satisfy the relationship:

$$T(E) + R(E) = 1 \tag{13.41}$$

It is interesting to note the similarity of various phenomena in physics. The above situation is analogous to the reflectivity and transmission of electromagnetic waves at the boundary between two media, which we met in the previous chapter, see Section 12.9.2. In fact, such a situation can be established in many other situations and is a characteristic of wave phenomena.

The energy spectrum due to quantum confinement in the lateral directions of the QPC will take the form of Equation (13.19) and for the situation we have described above can be expressed as:

$$E_{k_x}(m,n) = E_{m,n} + \frac{\hbar^2 k_x^2}{2m_e^*} \qquad (13.42)$$

In effect k_x, m, n are the quantum numbers that describe the system. The number of electrons in any particular state will be given by $2f_{FD}[E_{k_x}(m,n) + eV_a/2 - E_F]$, where the factor of 2 arises from spin degeneracy and $f_{FD}(E)$ is the Fermi - Dirac distribution function. We can express the left and right going currents as:

$$I_L = -\frac{2e}{L_c} \sum_{m,n} \sum_{k_x > 0} v_x T(E_\parallel) f_{FD}[E_{k_x}(m,n) + eV_a/2 - E_F] \qquad (13.43)$$

$$I_R = -\frac{2e}{L_c} \sum_{m,n} \sum_{k_x < 0} v_x T(E_\parallel) f_{FD}[E_{k_x}(m,n) - eV_a/2 - E_F] \qquad (13.44)$$

The total net current will be the difference between these two contributions:

$$\begin{aligned} I_{Tot} &= I_L - I_R \\ &= -\frac{2e}{L_c} \sum_{m,n} \sum_{k_x > 0} v_x T(E_\parallel) \{ f_{FD}[E_{k_x}(m,n) + eV_a/2 - E_F] \\ &\quad - f_{FD}[E_{k_x}(m,n) - eV_a/2 - E_F] \} \end{aligned} \qquad (13.45)$$

This is a rather complex function to evaluate due to the equilibrium conditions and can be greatly simplified for the low temperature approximation:

$$\lim_{T \to 0} f_{FD}(E - E_F) = \Theta(E - E_F) \qquad (13.46)$$

The difference in the two Fermi - Dirac distribution functions in Equation (13.45) can be expressed as $eV_a \delta(E_F - E)$. Since there is no physical sense in using the resistance as a characteristic property, it is customary to use the *conductance*, which is expressed as:

$$G = \frac{I}{V} \qquad (13.47)$$

It is now possible to express the low temperature conductance of the QPC in the form:

$$G = \frac{2e^2}{h} \sum_{m,n} T(E_F, m, n) \qquad (13.48)$$

where the sum over k_x is replaced by an integral and integrated over the energy. The summation above is over electron states (m, n), with energy $E < E_F$. The prefactor e^2/h is called the *quantum of conductance* and has a value of $G_0 = e^2/h = 39.6\mu S$,[5] with an inverse conductance equivalent to $1/G_0 = 25.2$ kΩ. Equation (13.48) is known as the *Landauer formula*. The lateral quantization effects that we discussed earlier, and given in Equation (13.42), mean that we can interpret the Landauer formula as having a discrete number of channels. We can write this as:

$$G = 2G_0 \sum_{m,n} T_{m,n} \qquad (13.49)$$

where each channel (m, n) will contribute $G_0 T_{m,n}$ to the conductance. In the ballistic regime the transmission coefficient $T_{m,n} = 1$. When further channels open, integral numbers of G_0 will pass, giving the conductance a step-like appearance. To experimentally observe this, it is necessary to control the width of the QPC. This can be done by defining a conduction channel in a 2DEG in a semiconductor with metallic contacts, which will form a depletion zone in the 2DEG and thus gating the output. The constricted region will have an energy spectrum determined by:

$$E_{k_x}^{QPC}(m,n) = \frac{\hbar^2 \pi^2}{2m_e^*} \left[\frac{k_x^2}{\pi^2} + \frac{m}{L_y^2(x, V_G)} + \frac{n}{L_z^2} \right] \qquad (13.50)$$

where the cross-section of the QPC is determined by the thickness of the 2DEG, L_z and the lateral dimension, $L_y^2(x, V_G)$ will vary in the x-direction and is controlled by the gate voltage, V_G. This means that the conductance will depend on V_G:

$$G = G(V_G) = G_0 N_{open}(V_G) \qquad (13.51)$$

where $N_{open}(V_G)$ is the number of open channels which can be evaluated as follows: Assuming $N_{open} = (m, 1)$, i.e., only one mode number in the z-direction, a new channel will open when the energy position at the top of the barrier passes the Fermi level as we alter the minimum width of the QPC:

$$E_F = \frac{\hbar^2 k_F^2}{2m_e^*} = \frac{\hbar^2 \pi^2}{2m_e^*[L_y(V_G)_{min}]^2} m^2 \qquad (13.52)$$

This gives the number of open channels as:

$$N_{open} = \left[\frac{k_F}{\pi} [L_y(V_G)_{min}] \right]_{int} \qquad (13.53)$$

where the subscript indicates that this must be an integer value, where the discrete jumps occur at the critical moments when the gate voltage is such that a new channel can be accommodated in the QPC structure. A more general approach to the scattering formalism was considered by Büttiker for any number of contact leads to a device, which is considered as the scattering center. In Figure 13.14, we show the conductance of the QPC as a function of the gate voltage. Also shown in the figure is an AFM image of the QPC where because of the electrostatic interaction with the AFM tip, the conduction channels can be visualized.

13.4.3 The Insulating Barrier and Tunnel Junctions

The above section showed that under reduced dimensions, conductors behave in a very different way to the bulk material. This is also true of insulating materials. In its normal bulk form, the insulator will not allow the passage of electrons through its bulk. However, under reduced dimensions, the conditions for transfer are fundamentally altered. The potential barrier is a well known problem in quantum mechanics and is one of the highlights that illustrate the fundamental new physics that emerged with the discovery of this area of physics. Indeed quantum tunneling effects are now very well studied, both theoretically and experimentally, having a number of very important applications, such as the scanning tunneling microscope (STM) and the Josephson junction (see Section 11.7).

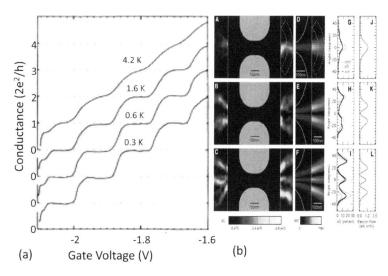

FIGURE 13.14: Conductance in a QPC as a function of the gating voltage. (a) The step-like variation is clear where the step height corresponds to G_0. The effect of sample temperature can be seen to smooth the step profile. (b) AFM image of a QPC where the bright regions correspond to the electron channels. Increasing the gate voltage reveals a discrete increase in the number of open channels (not shown). (A to C) Angular pattern of electron flow of individual modes of the QPC, comparing experiment with theory [(A) first mode, (B) second mode, (C) third mode]. (D to F) Calculated wave-function $-\psi|^2$ for electrons passing from (D) the first mode, (E) the second mode, and (F) the third mode of the QPC (the areas in each simulation corresponding to areas not scanned in the experiment are dimmed). (G to I) Measured angular distribution of electron flow from (G) the first mode, (H) the second mode, and (I) the third mode. (J to L) Angular distribution of the wave-function $-\psi|^2$ from (J) the first mode, (K) the second mode, and (L) the third mode. [Reprinted figure with permission from: B. J. van Wees et al., Phys. Rev. B., **43**, 12431 (1991). Copyright 1991 by the American Physical Society. Reprinted figure with permission from: M. A. Topinka, et al., Science, **289**, 2323 (2000). Copyright 2000 Science AAAS.]

The basic form of the potential barrier problem can be considered from the incidence of an electron at a potential barrier of height, V_0, where the tunnel effect will occur when the electron energy, $E < V_0$. The situation is illustrated in Figure 13.15. Three zones can be identified: (I) $x < 0$, where the wave function has two components consisting of the incident component, $\psi_i(x)$, and the portion reflected from the barrier, $\psi_r(x)$; (II) $0 < x < d$, the barrier region itself, where the wave function will exponentially decay (i.e., where $E < V_0$), and (III) $x > d$, where the emerging transmitted electrons will move to the right have a wave function, $\psi_t(x)$. The functional form the wave functions in the three regions can be expressed as:

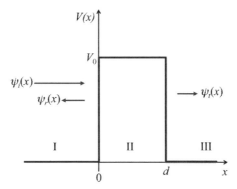

FIGURE 13.15: Potential barrier problem of an electron of energy, $E < V_0$, incident on a barrier.

$$\psi_I(x) = \psi_i(x) + \psi_r(x) = Ae^{ik_1x} + Be^{-ik_1x}; \quad x < 0 \quad (13.54)$$

$$\psi_{II}(x) = De^{k_2x} + Ee^{-k_2x}; \quad 0 < x < d \quad (13.55)$$

$$\psi_{III}(x) = \psi_t(x) = Fe^{ik_1x}; \quad x < 0 \quad (13.56)$$

where $k_1 = \sqrt{2m_e E/\hbar^2}$ and $k_2 = \sqrt{2m_e(V_0 - E)/\hbar^2}$.

There are many books on quantum mechanics that deal explicitly with this problem and we will just give the results of the calculations here. The root to solving the problem is a consideration of the boundary conditions between the three regions, where the wave functions and their first spatial derivatives are continuous. The transparency of a barrier to the electrons depends critically on the height, V_0 and width, d, of the barrier as well as the energy, E, of the incident electrons. The transmission probability (in the language of quantum mechanics), for the case considered here with $E < V_0$, takes the form:

$$T = \frac{|F|^2}{|A|^2} = \frac{4k_1^2 k_2^2}{4k_1^2 k_2^2 + (k_1^2 + k_2^2)^2 \sinh^2(k_2 d)}$$

$$= \frac{4E(V_0 - E)}{4E(V_0 - E) + V_0^2 \sinh^2(k_2 d)} \quad (13.57)$$

The reflection coefficient takes the form:

$$\mathcal{R} = \frac{|B|^2}{|A|^2} = \frac{(k_1^2+k_2^2)^2 \sinh^2(k_2 d)}{4k_1^2 k_2^2 + (k_1^2+k_2^2)^2 \sinh^2(k_2 d)}$$
$$= \frac{V_0^2 \sinh^2(k_2 d)}{4E(V_0-E) + V_0^2 \sinh^2(k_2 d)} \qquad (13.58)$$

It can be readily seen that these equation adhere to the general condition: $\mathcal{T} + \mathcal{R} = 1$. It is possible to consider the extremes, or limits of the problem, where $E \to 0$ and $E \to \infty$. In the former case, the tunnel probability will monotonically decrease with the energy. In the latter case, as the electron energy reaches the height of the barrier we can write:

$$\lim_{E \to V_0} \mathcal{T} \to \left(1 + \frac{m_e V_0 d^2}{2\hbar^2}\right)^{-1} \qquad (13.59)$$

where we have used the fact that $\sinh^2(k_2 d) \to (k_2 d)^2$ in this limit. The quantity $m_e V_0 d^2 / 2\hbar^2$ is sometimes referred to as the *opacity* of the barrier. In the classical limit, the opacity becomes very large such that the transmission probability becomes vanishingly small. When $k_2 d \gg 1$ we can approximate $\sinh(k_2 d) \simeq e^{k_2 d}/2$ and the tunnel probability can be expressed as:

$$\mathcal{T} \simeq \frac{16 E(V_0-E)}{V_0^2} e^{-2k_2 d} \qquad (13.60)$$

This result turns out to be very important for the workings of the STM, where the tunnel current is proportional to $\mathcal{T}$.

The problem as presented here should bear a resemblance to the scattering problem discussed above in Section 13.4.2. In fact, the reason that quantum tunneling can occur is due to the wave description of the electron. This type of problem can be seen in optics and acoustics as the general phenomenon of a wave passing from one medium to another, for example light passing through a window or sound passing through a wall.

In terms of the barrier phenomenon in electronic devices, the tunnel junction is a general type of device and the tunnel probabilities can be controlled by setting the thickness and height of the barrier. This component forms a part of many nanodevices, such as the Josephson junction, the resonant tunnel diode (RTD) and the magnetic tunnel junction (MTJ). In the latter, the spin of the electron adds a further degree of freedom to the problem and the tunnel probabilities will also depend on the spin states on either side of the barrier.

13.4.4 Single Electron Transport: Quantum Dots and Coulomb Blockade

In general, the interaction between electrons is neglected when considering the transport properties of solids. While the Coulomb interaction between electrons is important, it can for the most part be neglected since it constitutes only a small perturbation to the ground state of the system. However, in small nanometric structures this assumption is no longer valid as even a small perturbation will have an influence on the motion of small numbers of electrons or charge carriers. Charging in mesoscopic and nanoscopic devices can lead to the prevention of the passage of further charges. This effect is called *Coulomb blockade*. The control of voltages in a circuit can allow the motion of charges in specific amounts even down to the motion of a single electron. Such processes depend on charge quantization as we will discuss in the simple case of an isolated quantum dot, sometimes referred to as an *island*. Connection from the outside to the quantum dot can be made using a tunnel junction to which we add a voltage source to allow us to alter the energy levels of the system.

An important consideration in such a system is the capacitive coupling through the tunnel junction. We also need to distinguish whether the island is a metal or semiconductor, since this will influence the number of charge carriers stored on the island as well as the quantum confinement effects and allowed energy states. We briefly touched on some of the effects of spatial confinement in semiconductors in Section 9.6 and above in Section 13.4.1. Metallic systems differ mainly in that they have much larger electron densities, though both depend on the discrete nature of the tunneling process to couple to an external circuit.

In the system we will describe, the island will be coupled via two tunnel junctions, as illustrated in Figure 13.16. The circuit has a voltage source and ammeter to measure the current. The tunnel junctions (TJ_1 and TJ_2) are modelled with a resistance and a capacitance for the equivalent circuit. The charge on the quantum dot

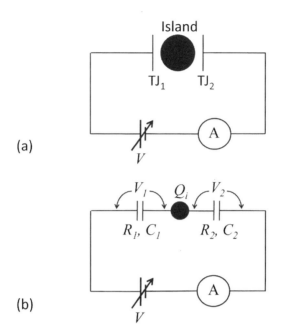

FIGURE 13.16: (a) Quantum dot coupled via two tunnel junctions (TJ) to an external circuit. (b) Equivalent circuit, where the TJs are characterized by a capacitance C and resistance, R.

can be expressed as an integer number of the elementary electron charge: $Q_i = -N_i e = Q_2 - Q_1$. Here $Q_{1,2} = V_{1,2} C_{1,2}$ is the charge on $TJ_{1,2}$ and we have that the applied potential, $V = V_1 + V_2$. The electrostatic potential energy can be expressed as:

$$E_p = \frac{1}{2} QV = \frac{Q^2}{2C} \tag{13.61}$$

So for the quantum dot, we have:

$$E_i = \frac{Q_i^2}{2C_{eq}} = \frac{N_i^2 e^2}{2C_{eq}} = E_C N_i^2 \tag{13.62}$$

where $E_C = e^2/2C_{eq}$, is the energy due to one electron and $C_{eq} = C_1 + C_2$. Since the potential across the two TJs must be equal to the applied voltage we find:

$$V_1 = \frac{VC_2 - Q_i}{C_{eq}} = \frac{VC_2 + N_i e}{C_{eq}} \tag{13.63}$$

and

$$V_2 = \frac{VC_1 - N_i e}{C_{eq}} \quad (13.64)$$

again $C_{eq} = C_1 + C_2$ and corresponds to the total island capacitance. The electrostatic energy stored in the two capacitors will be:

$$E_s = \frac{Q_1^2}{2C_1} + \frac{Q_2^2}{2C_2} = \frac{V^2 C_1 C_2 + Q_i^2}{2C_{eq}} \quad (13.65)$$

The work done by the voltage source in transferring electrons on to or off of the island will be: $W_s = V\Delta Q$, where ΔQ is the total charge transferred. For the transfer of a single electron we have: $V \to V - e/C_{eq}$, therefore $\Delta Q = -eC_{1,2}/C_{eq}$, where the $C_{1,2}$ indicates the dependence upon which TJ was crossed in the electron transfer. The work done can then be evaluated as:

$$W_s(N_{1,2}) = -N_{1,2} eV \frac{C_{2,1}}{C_{eq}} \quad (13.66)$$

The total energy of the circuit, including the voltage source will therefore be:

$$E(N_1, N_2) = E_s - W_s(N_{1,2}) = \frac{V^2 C_1 C_2 + Q_i^2}{2C_{eq}} + \frac{eV(N_1 C_2 + N_2 C_1)}{C_{eq}} \quad (13.67)$$

From this energy we can evaluate the condition for Coulomb blockade. This is done by considering the tunneling of a single electron via one of the tunnel junctions. The energy change due to an electron tunneling through junction 1 can be expressed as:

$$\Delta E_1^\pm = E(N_1, N_2) - E(N_1 \pm 1, N_2) = \frac{e}{C_{eq}} \left[-\frac{e}{2} \mp (N_i e + V C_2) \right] \quad (13.68)$$

The ± sign refers to the direction of charge transfer. The corresponding expression for junction 2 will be:

$$\Delta E_2^\pm = E(N_1, N_2) - E(N_1, N_2 \pm 1) = \frac{e}{C_{eq}} \left[-\frac{e}{2} \pm (N_i e - V C_1) \right] \quad (13.69)$$

For an initially neutral island ($N_i = 0$), we have

$$\Delta E_{1,2}^{\pm} = -\frac{e^2}{2C_{eq}} \mp \frac{eVC_{2,1}}{C_{eq}} \qquad (13.70)$$

This energy must be greater than zero, meaning that for all possible transitions to and from the island, the Coulomb energy causes the energy change to be negative until the applied potential exceeds the threshold, which depends on the lesser of the two capacitances. When these are equal, i.e., $C_1 = C_2 = C$, we obtain the condition:

$$|V| > \frac{e}{C_{eq}} \qquad (13.71)$$

Therefore, for tunneling to occur we must apply a certain voltage, below which no charge can flow. This is the Coulomb blockade region resulting from the Coulomb energy, $e^2/2C_{eq}$, due to an extra electron tunneling on to the island. This situation is schematically illustrated in an energy diagram in Figure 13.17. On the island the lowest energy states (at $T = 0$ K) will be occupied and all excited states empty. The energy difference (gap) between the levels is equal to the Coulomb energy above and below the Fermi level on the electrodes. At equilibrium, there are no states available for an electron to tunnel on to the island from either side. In the situation of an applied potential of $V > e/2C$, an electron can now pass from the right electrode to the island. Further passage is again blocked at this point until a potential of $V > 3e/2C (= e/2C + e/C)$ is applied. The next energy level above that shown will at e^2/C above this and at equal intervals beyond that. Between levels then we say the island is Coulomb blockaded. The current - voltage ($I - V$) characteristic for this device will also have a staircase structure, due to the discrete nature of the energy levels on the island.

For the effects of charge quantization to be observable, the energy between levels must be greater than the thermal energy: $e^2/C_{eq} > k_B T$. Considering the Heisenberg uncertainty relation: $\Delta E \Delta t \geq \hbar$, we can equate the energy with the Coulomb energy and the time with the characteristic time associated with an RC circuit, where here the resistance will be that associated with the junction, such that:

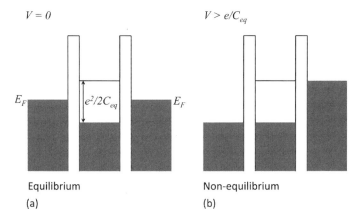

FIGURE 13.17: Quantum dot energy diagram. (a) Equilibrium condition ($V = 0$). The energy gap in the island region corresponds to the Coulomb energy for a single electron. (b) Under an applied voltage greater than the critical value for Coulomb blockade, such that an electron can now pass on to the island.

$$\frac{e^2}{C_{eq}} R_j C_{eq} > h \tag{13.72}$$

This gives the condition for the junction resistance for which Coulomb charging effects can be observed as:

$$R_j > \frac{h}{e^2} \tag{13.73}$$

which has a numerical value of 25.8 kΩ.

13.4.5 Resonant Tunneling

The concept of resonant tunneling effect is based on the a double junction structure, typically fabricated from semiconductor heterostructures. One important application of this effect is in the so-called *resonant tunneling diode* or RTD. The basic structure of the RTD is illustrated in Figure 13.18 (a), where two AlGaAs layers sandwich a GaAs central quantum well (QW). The outer layers are heavily doped n^+ - GaAs layers, such that the Fermi levels are in the conduction band. These will form the contacts with an external circuit and power supply. The schematic conduction band structure for the device is illustrated in Figure 13.18(c). The QW has one quasi-bound state illustrated with an energy E_1. Further bound states can

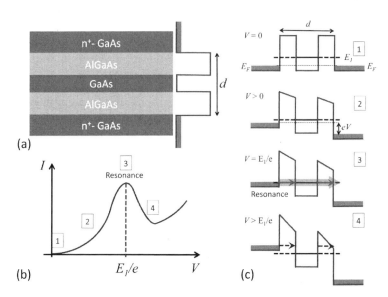

FIGURE 13.18: Resonant tunnel diode (RTD). (a) Basic design using semiconductors GaAs and AlGaAs. (b) Current - voltage characteristics and (c) Band diagrams (conduction band) for different applied voltages, with points indicated (1 - 4) in (b), see text.

occur, depending on the thickness of the QW and the band structures of the AlGaAs and GaAs layers. The current - voltage characteristics are shown in Figure 13.18 (b). For zero applied potential, the diode is in thermal equilibrium (1). Once a small voltage is applied the current will gradually increase (2) since the energy levels are shifted by eV, as shown. This will allow a small tunnel current to be established through the whole barrier structure. At the point where the applied potential is such that the quasi-bound state level lies between the Fermi level and conduction band edge, E_c, there will be a favorable transfer of charge since the effective barrier width becomes equal to the thickness of the AlGaAs layer and is much less than the whole barrier structure thickness, d. This corresponds to the resonant tunneling process, see Figure 13.18(b) point (3). Once the voltage passes the limit where the conduction band edge is above the quasi-bound state energy, the effect of *negative resistance* will be observed as the barrier width returns to being the entire barrier structure, d, point (4) in Figure 13.18 (b). The current will drop off and between points (3) and (4) we observe a negative slope in the

$I - V$ curve. In the case where the current - voltage characteristic is non-linear, it is customary to define the differential resistance:

$$R_d = \left(\frac{dI}{dV}\right)^{-1} \tag{13.74}$$

A negative differential resistance is used to denote $R_d < 0$, which corresponds to the unusual situation where an increase in applied voltage results in a decrease in the current. This situation is similar to that found in the Esaki diode. It is possible to evaluate the tunnel probabilities using a similar approach to that outlined in Section 13.4.3. We will not perform these calculations here, as they are somewhat academic.

13.4.6 Single Electron Transistor (SET)

The formation of the single electron transistor (SET) is based on the quantum dot (QD) structure we discussed in Section 13.4.4. The principal difference is that we add a third connection to the QD, which has a second voltage supply. This is called *gating*, where the gate potential, V_G, is used to control the transfer of charge between the other two connectors, which we now call the *source* (S) and the *drain* (D), in analogy to the field effect transistor (FET). The device is illustrated in Figure 13.19 (a) along with the equivalent circuit (b). The gate is also capacitively coupled to the island with a gate capacitance of C_G. The additional charge balance on the island will be such that:

$$Q_G = C_G(V_G - V_2) \tag{13.75}$$

and the island charge will now be calculated from:

$$Q_i = Q_2 - Q_1 - Q_G = -N_i e \tag{13.76}$$

Much of the analysis is similar to that which was evaluated in Section 13.4.4. The gate potential will modify the junction potentials with respect to Equations (13.63) and (13.64) to give:

$$V_1 = \frac{[V(C_G + C_2) - C_G V_G - Q_i]}{C_{eq}} \tag{13.77}$$

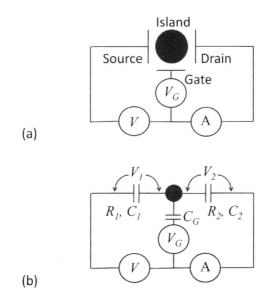

FIGURE 13.19: Single electron transistor (a) schematic outline. (b) Equivalent circuit for the SET.

and

$$V_2 = \frac{[VC_1 + C_G V_G + Q_i]}{C_{eq}} \tag{13.78}$$

where now $C_{eq} = C_1 + C_2 + C_G$ and the electrostatic energy will now include the contribution from the gate capacitance, $e^2/2C_G$, sich that:

$$E_s = \frac{[(V - V_G)^2 C_1 C_G + V^2 C_1 C_2 + V_G^2 C_G C_2 + Q_i^2]}{2C_{eq}} \tag{13.79}$$

The corresponding expressions for the work done in electron tunneling are now given by:

$$W_s(N_1) = -N_1 \left[\frac{eVC_2}{C_{eq}} + \frac{C_G}{C_{eq}} e(V - V_G) \right] \tag{13.80}$$

and

$$W_s(N_2) = -N_2 \left[\frac{eVC_1}{C_{eq}} + \frac{C_G}{C_{eq}} eV_G \right] \tag{13.81}$$

The energy change due to an electron tunneling through junction 1 will now take the form:

$$\Delta E_1^{\pm} = \frac{e}{C_{eq}}\left\{-\frac{e}{2} \mp [N_i e + V(C_G + C_2) - V_G C_G]\right\} \quad (13.82)$$

The corresponding expression for junction 2 will be:

$$\Delta E_2^{\pm} = \frac{e}{C_{eq}}\left\{-\frac{e}{2} \pm [N_i e - VC_1 - V_G C_G]\right\} \quad (13.83)$$

The role of the gate can now be seen to be that of controlling the effective charge on the island as it can be used to shift the Coulomb blockage condition with V_G. The application of the gate voltage effectively shifts the energy levels on the QD up or down, depending on the polarization of the applied bias. It is therefore possible to establish stable Coulomb blockade regimes with $N_i > 0$. As with the case for the simple quantum dot, the low temperature tunneling condition for the SET is given for: $\Delta E_{1,2} > 0$, such that the total energy will be reduced by the tunneling event. The case for forward and reverse tunneling can now be expressed as:

$$-\frac{e}{2} \mp [N_i e + V(C_G + C_2) - V_G C_G] > 0 \quad (13.84)$$

$$-\frac{e}{2} \pm [N_i e - VC_1 - V_G C_G] > 0 \quad (13.85)$$

These conditions refer to the four possible tunnel events between the contacts and the island. The Coulomb blockade will correspond to the case where the four conditions in Equations (13.84) and (13.85) are not satisfied. This means that they are blocked to electron transfer. Adjusting the source - drain or gate voltages can modify this situation and open one or more of the channels for tunneling. This will allow the transfer of electrons to and from the island, which will occur in a very controlled manner. In Figure 13.20, we show various situations, including (a) $V = 0$; $V_G = 0$: where no channels are open, i.e., Coulomb blockaded, (b) $V \geq e^2/C_{eq}$; $V_G = 0$: Left electrode to island channel allows tunneling to island, and from the island to the right electrode and electron can also tunnel, (c) $V = 0$; $V_G < 0$: the negative gate voltage will shift the energy states on the island down with respect to the electrodes opening up one way channels from the

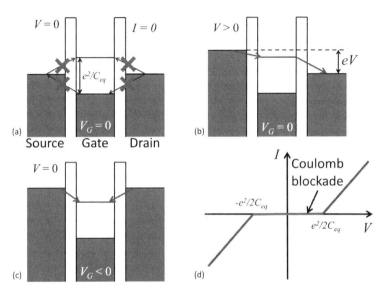

FIGURE 13.20: Variation of the band diagram for the single electron transistor for different applied potentials between source and drain, V and gate, V_G. (a) $V \leq e^2/C_{eq}$; $V_G = 0$; Coulomb blockade. (b) $V \geq e^2/C_{eq}$; $V_G = 0$: channels from left to right open for current flow. (c) $V = 0$; $V_G < 0$; Discrete number of electrons can transfer to the island, after which it will become Coulomb blockaded. (d) Current - voltage variation for the SET.

electrodes to the island. However, once the states are occupied the island becomes Coulomb blockaded again, thus closing all tunneling channels. In Figure 13.20 (d), we show the schematic form of the current - voltage characteristic. It is seen that for voltages between $\pm e/C_{eq}$, there will be no current since this corresponds to the Coulomb blockade regime. Once this potential is reached, the current will start to flow as charge can transfer via quantum tunneling. Further jumps will also occur at periodic voltages, giving a step-like form the the $I - V$ curve.

Using the above conditions, given by Equations (13.84) and (13.85), we can construct a stability diagram for electron transfer, which can be expressed in terms of the two potentials, V and V_G. To simplify the situation we will take the special case for $C_G = C_2 = C$; $C_1 = 2C$. The stability diagram for this situation is illustrated in Figure 13.21. The structure of the Coulomb blockade regions is also referred to as the *Coulomb diamond* formation. We note that the four solid lines correspond to the four critical conditions, where in

Equations (13.84) and (13.85), the $>$ symbol is substituted with an $=$ symbol, with $N_i = 0$ and are labelled with numbers (1) - (4). Further delimiting occurs for discrete numbers of charges on the QD; $N_i \neq 0$.

The periodic structure of the Coulomb diamond formation gives the SET the capacity to exhibit Coulomb oscillations, which occur as the gate voltage is swept with a small static potential between the source and drain, i.e., following a horizontal line along the stability diagram, as shown by the dashed lines in Figure 13.21. The periodicity of the peaks will be give by: $\Delta V_G = e/C_G$ and peak values occur when the energy levels on the island pass the Fermi energy of one of the contacts (depending on the voltage bias of V, which is a resonance condition of the structure).

Further complexity can be added by coupling quantum dots, all of which can be individually gated to control the passaged of electrons from the external circuit and between the quantum dots themselves. Since each of the gates can be controlled separately, a high degree of control can be achieved, where applications such as the *electron turnstile* and the *single electron pump* can be produced. By

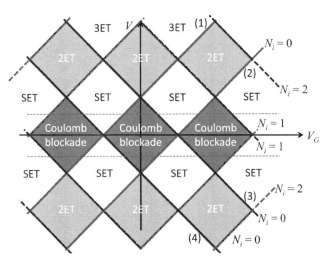

FIGURE 13.21: Stability diagram for a single electron transistor, which is expressed as the plot of $V - V_G$. Along the V_G axis we see a periodic array of Coulomb blockade conditions. These are separated by the conditions for single electron transfer (SET), which requires a small potential between source and drain, V. Further from the origin we see the additional conditions that can be found for the transfer of discrete packets of electrons, 2ET, 3ET etc., which occur for $N_i > 0$.

applying a periodic or alternating signal to the gate(s), the current can be controlled directly from the frequency of the gate signal:

$$I_{SET} = ef_G \tag{13.86}$$

where the current is simply the product of the electron charge and the frequency of the ac gate voltage, f_G.

13.4.7 Optical Properties of Nanostructures

Optical absorption in quantum well type structures occurs via the transfer of photon energy in the excitation of electrons between (filled states in the valence band and empty states in the conduction band) allowed transitions, which are governed by the *selection rules* of the system. In general, optical transitions of this type can be determined using *Fermi's golden rule*, where the conservation of energy is satisfied: $E_f = E_i + \hbar\omega$. The general form of Fermi's golden rule can be expressed as:

$$W_{i \to f} = \frac{2\pi}{\hbar} |M|^2 g(\hbar\omega) \tag{13.87}$$

which expresses the rate of transition in terms of the joint density of states at the photon energy, $g(\hbar\omega)$, and the *matrix elements*:

$$|M| = \langle \psi_f | \mathcal{H}' | \psi_i \rangle \tag{13.88}$$

where the interaction Hamiltonian, $\mathcal{H}'$, refers to the effect of the electric field of the photon, $\varepsilon_0 e^{\pm i\mathbf{k}\cdot\mathbf{r}}$, with the dipole moment, $\mathbf{p} = -e\mathbf{r}$. Using the Bloch functions for the electron states, the matrix elements take the form:

$$\begin{aligned}|M| &= \frac{e}{\Omega} \int u_f^*(\mathbf{r}) e^{-i\mathbf{k}_f \cdot \mathbf{r}} (\varepsilon_0 \cdot \mathbf{r} e^{\pm i\mathbf{k}\cdot\mathbf{r}}) u_i(\mathbf{r}) e^{i\mathbf{k}_i \cdot \mathbf{r}} d^3\mathbf{r} \\ &= \frac{e}{\Omega} \int u_f^*(\mathbf{r}) u_i(\mathbf{r}) (\varepsilon_0 \cdot \mathbf{r}) e^{i(\mathbf{k}_i - \mathbf{k}_f \pm \mathbf{k}) \cdot \mathbf{r}} d^3\mathbf{r} \end{aligned} \tag{13.89}$$

where Ω is the volume of the crystal, over which the integration is taken. The exponential shows the conservation of momentum condition of the phase factor, which must be zero if there is to be any absorption, otherwise the integral will be zero: $\mathbf{k}_f - \mathbf{k}_i = \pm \mathbf{k}$, where $\mathbf{k}$ is the photon momentum. The requirement is also inherent in the periodicity of the Bloch function. At optical frequencies, the wave vector for the photon will be much smaller than that for the electrons and we have: $\mathbf{k}_f \simeq \mathbf{k}_i$.

The factor $g(\hbar\omega)$ is the joint density of states to be evaluated at the photon energy. For the electrons in the bands we can write:

$$g(E)dE = 2g(k)dk \quad \text{or} \quad g(E) = \frac{2g(k)}{dE/dk} \tag{13.90}$$

where the factor 2 arises from spin degeneracy. The factor dE/dk is the gradient of the dispersion relation $(E - k)$ in the band structure. From previous chapters, we see that we can write:

$$g(E)dE = \frac{1}{2\pi^2}\left(\frac{2m_e^*}{\hbar^2}\right)^{3/2} E^{1/2} \tag{13.91}$$

and

$$g(k)dk = \frac{k^2}{2\pi^2}dk \tag{13.92}$$

The evaluation of the joint density of states for $g(E)$ at E_i and E_f, which are related to $\hbar\omega$ via the band structure, is performed as follows: Conservation of energy for photon absorption via the excitation of an electron from the valence to the conduction band gives:

$$\hbar\omega = \Delta + E_C(\mathbf{k}) + E_V(\mathbf{k}) = \Delta + \frac{\hbar^2 k^2}{2m_e^*} + \frac{\hbar^2 k^2}{2m_h^*} \tag{13.93}$$

where Δ is the band gap energy. Therefore, for absorption at $E = \hbar\omega$ we must have:

$$g(\hbar\omega) = 0; \quad \hbar\omega < \Delta \tag{13.94}$$

$$g(\hbar\omega) = \frac{1}{2\pi^2}\left(\frac{2\mu}{\hbar^2}\right)^{3/2}(\hbar\omega - \Delta)^{1/2}; \quad \hbar\omega \geq \Delta \tag{13.95}$$

μ is the reduced mass $(\mu = (1/m_e^* + 1/m_h^*)^{-1}$, where $m_h^* = m_{hh}^*$ or m_{lh}^*. We therefore see that there will be no absorption for $\hbar\omega < \Delta$, and the absorption coefficient, $\alpha(\hbar\omega) \propto (\hbar\omega - \Delta)^{1/2}$.

In the specific case of a QW structure, where the plane of the layer is perpendicular to the z-direction and with photon propagation also along z, such the polarization vector lies in the $x - y$ plane we have: $\langle \psi_f | x | \psi_i \rangle = \langle \psi_f | y | \psi_i \rangle \neq \langle \psi_f | z | \psi_i \rangle$. For the general case of a transition between the nth state in the valence band and the mth state in the conduction band of the QW, we can express the corresponding wave functions as:

$$\psi_i = \frac{1}{\sqrt{\Omega}} u_V(\mathbf{r}) \phi_{hn}(z) e^{i\mathbf{k}_{x,y} \cdot \mathbf{r}_{x,y}} \quad (13.96)$$

$$\psi_f = \frac{1}{\sqrt{\Omega}} u_C(\mathbf{r}) \phi_{em}(z) e^{i\mathbf{k}'_{x,y} \cdot \mathbf{r}_{x,y}} \quad (13.97)$$

in which $\phi_{hn}(z)$ and $\phi_{em}(z)$ denote the bound states in the QW in the z-direction. As stated previously, since the photon momentum is negligible, we can write: $\mathbf{k}_{x,y} = \mathbf{k}'_{x,y}$. The matrix elements for the QW can be written as the product:

$$M = M_{CV} M_{mn} \quad (13.98)$$

with

$$M_{CV} = \langle u_C(\mathbf{r}) | x | u_V(\mathbf{r}) \rangle \quad (13.99)$$

$$M_{mn} = \langle \phi_{em}(z) | \phi_{hn}(z) \rangle \quad (13.100)$$

This latter shows that the optical transitions in the QW are proportional to the overlap in the electron and hole states functions. Using this we can evaluate the selection rules $\Delta n = m - n$. For simplicity we consider the case of an infinite QW potential, which gives:

$$M_{mn} = \frac{2}{d} \int_{-d/2}^{d/2} \sin\left(k_n z + \frac{n\pi}{2}\right) \sin\left(k_m z + \frac{m\pi}{2}\right) dz \quad (13.101)$$

where d is the thickness of the QW. This will give unity for $n = m$ and is otherwise zero. Therefore we obtain the selection rule: $\Delta n = m - n = 0$. The allowed transitions are illustrated in Figure 13.22.

In Figure 13.23, we show the absorption spectrum for a multiple quantum well structure, which is compared with the corresponding case for a bulk material. We note that the step-like structure of the

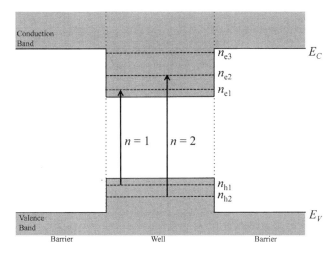

FIGURE 13.22: Allowed transitions for a QW structure for $\Delta n = 0$, with $n = 1$ and $n = 2$.

absorption is reminiscent of the shape of the density of states for the 1D confined structures. Heavy hole (hh) and light hole (lh) excitonic states are evident for the $n = 1, 2$ and 3 edges. The reason that these are observed as enhanced peaks at the edges of the transition energies is that the Coulomb interaction increases the absorption rate and hence produces the resonant peaks. The electron - hole pair formation and the Coulomb interaction between the two states produces a bound pair in the form of a quasi-particle (exciton). The excitonic states, since they have one electron bound to a positive

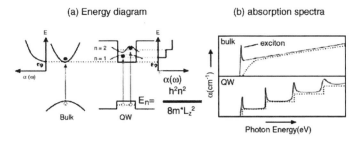

FIGURE 13.23: Absorption spectrum for a multilayered quantum well structure. (a) Energy diagrams for the excitation of an electron from the valence to the conduction band. (b) Absorption spectra for the bulk and QW systems. Note that the peaks at the absorption edges arise from the formation of excitons, see text. [Reprinted figure with permission from: H. Hosono, Thin Solid Films, **515**, 6000 (2007)]

(hole) charge, have energies which resemble the hydrogenic states of the Bohr model. The energy of the nth level relative to the ionization energy can be expressed in the form:

$$E_n^{ex} = -\frac{\mu R_H}{m_e \epsilon_r^2 n^2} = -\frac{R_X}{n^2} \quad (13.102)$$

In analogy with the Bohr model we can define the excitonic radius for the electron - hole orbit, which for the nth state will be:

$$r_n^{ex} = -\frac{m_e \epsilon_r n^2}{\mu} a_0 = -\frac{n^2}{a_X} \quad (13.103)$$

where a_0 is the Bohr radius and a_X the excitonic Bohr radius. For the $n = 1$ state in GaAs we have $R_X = 4.2$ meV and $a_X = 13$ nm. In QW and confined systems, the excitonic states are further enhanced since the electron - hole pair are forced to be even more closely bound, thus increasing the attractive potential.

Further to the absorption behavior, QW structures have important applications in electroluminescence. In fact, this is one of the major applications of QW systems. The use of QW structures provides a significant improvement on device performance and efficiencies and have the advantage of allowing the active wavelengths to be tuned via the thickness of the active region. The generation of light is provided by the recombination of electrons with holes, giving a luminescent spectrum which peaks at the energy of the gap. The width of the peak is related to the carrier density and temperature. As we noted in Chapter 9, the energy of the QW peak is up-shifted with respect to the bulk semiconductor due to the confinement effects discussed and we can tune the energy to the desired wavelength in the fabrication of the device. Such flexibility means that we can fabricate devices with better output characteristics. This is exploited in the manufacture of LEDs and semiconductor lasers, for example.

The confinement of electronic states in QW like structures can lead to multiple states where transitions can occur between the different bound states in the same band. Such processes are referred to as *intersubband transitions*, e.g. between states n_{e1} and n_{e2}, see Figure 13.22. In the valence band this will occur for hole states, while in the conduction band for electrons. Since the energies involved are much lower the corresponding wavelengths will be longer. Such transitions typically occur in the infrared region of the electromagnetic spectrum, (infrared) heat sensors exploit such processes.

Quantum dots and nanoparticles have confinement effects in all three spatial dimensions, but the basic principles are essentially no different from the QW structures we have considered above.

In addition to the confined electronic states, structured materials can be used to produce some rather interesting optical properties. Of particular importance are periodically structured materials with differing refractive indices. These provide quasi-crystal structures which function in a similar way to the periodic nature of real crystals. However, in this case it is not the wave-like nature of electrons that are of importance, but that of photons. In analogy, such periodic materials are referred to as *photonic crystals*. In fact, *photonics* has become a major area of research and has some important applications. The color of some insects and the wings of butterflies are actually due to the periodic structures that naturally occur and not from pigments. This rather graphically illustrates the relevance of photonic systems and the active wavelengths are only related to the periodicity and relative dielectric constants of the materials used. The model used to calculate the optical properties is in fact very similar to the Kronig - Penney model used in idealized periodic potentials for electronic systems. The calculations for photonic crystals are based on the Maxwell equations rather than the Schrödinger equation. Furthermore, well defined defects, and arrays of defects, in the periodic structures can be used in ways similar to the quantum confined structures of semiconductors. These use confinement effects of the photons for very specific applications, such as high - Q cavities, waveguides, and cladding for optical fibers.

13.5 ASPECTS OF NANOMAGNETISM

As with the case of electronic properties, the size reduction of magnetic entities also has a number of important effects on their magnetic properties. One of the principal properties that govern the behavior of a magnetic body are the magnetic anisotropies at play. We have outlined the principal sources of anisotropy that occur in magnetic systems, these are: magnetocrystalline anisotropy, shape anisotropy, magnetoelastic anisotropy, and surface anisotropy. The other main consideration for an ordered magnetic material is the exchange energy of the system.

In terms of the reduced dimensions in nanosystems, there are a number of effects that are important. Firstly, the large amount of surface will mean that surface anisotropy effects will become crucial. This can dominate for very small nanoparticles, where a significant proportion of the atoms in the object are located on the surface. Of course, as with other properties, the length scales for physical properties are important and we will outline these in the next section.

13.5.1 Magnetic Length Scales

The principal magnetic length scales arise from the energy considerations we met in Chapter 10. There are two main characteristic length scales that are of importance, these are:

1) The *exchange length*, which is defined as:

$$\Lambda_{exch} = \sqrt{\frac{2A}{\mu_0 M_s^2}} \tag{13.104}$$

2) The domain (Bloch) wall width:

$$\delta_{DW} = 2\sqrt{\frac{A}{K}} \tag{13.105}$$

In both of these expressions we see that the exchange stiffness constant, A, appears. These expressions give an indication of the length over which the exchange energy dominates the magnetostatic energy, Equation (13.104), and over the magnetic anisotropy energy, Equation (13.105).

The energy considerations are central to much of our thinking with regards to the size effects in magnetic objects of reduced dimensions. For example, the largest size that a magnetic particle may support a single magnetic domain will depend on the magnetic energies involved. This essentially means that we need to consider how much energy is saved by having a single domain rather than two domains, which will have domain wall energy, but reduces the magnetostatic energy. Such considerations lead to the critical diameter, for a spherical particle, of:

$$D_{cr} = \frac{9\pi\sqrt{AK}}{\mu_0 M_s^2} \qquad (13.106)$$

This clearly reflects the magnetic energies at play. In Table 13.5, we show some typical values of these critical length scales selected ferromagnetic materials. The calculated values are based on the bulk parameters for these materials

We note that a ferromagnetic (or ferrimagnetic) particle below the critical dimensions, given in Equation (13.106), will have a single magnetic domain and is thus referred to as a *single domain particle* or sdp. We further note that Equation (13.106) is based on a spherical particle which has no shape anisotropy. Clearly elongated particles will have an additional anisotropy contribution and, depending on the relative orientations of the magnetocrystalline and shape anisotropies, can have very different critical dimensions. The shape anisotropy can be used as a "tuning" parameter for magnetic behavior in nanostructures since it is an extrinsic parameter.

13.5.2 The Stoner - Wohlfarth Model

The Stoner - Wohlfarth model, which dates back to 1948, is a simple attempt to evaluate the hysteretic behavior of ferromagnetic systems. It has subsequently become a very important first approximation method for the study of ferromagnetic nanoparticle systems. In the model we assume the particle to be a single magnetic domain with uniaxial anisotropy. This assumption is based on

TABLE 13.5: Magnetic length scales for selected ferromagnetic materials.

Material	Anisotropy $\times 10^4$ (J m^{-1})	Λ_{exch} (nm)	δ_{DW} (nm)	D_{cr} (nm)
bcc Fe	4.81	3.3	20.3	6.9
fcc Co	−12	4.8	15.8	10.1
hcp Co	41.2	4.7	8.3	13.7
fcc Ni	−0.56	7.6	39.2	16.2
fcc Ni$_{80}$Fe$_{20}$	0.027	5.1	199	10.8

the original premise of the model that the magnetization remains constant throughout the sample. This has the consequence of maintaining the total exchange energy constant during the magnetization reversal process. Such behavior is termed *uniform* or *coherent rotation* and is also appropriate for weakly or non-interacting magnetic nanoparticle assemblies.

The magnetic anisotropy of the particle, which we will consider as uniaxial, can be of magnetocrystalline or magnetostatic origin, since it has no imprtant bearing on the outcome of the model. We will use just an effective anisotropy for this reason. As such, we can express the free energy of the system as:

$$E_{SW} = K_{eff} \sin^2 \theta - \mu_0 M_s H \cos(\theta - \phi) \qquad (13.107)$$

where angles are defined in Figure 13.24. For our definition we have $K_{eff} > 0$, so that an energy minimum occurs for $\theta = 0$, i.e., with the magnetization along the easy axis. For further discussion of such considerations, see Skomski (2008).

The equilibrium orientation of the system is defined by the configuration which gives a minimum energy, indicated as $\theta = \theta^*$. The condition for minimizing the free energy is given by:

$$\left(\frac{\partial E_{SW}}{\partial \theta}\right)_{\theta=\theta^*} = 0 \quad \text{and} \quad \left(\frac{\partial E_{SW}^2}{\partial \theta^2}\right)_{\theta=\theta^*} > 0 \qquad (13.108)$$

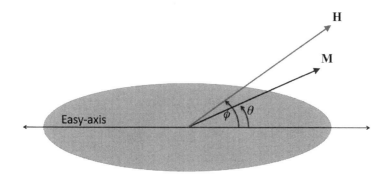

FIGURE 13.24: Schematic diagram of a magnetic nanoparticle with uniaxial anisotropy. Definitions of orientations of the applied magnetic field and magnetization with respect to the easy axis are shown.

From this we obtain:

$$[\sin\theta\cos\theta - h\sin(\theta-\phi)]_{\theta=\theta^*} = 0 \qquad (13.109)$$

and

$$[\cos 2\theta + h\cos(\theta-\phi)]_{\theta=\theta^*} \geq 0 \qquad (13.110)$$

where we have used the following normalizations: $h = H/H_K$; $m = M/M_s$ and $H_k = 2K_{\mathit{eff}}/\mu_0 M_s$ is called the *anisotropy field*. The calculations of the magnetization state are made by varying the strength and direction of the applied magnetic field, where the magnetization will respond by aligning in that field due to the relative strength and direction of the magnetic anisotropy as well as that of any applied magnetic field. It is customary to decompose the magnetization into parallel and perpendicular components, which we can express as: $m_\parallel = \cos(\theta-\phi)$ and $m_\perp = \sin(\theta-\phi)$. The variation of these transverse and longitudinal components of the magnetization are illustrated in Figure 13.25 for different orientations of the applied magnetic field.

In the Stoner - Wohlfarth model we see that the hysteretic magnetic behavior arises from the magnetic anisotropy, where the coercive field, H_c, is related to the anisotropy field. The model is also used to show the magnetization reversal or switching behavior in single domain particle systems. This is achieved by evaluating the boundaries of the hysteresis behavior via the first and second derivatives of the free energy. In fact, the entire model is based on the variation of the energy landscape, in which the minima determine the orientation of the magnetization which is affected by the strength and orientation of an applied magnetic field with respect to those of the magnetic anisotropies. The switching field dependence can be expressed in terms of the anisotropy and orientation of the applied magnetic field in the form:

$$H_{SW} = \frac{H_K}{(\sin^{2/3}\phi + \cos^{2/3}\phi)^{3/2}} \qquad (13.111)$$

This is displayed for the applied fields in the parallel and perpendicular directions. The form of the switching field curve is commonly

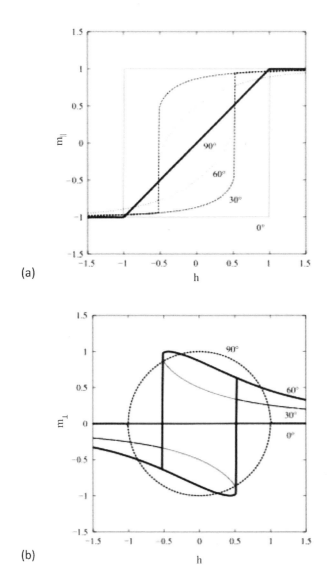

FIGURE 13.25: (a) Transverse, $m_\perp$, and (b) longitudinal, $m_\parallel$, components of the magnetization. The angles refer to the direction of the applied magnetic field for $\phi = 0°, 30°, 60°$ and $90°$. [©IOP Publishing. Reproduced with permission. All rights reserved. C. Tannous and J. Gieraltowski, Eur. J. Phys. **29**, 475487 (2008)].

known as the *Stoner - Wohlfarth astroid*. This is illustrated in Figure 13.26, where we show the theoretical curve and a comparison with experiment as measured using a micro-SQUID device.

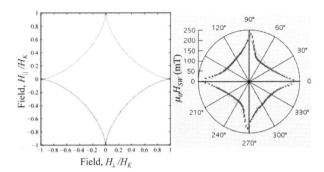

FIGURE 13.26: Switching fields for a single domain particle with uniaxial anisotropy. Due to the shape of the curve, this is popularly known as the Stoner - Wohlfarth astroid. Inside the astroid, the particle is capable of switching, while outside it is not. The right hand image shows an experimental micro-SQUID measurement of a single ferromagnetic nanoparticle, showing good agreement with the theory. [©IOP Publishing. Reproduced with permission. All rights reserved. C. Tannous and J. Gieraltowski, Eur. J. Phys. **29**, 475487 (2008). Reprinted figure with permission from: W. Wernsdorfer et al., J. Appl. Phys., **81**, 5543 (1997). Copyright 1997 by the American Institute of Physics.]

While the Stoner - Wohlfarth model is a good first approximation, it does assume that the magnetization is always homogeneous. This situation is generally favored only in small magnets which has the form of an *ellipsoid of rotation*. This means that the particle acts as a single macrospin and the exchange energy is a dominant energy in the system. Deviations from this model can occur for a number of reasons, such as edge and surface anisotropies, which can cause some form of non-collinear alignment of the magnetization at the surface, meaning $\nabla \mathbf{M} \neq 0$. Defects and other magnetic inhomogeneities will also bring about deviations from the model and add further contributions to the coercive field, H_c, which in the Stoner - Wohlfarth model arise only from the anisotropy. Non-coherent reversal of the magnetization can also occur in elongated single domain particles and therefore isn't well described by the SW model. Such processes are usually described by a *curling* of the magnetic moments in the particle.

13.5.3 Superparamagnetism and Ferromagnetic Nanoparticles

One of the fundamental problems of the reduced dimensions of ordered magnetic structures is the thermal instability that results from a reduction in the magnetic anisotropy energy with respect to the thermal energy. This leads to thermal fluctuations in the orientation of the magnetization state and ultimately to the loss of a well defined global magnetization state of the system. This gives rise to the so-called *superparamagnetic limit*, which fundamentally depends on the magnetic anisotropy energy and size of the magnetic entity. Also of great importance is the method of observation of the magnetic state, which has a characteristic measurement time. We will outline these concepts below.

The question of thermal agitation of the magnetic state of a sample was addressed by Nobel laureate Louis Néel as early as 1949 and was further considered by W. F. Brown in the early 1960s. The subsequent *Néel - Brown* model considers the probability per unit time of the magnetization of a single domain (nano)particle not being reversed by thermal agitation after a certain time, t, which is given by the exponential relation:

$$P = \nu_0 e^{-\Delta \varepsilon V / k_B T} \qquad (13.112)$$

This result is frequently quoted in a slightly different form, which considers the reversal or relaxation time, which is expressed as:

$$\tau = \tau_0 e^{\Delta \varepsilon V / k_B T} \qquad (13.113)$$

where the attempt frequency is given by; $\nu_0 = \tau_0^{-1}$, and has a value of the order of GHz and is in the range of ferromagnetic resonance measurements. The factor $\Delta \varepsilon V$ is the energy barrier between the different energy minima which describe the most stable state (orientation) of the magnetization. Here $\Delta \varepsilon$ can be considered as the anisotropy free energy density, given by K_u for uniaxial magnetocrystalline anisotropy or $\Delta \mathcal{N} \mu_0 M_s^2$ for strong shape anisotropy, with the factor $\Delta \mathcal{N}$ being the difference of the demagnetizing factors along the principal axes of the particle and depend on its shape. The quantity V is the particle volume. In Figure 13.27, we illustrate the energy landscape for the case of uniaxial anisotropy in the particle, with easy directions being along 0° and 180°. The barrier height corresponds to the uniaxial anisotropy constant multiplied by the particle volume, $\Delta \varepsilon V = K_u V$ It is thus immediately evident that as the particle reduces in size, the energy barrier will follow suit. Once this energy barrier height becomes comparable to the thermal energy, $k_B T$, the system becomes very unstable, with the thermal fluctuation being sufficient

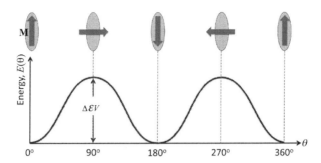

FIGURE 13.27: Energy due to the uniaxial anisotropy in a nanoparticle, showing the easy directions for 0° and 180°. The energy will be a minimum when the magnetization points along these directions. The application of a magnetic field will alter the energy landscape and can shift the position of the energy minimum, depending on its strength and orientation with respect to the easy axes.

to flip the magnetization direction along the easy axes of the magnetic system. (This is a similar situation to the original Brownian motion experiment, in which small particle are buffeted around by random thermal collisions with other particles.) In this situation, in which the particle can still be considered to be in a coherent magnetic state, where switching involves all spin inverting at the same time, the global state is not stable and looks to be paramagnetic if the time of measurement is greater than the inverse attempt frequency. Hence the name *superparamagnetic* is used to describe such a situation. Note that we have introduced the concept of measuring time. This can be thought of with analogy to taking a photograph of a moving object with different shutter speeds: when the shutter speed is slow with respect to the motion of the object, it will appear blurred in the image. If we increase the shutter speed so that it is faster than the movement of the object, then it will appear fixed. In the case of our magnetic particle, we have a similar situation. For slow measurement times, i.e., slower than the switching speed, the sample has no well defined magnetic orientation and the sample appears to the paramagnetic ($M_s = 0$). However, if we use a fast measurement time, i.e., quicker than the reversal time, then the sample appears to be in a fixed ferromagnetic state. Hence the importance of the method of observation.

A crucial consideration in the magnetism of nanostructures is the temperature at which the ferromagnetic state stabilizes with respect to the thermal energy. We define the *blocking temperature*,

T_B, as that temperature at which the system appears to be ferromagnetic, i.e., where the measurement of the magnetization gives a ferromagnetic like response. This will of course depend on the measurement technique employed. It is thus possible to use Equation (13.113) to express the blocking temperature in an Arrhenius type law of the form:

$$T_B = \frac{\Delta \varepsilon V}{k_B} \frac{1}{\ln(\tau_m/\tau_0)} \qquad (13.114)$$

where τ_m is the measurement time. For room temperature using ferromagnetic resonance techniques we find: $T_B \simeq 0.43(\Delta \varepsilon V/k_B)$. From this relation we see that for short measurement times the blocking temperature will be higher than for longer measurements, such as magnetometry methods, which can have measurement times on the scale of minutes.

The superparamagnetic limit is a very important consideration in the workings of nanomagnetic devices and applications, such as magnetic bit used in magnetic data storage technologies. The recording media industry is a huge global market and therefore these questions are worth a lot of money. So while nanotechnologies have been based on reducing the size of components, there are some physical limits that must be overcome. The superparamagnetic limit is a very important one. Clearly, the blocking temperature relies on the energy barrier, $\Delta \varepsilon V$. So if we reduce the volume of our object we need to compensate with an increased anisotropy. The principal solutions have been to use ferromagnetic materials with very high anisotropies, such as CoPt or FePt alloys, or we can use pillar like objects which have large shape anisotropies.

The exponential law governing the superparamagnetic regime means that there is a very fine separation between the sizes of objects which can be considered as stable. This is best demonstrated with an example. Consider a spherical nanoparticle of radius, r_{NP}, and uniaxial anisotropy, K_u. We now define the radius of a particle which is stable over 1 second and 1 year. These can be expressed using:

$$r_{NP} = \left[\frac{3k_B T}{4\pi K_u} \ln\left(\frac{\tau_0}{\tau_m}\right) \right]^{1/3} \qquad (13.115)$$

For an Fe nanoparticle ($K_u = 5 \times 10^4$ Jm^{-1}) at room temperature, we find the stable radius for 1 second to be about 7.5 nm, while for a year long stability the radius only increases to 9.4 nm. We can think of this as another magnetic length scale.

Magnetic interactions between particles can help to stabilize the magnetization and lead to the modification of the Arrhenius law, giving:

$$\tau = \tau_0 e^{\Delta \varepsilon V / k_B (T_B - T_0)} \tag{13.116}$$

This is referred to as the *Volger - Fulcher law*, where T_0 is an effective temperature which is proportional to H_i^2, the interaction field strength. Furthermore, an applied magnetic field will also stabilize the magnetic particle by the modification of the magnetic free energy, as given for example in the Stoner - Wohlfarth model. In this case the Arrhenius law has the modified form:

$$\tau = \tau_0 e^{\Delta \varepsilon V (1 - H/H_K)/k_B T} \tag{13.117}$$

where H_K is the anisotropy field and H the applied field strength.

It is worth mentioning that length scales in particle dimensions are important not only from the point of view of the properties, but also because their behavior can change in very specific ways and this means that we need to consider the best way to model their behavior. For large particles, i.e., above the critical radius, the particles can have a multidomain state. Once we are below this threshold most researchers model their behavior as a macrospin, which means that the magnetization is always considered to have a single value throughout the sample. Even in the reversal process, the magnetization has a fixed value in the homogeneous sample. This is one of the premises of the Stoner - Wohlfarth model. However, there is a further change that should be considered for nanostructures in which there is a significant proportion of the atoms in surface positions. In this case the local magnetic anisotropy can have a very different value and easy axes from that of the bulk or core atoms. In such cases the only correct way to model the behavior is as a collection of coupled spins. Modelling in this way is far from simple and some effective models are used which distinguish between bulk spins and an effective surface layer.

13.5.4 Magnetic Thin Films and Multilayers

Thin magnetic films represent an important category of system and have a long history of important applications and discoveries. As we have discussed earlier, such one dimensional size reduction illustrates the importance of surface effects and also has implications in the anisotropy of the sample. The difference in magnetic freedom at a surface can produce confinement effects in the generation of standing spin wave modes across the sample thickness. Additionally, the surface roughness of a film is an important parameter, since it can influence the surface magnetic properties, especially for ultrathin samples. The reduced coordination will obviously affect the local magnetic anisotropy and can give out-of-plane components to the magnetization and this is observed for example in Fe thin films for thicknesses between 2 and 5 monolayers (ML). Below 2 MLs there is no observed magnetization, which is attributed to a reduced Curie temperature, which again arises due to the reduced coordination, where for 2MLs there should be no atoms with full coordination. This reduces the exchange energy and hence the Curie temperature. Between 5 and 7 ML, there is a spin reorientation transition (SRT), which shifts the easy axis from the perpendicular to the in-plane direction. Changes in structural properties can also significantly alter the magnetic state, since changes in strain and hence interatomic separation will have a strong influence on the exchange interaction between neighboring spins. The deposition of any thin film will depend critically on the ambient conditions of substrate temperature and rate of deposition, which will affect the roughness of the surface and interface, while we should not forget the importance of the type of substrate itself, both in terms of its intrinsic nature, crystalline properties as well as imperfections and surface roughness.

The anisotropy of a thin film is often expressed as an effective anisotropy, taking into account various contributions. The effective anisotropy due to magnetocrystalline effects will have a thickness dependence since the surface contribution should be dominant for very thin films. For a thin film with asymmetric interfaces (s1 and s2) we can express the effective anisotropy as:

$$K_{eff}(t) = \left(K_{bulk} + \frac{K_{s1} + K_{s2}}{t}\right)\sin^2\theta \quad (13.118)$$

where we have considered the case of a uniaxial bulk anisotropy that is collinear with the two surface contributions.

The effects of sample temperature can reveal much about the magnetic properties of thin films. With increasing temperature, for example, we can expect the spectrum of thermally excited spin waves (or magnons) to increase in terms of the frequencies of the excited modes, much as we saw for the case of phonons. Furthermore, Stoner excitations (thermal reversal of electron spins) will also contribute to the general destruction of the magnetic order. The thickness at which the Curie temperature departs from the bulk value gives a measure of the length scale for the spin-spin coherence length. Such effects have been observed in thin films of a few tens of nm.

Magnetic multilayer systems, consisting of alternating ferromagnetic and non-magnetic layers, are a natural extension to the study of thin film systems. In addition to the magnetic surfaces and interfaces of 2D structures, the magnetic multilayers also depend on the interlayer magnetic interactions between adjacent magnetic films. The form of the interaction between the magnetic layers depends on a number of parameters, including thickness of intervening non-magnetic layer, its conductivity and crystallographic orientation. Also of importance is the interfacial roughness between the magnetic and non-magnetic layers. The magnetic properties of the individual layers also plays a central role. It should be clear that the coupling between magnetic layers can indeed be a complex problem. Much progress was made from the mid to late 1980s, when a refinement of deposition techniques and analytical tools allowed for high quality multilayers to be produced and characterized. One of the most important results to be found was the discovery of the oscillatory nature of magnetic coupling in metallic systems as a function of the thickness of the non-magnetic layer, in which the coupling is found to alternate between ferromagnetic and antiferromagnetic alignment and gradually decaying with the interlayer thickness. There are several mechanisms that can be produce such behavior, with the RKKY mechanism being one of the most popularly studied. This effect was first observed from the measurement of the giant magnetoresistance (GMR) effect. Ultimately this work led to the development of the study of the spin dependent electronic transport properties in devices such as magnetic tunnel junctions and spin valves. We have briefly introduced the topic of spintronics in Chapter 10.

Magnetic multilayers can be used in a number of applications, with the basic GMR effect along with the TMR (tunnel magnetoresistance) being the principle physical quantities exploited for magnetic sensor applications. These magneto-transport properties lend themselves to device applications, where for example the read heads in hard disk drives are based on the measurement of the magnetoresistance in a *spin valve* structure. This consists of a multilayer system where one of the ferromagnetic layers is fixed via the *exchange bias effect* through the coupling with an antiferromagnetic layer. The exchange bias causes the reference point of the hysteresis loop of the ferromagnetic layer to be shifted, along the magnetic field axis, by an amount, called the exchange bias field, which can be expressed as:

$$H_{EB} = \frac{nJ\mathbf{S}_{FM} \cdot \mathbf{S}_{AFM}}{M_{FM} t_{FM}} \qquad (13.119)$$

where n is the density of spins at the interface between the ferromagnetic and antiferromagnetic layers, which have spins, $\mathbf{S}_{FM}$ and $\mathbf{S}_{AFM}$, respectively. M_{FM} is the magnetization of the ferromagnetic layer, which has a thickness of t_{FM}. The only parameter that can be adjusted is the thickness of the ferromagnetic layer, which can be set such that the resistance state changes near $H = 0$. This will give the device a very sensitive response for low fields. This is sufficient to measure the magnetic fields of the magnetic bits on a hard disk and hence read its state. Indeed the active device used as the read head in magnetic disk drives typically has magnetic and non-magnetic layers of thicknesses as little as 1 to a few nm.

As we saw in Section 10.13, the spin wave spectrum depends on the intrinsic properties of the material in question, these are: the g-factor, the magnetization, magnetic anisotropies and exchange constants. Besides the thermal excitation of spin waves, we can use specific excitations of given energies if they are in the appropriate region of the frequency spectrum. Typically experiments are performed using photons in the visible region or alternating fields in the microwave to radio frequency range. In the Brillouin light scattering (BLS) type measurement, the shift in frequency from the incident beam via Stokes and anti- Stokes inelastic scattering processes, allow us to assess the energies associated with the generated spin waves. With the use of waveguide systems we can measure the absorption of radiation due to the excitation of magnetic resonances, which essentially gives us the same information. In addition to the uniform precession of the ferromagnetic resonance observed in bulk materials, low dimensional systems permit the excitation of higher order resonance modes due to the existence of magnetic boundaries, which can act as pinning centers. The resulting standing spin wave modes can be understood from the dimensions and pinning conditions of the magnetic boundaries.

In the simples case of a thin magnetic film with perfect pinning at the surfaces, i.e., the surface spins are rigidly fixed due to the surface anisotropy. The situation is rather like the vibrations on a string which is driven at its various resonance frequencies. The additional term in the resonance equation takes the form of Dk_n^2, where $D = 2A/M_s$ is the spin wave constant and for this case of perfect pinning the spin wave wave vectors take the form:

$$k_n = \frac{n\pi}{t} \qquad (13.120)$$

where n is an integer and indicates the excitation mode and t the thickness of the magnetic layer. The observation of multi-peaked absorption spectra is usually a sign of the existence of standing spin wave modes in thin films, and the pattern of resonance field and spin wave modal number provides us with an experimental route to the measurement of the spin wave constant and hence the exchange interaction in the ferromagnet. Furthermore, the measurement of the spin wave spectrum in magnetic multilayers allows us to compare the spectra with isolated thin films from which we can measure the frequency shifts which will allow us to evaluate the strength and sign of the magnetic exchange coupling between adjacent magnetic layers. For further discussion of ferromagnetic and spin wave resonance, see Section 10.13.

13.5.5 Magnetic Nanostructures

Size reduction in two and three dimensions will add more complexity to the general magnetic behavior of the object. This is because we are adding the surface component of the ferromagnetic body and may produce additional confinement effects. In this section, we will outline some of the basic effects due to size considerations and review some of the principal types of objects, including ferromagnetic wires, rings, magnetic dots, and antidot systems. The constraints imposed on the magnetic moments in a system of reduced dimensions can produce some very specific types of behavior, especially with respect to domain walls and their motion. Some of the principal forms of nanostructure are illustrated in Figure 13.28.

Magnetic nanowires can produce spin configurations that are very different from the usual magnetic behavior, resulting from the effect of the surface and the interaction of domain walls with the surface of the wire. Such nanowire systems have been proposed as an alternative method for magnetic data storage systems or *racetrack memory*. In such an application, magnetic domains, with orientations along the axis of the wire, are produced and moved along the wire via the application of a magnetic field or through the passage of a spin-polarized electrical current in the wire. An electrical current can become polarized in a ferromagnetic material and as the

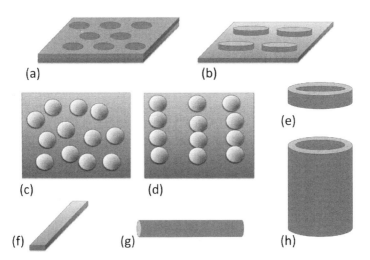

FIGURE 13.28: Examples of the types of nanostructures commonly studied. (a) Antidots, (b) nanodots, (c) assemblies of randomly spaced and oriented nanoparticles, (d) chains of nanoparticles joined due to dipolar interactions, (e) nanorings, (f) flat nanowires, (g) nanowires with circular cross-section, (h) nanotubes.

electrons pass from one domain to the next they impart their angular momentum to the domain wall and shift its position. By modifying the surface structure, such as with a notch, it is possible to produce pinning sites for domain walls.

In the case of nanoparticles, we have outlined some of the more important physics at play in the previous two sections. However, in addition to this we need also to underline the importance of inter-particle interactions, typically due to dipolar forces between the magnetic particles. This can bring about the alignment and formation of linear chains, if the particles are free to move, say in a liquid suspension. There are many methods for the production of nanoparticles and any description goes beyond the scope of this book. However, in most cases the production method produces an assembly of nanoparticles with a distribution of sizes. This most often takes the form of a *log - normal distribution*, given by:

$$P(V) = \frac{1}{\sqrt{2\pi}\sigma V} e^{-\frac{[\ln(V/V_0)]^2}{2\sigma^2}} \qquad (13.121)$$

where V is the particle volume, which has an averages size of V_0 and σ is the standard deviation of $\ln V$. This appears as a skewed distribution, with a longer tail at the high side of the peak value. Ferromagnetic nanoparticle assemblies have been extensively studied and find a number of applications, and notably in biological and medical areas. The surfaces of certain nanoparticles can be coated or functionalized, which allows them to become biocompatible. Inside the body they can be used for labelling, selective MRI enhancement, hyperthermia treatments, magnetic filtering among other biomedical applications[6].

The magnetic dots structure, which can be square, circular, elliptical or rectangular, is usually made by some form of electron or photolithography process. In square and rectangular structures, it is common that the magnetic configuration has some form of domain closure, meaning that in the absence of an applied magnetic field, the object has no net magnetization. In the case of circular dots, the ground state is called the vortex state and again has no net magnetization in the sample plane, see Figure 13.29. In fact, the region at the center of a square dot can also be see to be a vortex structure.

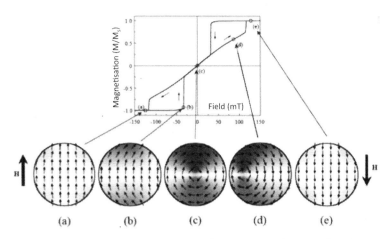

FIGURE 13.29: Magnetic vortex state with corresponding hysteresis loop for a magnetic disk or dot. (This simulation was carried out using the OOMMF software, see *http://math.nist.gov/oommf/*.)

The center of the vortex avoids being a true singularity by having the magnetic moment pointing out of the plane of the disk. The vortex state is characterized by its chirality and direction of the out of plane moment. It is possible to transit between these states at the cost of some magnetic energy. In Figure 13.29, we show the hysteretic behavior for a magnetic disk with a vortex as the ground state. As we stated previously, the ground state of the disk is a vortex with zero net magnetization in the plane. An applied field in the plane of the sample will cause the vortex core to shift to one side, favoring the net magnetization in the direction of the applied field. Beyond a critical field the core will be expelled from the magnetic disk. We can see from the hysteresis loop, that if we maintain the magnetic field to a low enough value, so as to not remove the vortex core, there is effectively no hysteresis, with the magnetization being reversible. Once the core is eliminated, hysteretic behavior will resume. The loops in the M-H cycle arise from this ferromagnetic character where the motion of the magnetization is irreversible. Applying ac magnetic fields can cause the vortex core to precess in *gyrotropic motion*. The motion has a natural resonance frequency, which typically lies in the range of a few hundred MHz. It is possible in some nanostructures to have more than one vortex core, for example in an elliptical magnetic particle. Since the vortex cores can have different states, they can be made to interact, with attractive and repulsive forces being possible, depending on the chirality and orientation of the central vortex core.

It is possible to imagine removing the central core from a magnetic vortex in a magnetic dot. The result is a nanoring structure, see Figure 13.28 (e). The ring structure can support a vortex state and will have chirality. However, the application of a magnetic field can only lead to the introduction of domain structures and hence hysteresis. From the zero field, or ground state, with zero magnetization in the vortex state (closure state), a magnetic field will lead to saturation when the magnetization is aligned with the field in the so-called *onion state*, see Figure 13.30. Increasing the thickness of a nanoring

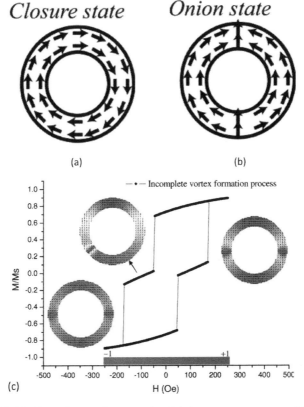

FIGURE 13.30: Magnetic states for a nanoring: (a) closure or vortex state and (b) onion state. (c) Magnetic hysteresis loop for the nanoring, showing the different magnetic configurations as a function of the applied magnetic field. [Reprinted figure with permission from: C. C. Chen et al., IEEE Trans. Magn., **45**, 3546 (2009).]

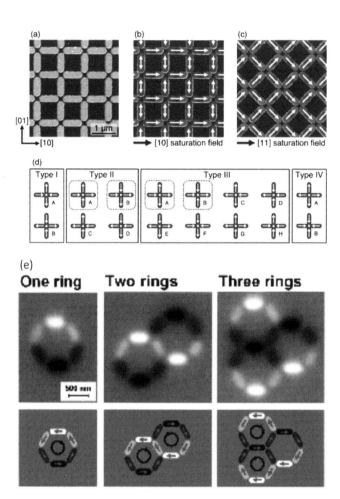

FIGURE 13.31: Artificial spin ices. (a) Image of a square array of nanodots. The major symmetry axes are indicated in the lower left corner. (b) Schematic illustration of the remanent magnetic configuration of the spin ice array after applying a saturation field parallel to the [10] direction and (c) parallel to the [11] direction. (d) The 16 possible remanent magnetic configurations for the artificial square spin ice vertices. After applying a saturation field along the [10] direction there is a fourfold degeneracy of the possible remanent vertex configurations. In comparison, after removing the [11] field, there is only one possible remanent vertex configuration. (e) Artificial kagome lattice with one, two and three rings. The different configurations are illustrated. [(a) - (d)©IOP Publishing. V. Kapaklis et al., New J. Phys., **14**, 035009 (2012). Reproduced with permission. All rights reserved. (e) Reprinted figure with permission from: E. Mengotti et al., Phys. Rev B, **78**, 144402 (2008). Copyright 2008 by the American Institute of Physics.]

leads to the formation of a magnetic nanotube, where the magnetization can have similar magnetic configurations to the nanoring. However, the extra length can allow more complex magnetic configurations to exist. The hysteresis loop for the ring structure has a plateau regions on either side near the coercive field and arises from the switch between the onion and vortex metastable states, see Figure 13.30 (c).

A magnetic films with regular shapes removed, for example by lithographic and etching methods, can lead to another type of structure called antidots. These can be of any shape, but most studies have used circular or square antidot structures in regular arrays. The periodic structures give rise to a periodic demagnetizing distribution, which is related to the dipolar interactions in the body. Depending on the spatial distribution and size of the antidots, specific domain structures, such as stripes can be formed. These will be modified by a magnetic field, depending on its strength and direction.

One final structure we can mention is that of arrays of elongated magnetic dots which can be arranged in specific configurations where competing interactions and anisotropies act to produce a *frustrated magnetic state*. This basically means that there is no one unique ground state. Such systems are generally referred to as *spin ices*. In Figure 13.31, we show the example of a square and a hexagonal artificial spin ice, also known as an artificial *kagome lattice*. It is notoriously difficult to coax the system into a low energy state and the ground state is practically impossible to achieve. This is due to the frustration introduced by the interdot interactions.

There are many variants of the structures we have outlined above and at the time of writing, there is intense research activity to study these types of materials. Here we have only given a brief overview of some of the more important results.

13.6 SUMMARY

Nanoscience is a broad area of study that comprises many areas of modern science. These include all the major natural sciences of Physics, Chemistry, and Biology. The principal theme is that these subjects are brought together by the common aspect that we are dealing with the properties and effects brought about by using objects of reduced physical dimensions, which as the name suggests is of the order of nanometers. The specific region of interest depending on the physical effect that is being measured. In this chapter, we have considered only those subjects with a direct connection to the physical properties of materials with size reduction in one or more dimensions.

One of the main reasons why objects show very large departures of their normal bulk properties at the nanoscale is that they have a significant proportion of their constituent atoms in surface positions. This means they are subject to less than full atomic coordination or some modification of their normal bulk crystalline symmetry. Such modifications of the local structure will be accompanied by alterations in most if not all of their normal bulk physical properties. Furthermore, the confinement effects produced by limiting the spatial extent of electronic wave functions can radically alter the electronic and optical properties of solids.

Surface modifications of crystalline order arise from the fact that the reduced symmetry of atomic coordination means that the normal bonding which occurs in the bulk of a crystal is incomplete. This will leave free or dangling bonds, which can be accommodated by a re-arragement of the surface atomic order and a relaxation of the normal interatomic separation between atoms. Such a modification of surface order is called surface reconstruction. There are a number of specialized characterization techniques that are only concerned with surfaces. Electron beams are of particular importance since their energies can be well controlled and from the fact that electrons in a certain energy range (20 - 1000 eV) have a very limited mean free path in solids, of the order of 1nm or less, meaning that they only probe the surface atomic layers of the solid. Such electrons can be used for diffraction purposes to evaluate surface crystallography and reconstructions as well as the electronic environment of the surface via electron spectroscopies. Actually higher energy electrons can be used to probe surfaces if their angle of incidence is reduced to a grazing incidence of a few degrees, such as in the RHEED technique. Furthermore, surface microscopy techniques such as scanning electron and tunneling microscopies and other surface probe methods have been developed over the last two to three decades. These are used in a number of ways to probe surface atomic order and morphologies as well as surface chemical mapping.

There are a large number of techniques that can be used to produce nanostructures. These go from very sophisticated methods for controlling electron and ion beams to the very basic in the form of chemical and electrochemical methods for making agglomerates of atoms and nano-sized templates as well as self-assembly techniques. These technologies are broadly based on the top-down or bottom-up approaches to the fabrication of nanosystems. At the more advanced end of the spectrum, electron beams can be controlled to form very specific patterns in a photoresist which will then be used to manufacture precisely architectured objects. Focused ion beams, or FIB, can also be used in a similar fashion to etch and deposit materials in almost exact constructions at the nanoscale. Such advanced tools are mainly used as a research technique and industrial production usually requires more large scale methods for producing batches of devices. Replication technologies are showing themselves to be of great interest as are important developments in chemical methods of production of nanoparticles. Other forms of low dimensional structure, such as carbon nanotubes and graphene are also of enormous interest since they involve a single or multiple of single atomic layers. These materials have a range of exceptional physical properties and are strong candidates for many electronic and nano-mechanical devices and applications.

There has been a large amount of interest in microelectromechanical systems or MEMS over recent decades. These are basically small mechanical devices that can be controlled or adapted to measure physical properties on the micron scale. These are very popular devices since they can be built on microchip systems and incorporated in the the usual CMOS microchip technologies. Such devices are common in the market place, with applications such as accelerometers and gyroscopes as well as an enormous range of sensors and actuators. Such technologies are commonly found in smart phones and gaming devices which require position and motion sensors. The extension of technologies to the nanoscale has lead to interest in nanoelectromechanical systems or NEMS. While the range of applications is still limited at the current time, advances in production and characterization are progressing. It should be remembered that the reduction of size generally leads to faster operation or higher frequency processing. Typical resonance frequencies of mechanical objects in this range of sizes can be well in to the GHz region, which is the operational frequency used in communications systems for example.

In the area of electronics, we have seen that there is enormous scope for development. Clearly the electronics industry is one of the most important global industries and developments in nanosystems are leading to smaller and faster products. At the time of writing, the active components in transistors are reaching the scale of a few 10s of nm in size. With developments in nanotube technologies and molecular electronics, this trend can be expected to continue for the next few decades at least. At the root of these advances in electronics are the radical changes to the electronic properties that occur on the nanoscale. For example, with nanometric dimensions, insulators can become transparent to electrons and conductors can become non-conducting if there are charging effects that lead to Coulomb blockade. However, understanding the basic principles of these limitations allows us to manipulate the devices to our advantage. We can use our knowledge of this behavior to produce a new range and class of device that is based on the control of the passage of single electrons. Since such devices require such small currents, there will be much less dissipation of energy, which is also an important consideration.

The energy quantization that we observe for objects of reduced dimensions has been a property that has been well studied since the inception of quantum well technologies from the 1980s. We can manipulate the dimensions of an object to produce specific well defined energy levels for the electronic states of a system. This allows us to control and design devices for specific applications and will have electronic and optical properties to suit specific functions. This means we can control the color of light emitted from semiconductor lasers and LEDs for example. Further to these, the optical properties of solids can be adjusted by making periodic structures which will limit the propagation of light at specific wavelengths. Such considerations have lead to the development of photonic crystals, which have a large range of applications even beyond the visible spectrum and can be used for camouflage at say radar frequencies.

As with electronic properties, the magnetic behavior of nanostructures can alter significantly from bulk materials. The magnetic characteristic length scales also fall within the nanometer region. For magnetic structures of this dimension, there is generally not enough room to form magnetic domains and a first approximation of the nanosystem can be envisaged as a coherent macrospin. The behavior of the magnetization then depends on the effective anisotropy of the structure and the applied magnetic field strength and direction. The Stoner - Wohlfarth model appears to be a good starting point for modelling this type of structure. However, the thermal agitation can lead to a fast switching of the magnetic state between the various energy minima of the anisotropy and give rise to the so called superparamagnetic behavior. Here there is no well defined orientation of the magnetization in the system. However, the speed at which we measure is an important factor in how the magnetic response appears. Cooling to low temperatures leads to a freezing of the magnetic state, with the transition occurring at the so-called blocking temperature. These properties depends critically on the strength of the magnetic anisotropy and the size of the nano-object. Developments in thin film technologies and structuring have followed to a certain extent that of the electronic properties. Multilayer systems also allow the magnetic properties to be tailored to a degree via the control of interactions between ferromagnetic layers and structures. In elongated and unidimensional structures, magnetic domain effects can be observed and exploited to make specific device applications. Of great interest are magnetic data storage technologies, much of which are based on magnetics and magnetic materials both for storage purposes and reading applications.

While we have only considered the physical response and behavior of nano-objects, there are a large number of applications that go well beyond the physical nature of materials and utilize these properties for other purposes. Of particular importance are the many applications that are emerging in the areas of medicine and biology. Much of this is related to the use of nanoparticles which can be exploited to label and identify regions in biological systems. For example, magnetic nanoparticles can by functionalized to attach themselves to cancer cells. Hyperthermia treatments can be then applied locally to heat and kill the cancer cells using radio-frequency signals. Also there is much interest in using magnetic particle as a delivery system for drug therapies. Since the magnetic particles can be directed to the region where treatment is required, dosage levels can be significantly reduced, thus limiting any undesirable side effects, such as those experienced in chemotherapies.

We have only been able to really see the tip of the iceberg and nanotechnologies has enormous prospects for future science and technology in the foreseeable future. We can expect devices to become smaller and faster and even cheaper. However, there are physical size limits and we appear to be coming close to those in a number of areas, such as in electronic and magnetic systems. This will require new thinking and future scientists will be faced with these challenges. However, this is the role of science in general.

REFERENCES AND FURTHER READING

Basic Texts

- A. Chambers, R. K. Fitch and B. S. Halliday, *Basic Vacuum Technology*, Adam Hilger, Bristol (1989)

- H. P. Myers, *Introductory Solid State Physics*, Taylor and Francis, London (1998)

- M. Prutton, *Surface Physics*, 2nd Edition, Oxford University Press, Oxford (1983)

- S. M. Lindsay, *Introduction to Nanoscience*, Oxford University Press, Oxford (2010)

Advanced Texts

- D. P. Woodruff and T. A. Delchar, *Modern Techniques of Surface Science*, Cambridge University Press, Cambridge (1986)

- V. V. Mitin, V. A. Kochelap and M. A. Stroscio, *Introduction to Nanoelectronics: Science, Nanotechnology, Engineering and Applications*, Cambridge University Press, Cambridge (2008)

- S. Bandyopadhyay, *Physics of Nanostructured Solid State Devices*, Springer, New York (2012)

- D. K. Ferry, S. M. Goodnick and J. Bird, *Transport in Nanostructures*, Second edition, Cambridge University Press, Cambridge (2009)

- E. L. Wolf, *Nanophysics and Nanotechnology: An Introduction to Modern Concepts in Nanoscience*, Second edition, Wiley - VCH (2006)

- S. Datta, *Electronic Transport in Mesoscopic Systems*, Cambridge University Press, Cambridge (1995)

- R. C. O'Handley, *Modern Magnetic Materials: Principles and Applications*, Wiley - Interscience, J. Wiley and Sons, New York (2000)

- M. J. Madou, *"Fundamentals of Microfabrication and Nanotechnology"*, CRC Press, Taylor and Francis (2011).

- R. Skomski, *Simple Models of Magnetism*, Oxford University Press, Oxford (2008)

EXERCISES

Q1. Illustrate the form of the following surface reconstructions:

a) bcc (110)c(4×2)

b) fcc (110) (3×2)

c) zinc-blende (111) c(4×2)

Q2. Compare and contrast the techniques of LEED and RHEED, giving advantages and disadvantages of each.

Q3. Check the validity of the plasma frequency for Si, which has four valence electrons per atom and has a plasmon loss peak at 16.9 eV. The lattice constant for Si is 5.43 Å. What is the corresponding surface plasmon loss energy?

Q4. The surface concentration of a particular GaAs dopant is temperature independent, has an activation energy of

3.5 eV and a value of diffusion coefficient at 700° C of 10^{-16} cm²s^{-1}. The diffusion length of all processing should be 10^{-6} cm. Calculate:

a) the process time at 700°C,

b) the area density of impurities in the diffused layer at the same temperature,

c) the change in process time if an elevated temperature of 800°C is used,

d) the flux of atoms after 30 minutes at 800°C.

Q5. Show that gravitational forces scale as S4.

Q6. The semiconductor CdSe has a melting temperature of 1678 K, however, in this material is formed into nanoparticles of 3 nm diameter this melting temperature significantly reduces to 700 K. Explain this empirical observation. Estimate the melting temperature for 2 nm diameter particles.

Q7. a) How does the frequency of an electronic device scale with its size?

b) How large would a device need to be for it to function at microwave frequencies at room temperature? Consider only the thermal velocity of the electron.

Q8. Consider a sample of crystalline GaAs with an electronic effective mass of $m* = 0.067 m_e$ and a mobility of $\mu = 10^5$ cm²V^{-1}s^{-1} at liquid nitrogen temperatures. Calculate the following parameters:

a) de Broglie wavelength, λ;

b) Scattering time, τ_e;

c) Thermal electron velocity, v_T;

d) Mean free path, λ_e;

e) Diffusion coefficient, $D(\alpha = 3)$.

Determine the transport regime for devices with feature sizes of $L_x = 0.05, 0.5$ and 5 mm.

Q9. Estimate the emission wavelength of a 15 nm GaAs quantum well laser at room temperature. (Use the following data for GaAs: $m^* = 0.067 m_e$; $m^*_{hh1} = 0.5 m_e$; $\Delta = 1.424$ eV at 300 K.)

Q10. Explain what is meant by Coulomb blockade and explain the origin of the formation of the Coulomb Diamond structure experimentally observed in single electron transfer devices. Further show that a potential of e/C is required to transfer a single electron across a tunnel junction.

Q11. Determine the energy of the Coulomb barrier for islands of 2, 5, and 10 nm diameter. Assume that the capacitance of the island contact junctions to be 1fF.

Q12. Consider a semiconductor of width W and length L connected to two regions (contacts) of 2DEGs of the same material, with an effective mass of 0.07 me. In the narrow region of the device it is necessary to take account of the discreteness of transversal modes. Plot the electron density as a function of the Fermi energy for W = 100 nm assuming a hard wall potential.

Q13. A typical tunnel junction is formed using an oxide layer of thickness 5 nm and dielectric constant, $\epsilon_r = 5$. Estimate the maximum area of the capacitor plates for Coulomb blockade to be observed at temperatures of:

a) 4.2 K

b) 300K

Q14. Consider the energy associated with electrons in a quantum dot. Show that the degenerate level (at k = 0) in a semiconductor between the heavy and light hole bands is lifted when a quantum dot is formed. (For the purposes of argument use a cubic form of quantum dot, of side a.) Show that the energy difference in the ground state is given by:

$$\Delta^0_{VB} = \frac{3\hbar^2 \pi^2}{2a^2} \left(\frac{m^*_{hh} - m^*_{lh}}{m^*_{hh} m^*_{lh}} \right) \tag{13.122}$$

Q15. Show that quantum effects become observable in the condition:

$$\Delta x = \frac{h}{\sqrt{mk_B T}} \tag{13.123}$$

Q16. Evaluate the ground state of an Fe^{2+} ion. In crystalline Fe, the orbital angular momentum is quenched. Taking this into account, evaluate the magnetic moment of Fe, assuming it forms Fe^{2+} ions in solid. What is the corresponding magnetization of Fe? Note that Fe has a bcc structure and lattice parameter of 2.87 Å.

Q17. Consider the energies involved in a Bloch domain wall for a ferromagnetic particle with uniaxial magnetic anisotropy. Find an expression for the critical diameter of a spherically shaped particle in terms of the anisotropy strength and the exchange stiffness constant. Determine a value for the size of an iron single domain particle. (Look up any constants that you may require for this estimate.)

Q18. A magnetic measurement is made using a SQUID magnetometer on a monodisperse assembly of Fe nanoparticles with diameter 8 nm. The results indicate a blocking temperature of $T_B = 33$ K, where it was assumed that the characteristic measuring time is 100 s. Evaluate the corrresponding blocking temperature for these nanoparticles using ferromagnetic resonance, where the measuring time can be approximated as around 10^{-10} s. The anisotropy constant for Fe is about 0.48×10^6 erg cm^{-3}. State any assumptions made in the calculation.

Q19. Consider the spin wave spectrum for a 50 Å Fe thin film. Evaluate the expected frequency for the first spin wave mode. Use $D = 280$ meVÅ^2.

Q20. Spin injection is a crucial process in the electronics of spin. Describe how this can be achieved optically in a semiconducting device.

Q21. Explain why the existence of a spin current does not necessary have to be accompanied by an electrical current.

NOTES

[1] The basic units of pressure have the following conversion factors: 1 atm = 760 torr = 1013 mbar = 1.013×10^5 Pa (Nm^{-2})

[2] This was first demonstrated in 1989 by Don Eigler, who worked at IBM. He wrote the letters IBM with 35 Xe atoms. A great publicity coup for IBM!

[3] The Nobel Prize for 1986 was also shared with Ernst Ruska, who was the inventor of the electron microscope, which was first developed in 1931.

[4] M. J. Madou, *"Fundamentals of Microfabrication and Nanotechnology"*, CRC Press, Taylor and Francis (2011).

[5] The unit of conductance is siemens: $1\,S = 1AV^{-1} = 1\Omega^{-1}$.

[6] see for example Q A Pankhurst *et al.*, J. Phys. D: Appl. Phys. **36**, R167 (2003) and J. Phys. D: Appl. Phys. **42**, 224001 (2009).

APPENDIX A

SPACE GROUPS

TABLE A.1: Crystallographic space groups in 3 dimensions

Number	Crystal System	Point Group		Space Groups (International Short Symbol)
		International	Schönflies	
1	Triclinic (2)	1	C_1 Chiral	P1
2		$\bar{1}$	C_i	P$\bar{1}$
3–5	Monoclinic (13)	2	C_2 Chiral	P2, P2$_1$, C2
6–9		m	C_s	Pm, Pc, Cm, Cc
10–15		2/m	C_{2h}	P2/m, P2$_1$/m, C2/m, P2/c, P2$_1$/c, C2/c
16–24	Orthorhombic (59)	222	D_2 Chiral	P222, P222$_1$, P2$_1$2$_1$2, P2$_1$2$_1$2$_1$, C222$_1$, C222, F222, I222, I2$_1$2$_1$2$_1$
25–46		mm2	C_{2v}	Pmm2, Pmc2$_1$, Pcc2, Pma2, Pca2$_1$, Pnc2, Pmn2$_1$, Pba2, Pna2$_1$, Pnn2, Cmm2, Cmc2$_1$, Ccc2, Amm2, Aem2, Ama2, Aea2, Fmm2, Fdd2, Imm2, Iba2, Ima2
47–74		mmm	D_{2h}	Pmmm, Pnnn, Pccm, Pban, Pmma, Pnna, Pmna, Pcca, Pbam, Pccn, Pbcm, Pnnm, Pmmn, Pbcn, Pbca, Pnma, Cmcm, Cmce, Cmmm, Cccm, Cmme, Ccce, Fmmm, Fddd, Immm, Ibam, Ibca, Imma

TABLE A.2: Crystallographic space groups in 3 dimensions (Cont.)

Number	Crystal System	Point Group		Space Groups (Intl. Short Symbol)
		International	Schönflies	
75–80	Tetragonal (68)	4	C_4 Chiral	P4, P 4_1, P42, P43, I4, $I4_1$
81–82		$\bar{4}$	S_4	$P\bar{4}$, $I\bar{4}$
83–88		4/m	C_{4h}	P4/m, $P4_2/m$, P4/n, $P4_2/n$, I4/m, $I4_1/a$
89–98		422	D_4 Chiral	P422, $P42_12$, $P4_122$, $P4_12_12$, $P4_222$, $P4_22_12$, $P4_322$, $P4_32_12$, I422, $I4_122$
99–110		4mm	C_{4v}	P4mm, P4bm, $P4_2cm$, $P4_2nm$, P4cc, P4nc, $P4_2mc$, $P4_2bc$, I4mm, I4cm, $I4_1md$, $I4_1cd$
111–122		$\bar{4}2m$	D_{2d}	$P\bar{4}2m$, $P\bar{4}2c$, $P\bar{4}2_1m$, $P\bar{4}2_1c$, $P\bar{4}m2$, $P\bar{4}c2$, $P\bar{4}b2$, $P\bar{4}n2$, $I\bar{4}m2$, $I\bar{4}c2$, $I\bar{4}2m$, $I\bar{4}2d$
123–142		4/mmm	D_{4h}	P4/mmm, P4/mcc, P4/nbm, P4/nnc, P4/mbm, P4/mnc, P4/nmm, P4/ncc, $P4_2/mmc$, $P4_2/mcm$, $P4_2/nbc$, $P4_2/nnm$, $P4_2/mbc$, $P4_2/mnm$, $P4_2/nmc$, $P4_2/ncm$, I4/mmm, I4/mcm, $I4_1/amd$, $I4_1/acd$,
143–146	Trigonal (25)	3	C_3 Chiral	P3, $P3_1$, $P3_2$, R3
147–148		$\bar{3}$	S_6	$P\bar{3}$, $R\bar{3}$
149–155		32	D_3 Chiral	P312, P321, $P3_112$, $P3_121$, $P3_212$, $P3_221$, R32
156–161		3m	C_{3v}	P3m1, P31m, P3c1, P31c, R3m, R3c
162–167		$\bar{3}m$	D_{3d}	$P\bar{3}1m$, $P\bar{3}1c$, $P\bar{3}m1$, $P\bar{3}c1$, $R\bar{3}m$, $R\bar{3}c$

TABLE A.3: Crystallographic space groups in 3 dimensions (Cont.)

Number	Crystal System	Point Group		Space Groups (Intl. Short Symbol)
		International	Schönflies	
168–173	Hexagonal (27)	6	C_6 Chiral	$P6, P6_1, P6_5, P6_2, P6_4, P6_3$
174		$\bar{6}$	C_{3h}	$P\bar{6}$
175–176		6/m	C_{6h}	$P6/m, P6_3/m$
177–182		622	D_6 Chiral	$P622, P6_122, P6_522, P6_222, P6_422, P6_322$
183–186		6mm	C_{6v}	$P6mm, P6cc, P6_3cm, P6_3mc$
187–190		$\bar{6}m2$	D_{3h}	$P\bar{6}m2, P\bar{6}c2, P\bar{6}2m, P\bar{6}2c$
191–194		6/mmm	D_{6h}	$P6/mmm, P6/mcc, P6_3/mcm, P6_3/mmc$
195–199	Cubic (36)	23	T Chiral	$P23, F23, I23, P2_13, I2_13$
200–206		$m\bar{3}$	T_h	$Pm\bar{3}, Pn\bar{3}, Fm\bar{3}, Fd\bar{3}, Im\bar{3}, Pa\bar{3}, Ia\bar{3}$
207–214		432	O Chiral	$P432, P4_232, F432, F4_132, I432, P4_332, P4_132, I4_132$
215–220		$\bar{4}3m$	T_d	$P\bar{4}3m, F\bar{4}3m, I\bar{4}3m, P\bar{4}3n, F\bar{4}3c, I\bar{4}3d$
221–230		$m\bar{3}m$	O_h	$Pm\bar{3}m, Pn\bar{3}n, Pm\bar{3}n, Pn\bar{3}m, Fm\bar{3}m, Fm\bar{3}c, Fd\bar{3}m, Fd\bar{3}c, Im\bar{3}m, Ia\bar{3}d$

APPENDIX B

THE FOURIER TRANSFORM

The Fourier transform, named after Jean Baptiste Joseph Fourier (1768–1830), is a mathematical tool for the study of periodic properties and has an enormous range of applications in Physics and Engineering. The transform is used to analyze the periodicity of a function and in performing a transform provides a suitable change of units. For example, the Fourier transform of a temporal function (with units in seconds), i.e., time domain, will return a function with the temporal variation or frequency (with units Hz or s^{-1}), i.e., frequency domain. Such a transform provides a frequency spectrum of the original function.

The mathematical expression of the Fourier transform can be given as follows. Consider a function $f(x)$, the Fourier transform of this function is generally expressed as:

$$\mathcal{F}(u) = \int_{-\infty}^{\infty} f(x) e^{-2\pi i u x} \, du \qquad (B.1)$$

which is valid for any real number u. The inverse transform is expressed as:

$$f(x) = \int_{-\infty}^{\infty} \mathcal{F}(u) e^{-2\pi i u x} \, dx \qquad (B.2)$$

and is valid for any real number x.

Apart from the spectral analysis of time dependent properties, the Fourier transform is also frequently used in the analysis of spatially dependent functions and properties. The periodicity of crystals and the transformation from real to reciprocal space can also be analyzed with the aid of Fourier analysis.

REFERENCES AND FURTHER READING

- D. W. Jordan and P. Smith, *Mathematical Techniques: An Introduction for the Engineering, Physical and Mathematical Sciences*, (3e) Oxford University Press, Oxford (2002)

- M. L. Boas, *Mathematical Methods in the Physical Sciences*, (3e) J. Wiley and Sons, New Jersey (2006)

- K. A. Stroud, *Fourier Series and Harmonic Analysis*, Stanley Thornes (Publishers) Ltd, Cheltenham (1984)

- E. Kreyszig, *Advanced Engineering Mathematics*, (5e) J. Wiley and Sons, New York (1983)

APPENDIX C

FUNDAMENTAL CONSTANTS

TABLE C.1: Fundamental Constants in Physics

Quantity	Symbol	Value
Electron charge	e	1.60219×10^{-19} C
Electron rest mass	m_e	9.1095×10^{-31} kg
Proton rest mass	m_p	1.6726×10^{-27} kg
Neutron rest mass	m_n	1.6749×10^{-27} kg
Planck's constant	h	6.6262×10^{-34} Js
Planck's constant	$\hbar = h/2\pi$	1.05459×10^{-34} Js
Speed of light	c	2.997925×10^{8} ms^{-1}
Permittivity of free space	ϵ_0	8.8542×10^{-12} Fm^{-1}
Permeability of free space	μ_0	$4\pi \times 10^{-7}$ Hm^{-1} or 1.2566×10^{-6} Hm^{-1}
Boltzmann's constant	k_B	1.3807×10^{23} mol^{-1}
Avogadro's number	N_A	6.022×10^{-23} JK^{-1}
Gas constant	$R = k_B N_A$	8.314 JK^{-1} mol^{-1}
Bohr magneton	μ_B	9.2741×10^{-24} JT^{-1}
Nuclear magneton	μ_N	5.0508×10^{-27} JT^{-1}
Reciprocal fine structure constant	$1/\alpha$	$1/137.036$
Electron radius	r_e	2.81794×10^{-15} m
Bohr radius	$a_0 = 4\pi\epsilon_0 \hbar^2 / m_e e^2$	0.529177×10^{-10} m
Rydberg constant	$\text{Ry} = \hbar^2 / 2m_e a_0^2$	2.17991×10^{-18} J

(*Continued*)

TABLE C.1: Continued

Quantity	Symbol	Value
Free electron g-factor	g	2.002319304386
Magnetogyric (gyromagnetic) ratio	$\gamma = g\mu_B/\hbar$	$1.7608592 \times 10^{11}\ s^{-1}T^{-1}$
Electron volt	eV	$1.60219 \times 10^{-19}\ J$

INDEX

A

Amorphous, 58, 124
Angular momentum quantum
  number, 13–14
Antiferromagnetic order
  Curie-Weiss law, 355
  Néel temperature, 352, 353
  properties of, 355t
  spin-flip transition, 354
  sublattice, 352
Atomic force microscopy (AFM), 371
Atomic magnetic moment
  Hund's rules and the ground
    state, 326–330
  moments and energies, 330–331
  orbital and spin angular
    momenta, 324–326
Atomic number, 1
Atomic orbital, 12–13

B

Band theory
  Bloch theorem and functions,
    201–203
  Brillouin zones and Fermi
    surface
    2D square array, 205, 205f
    band gaps, 207
    three-dimensional FCC and
      BCC structures, 206, 206f
    Wigner-Seitz cell, 205
  electronic properties
    conduction band, 225
    conductor/metal, 225
    direct band gap, 226
    indirect band gap, 226
    insulator, 225
    semiconductor, 225, 227
    silicon, band structure for,
      225, 226f
    valence band, 225
  electronic states, 198
  Kronig-Penney model
    Bloch functions, 209
    Schrödinger equation, 208, 210
  nearly free electron model,
    216–218
  periodic potential, 199–200
    free electrons, 211–216
    Schrödinger equation, 203–205
  potentials and wave-functions,
    222–224
  tight-binding model
    electronic wave functions, 218
    energy bands, 222, 222f
    FCC structure, 221, 221f
    Hamiltonian, 219
    Wannier function, 219, 220
Bardeen Cooper and Schrieffer
  (BCS) theory
  Cooper pair, 417, 418, 421
  electrons, 429
  Fermi surface, 420
  ground state, 422–425
  isotope effect, 416
  outcomes of, 425–430
  phonon coupling, 419, 419f
  two-electron wave-function, 422
Barkhausen effect, 370
body-centered cubic (BCC), 52
Bohr magneton, 185

Bohr model
  de Broglie's hypothesis, 8
  electromagnetic radiation, 6
  negative sign, 8
  wave interference, 9
Bose-Einstein condensation(BEC), 395
Bragg's law, 68–70
Bravais lattice, 43–44
Brittle fracture, 110
Brillouin light scattering (BLS), 552
Burgers vector, 105–108

C

Color centers, 103–104
Colossal magnetoresistance (CMR), 343
Continuity equations, 298–300
Coulomb blockade, 523
Covalent bonding
  anti-bonding state, 24
  bonding state, 23
  Pauli exclusion principle, 24
Crystal structure determination
  atomic form factor, 76–77
  diffraction methods
    electron diffraction, 85–88
    neutron diffraction, 88–90
    X-ray diffraction, 80–85
  reciprocal lattice
    Bravais lattice, 64
    Fourier components, 63
    interplanar spacing, 65, 66f
    Kronecker delta, 64
    Miller indices, 65
    primitive lattice, 63
  structure factor
    BCC structure, 78
    FCC structure, 79
    phase problem, 80
  waves, diffraction of
    Bragg's law, 68–70
    elastic scattering, 68
    electromagnetic spectrum, 67
    Ewald sphere construction, 74–76
    inelastic scattering, 68
    reconciling, Bragg and von Laue approaches, 72–74
    Von Laue approach, 70–72
Crystalline structure
  atomic packing
    atomic planes, 56
    conventional unit cell, 55
    crystal types, 57t
    hard-sphere model, 55, 56f
    packing fraction, 55
  BCC structures, 52, 52f, 53
  Bravais lattice, 43–44
  crystal planes and axes, Miller indices
    analogous triangles, 50
    anisotropic, 45
    brackets, convention for, 48, 48t
    cosine rule, 50
    cubic lattices, low index planes for, 46, 47f
    physical properties, 45
    trigonal and hexagonal crystal classes, 49
  CsCl structure, 53, 53f
  diamond crystal structure, 53, 54
  FCC structures, 51, 52
  HCP structures, 51, 51f, 52
  symmetry, aspects of
    basis, 36, 38–41
    building block, 36
    elements of, 41–43
    translational symmetry, 37–38
    unit cell, 38–41

D

Debye model, 152–155
Degenerate electron gas, 184
Density functional theory, 224
Density of states (DOS), 155

Depletion zone, 304–306
Device under test (DUT), 514
Dielectric materials
  basic properties of
    dielectric breakdown, 450–451
    electrical conductivity, 449
    ionic conduction, 449–450
  electrostatics and Maxwell equations
    capacitance, 454, 455
    electric dipole moment, 452
    electric susceptibility, 452
    metallic plates, 453
  ferroelectrics
    $BaTiO_3$, cubic perovskite structure, 469, 470
    Curie–Weiss law, 470, 471t
    properties of, 469, 469t
    pyroelectrics, 469
  function
    electric field, application of, 458, 458f
    electronic polarization, 460–462
    ionic polarization, 462–467
    Kramers-Kronig relations, 460
    total dielectric function, 467–468
  local field, 455–458
  multiferroic materials, 472–473
  piezoelectrics, 472
  solids, optical properties of
    anti-Stokes scattering, 481
    cubic crystals, 480
    electromagnetic waves, absorption of, 477–478
    exciton, 480
    photon energy, 479
    Raman scattering, 481
    Stokes scattering, 481
    transmission and reflection coefficients, 475–477
    wave equation, 474–475
Diffusion current, 249
3 Dimensions, crystallographic
  space group in, 572t–574t
Dislocations
  Burgers vector, 105–108
  edge dislocations, 104–105
  energy, 115–117
  interactions
    Burgers vectors, 119
    dislocation reaction, 118
    Frank-Read source, 118, 119f
    jogs, 117, 117f, 118
    partial dislocations, 119–120
    sessile, 118
    slip system, 118
    stacking fault, 120, 121
  screw dislocations, 105
  solids, and mechanical properties of
    climb, 114
    cross-slip, 115
    plastic deformation, 109
    shear force, 113
    slip planes, 110, 111, 111f, 112, 114–115
    stress-strain curve, 108, 109
    work hardening, 109
Doping, 284
Ductile, 110
Dulong and Petit's law, 150
Dzyaloshinski-Moriya (DM) interactions, 343

E

Edge dislocations, 104–105
Eigenvalue equation, 12
Einstein relation, 250
Einstein's model, 150–152
Electrical conductivity, 449
Electron diffraction
  LEED experiment, 88
  RHEED experiment, 86, 87
  TEM, 86
Electron dynamics and transport phenomena

Bloch function, 233, 234
charge carriers, drift and
   diffusion of
   diffusivity and mobility, 249
   Einstein relation, 250
effective mass
   constant energy surfaces, 238, 238f
   energy bands, 239
   inverse effective mass tensor, 237
electron scattering, 251–253
Fermi surface
   Bragg plane, 241, 241f
   Brillouin zones, 240, 242, 242f
   group and phase velocities, 235, 235f
   k-space, 236
   magnetic field effects
      cyclotron resonance, 256–260
      Hall effect, 254–256
      magnetic sub-bands and oscillatory phenomena, 263–265
      magnetoresistance, 260–263
      quantum and fractional quantum Hall effects, 266–269
   positive charge carriers
      conduction band, 246, 247
      hole states, 245
      valence band, 244,
   wave vector, 233
Electron dynamics
   Bloch function, 233, 234
   effective mass
   group and phase velocities, 235, 235f
   k-space, 236
   wave vector, 233
Electron energy loss spectroscopy (EELS), 501

Electronic polarization, 460–462
Extrinsic semiconductors
   Fermi level, 289
   freeze-out range, 287
   intrinsic range, 288
   ionized donors, 286
   n-type, 286, 290
   p-type, 285, 290
   quadratic equation 287
   saturation range, 287–288

F

Facecentered cubic (FCC), 51, 52
Ferrimagnetic order
   Brillouin function, 356
   classes of, 359
   inverse magnetic susceptibility, 358, 358f
   octahedral and tetrahedral sites, 360
   properties of, 360t
Ferroelectrics
   $BaTiO_3$, cubic perovskite structure, 469, 470
   Curie–Weiss law, 470, 471t
   properties of, 469, 469t
   pyroelectrics, 469
Ferromagnetic order
   itinerant ferromagnetism, 349–351
   mean field theory
      Brillouin function, 347
      Curie temperature, 346, 347
      Hamiltonian, 345
      properties of, 349t
Ferromagnetic resonance (FMR), 379
Field effect transistor (FET), 529
Fourier transform, 575
Fractional quantum Hall effect (FQHE), 256, 268
Free electrons in metals

Drude theory, 172–174
electron gas
  Fermi-Dirac statistics of, 174–177
  specific heat of, 182–184
high frequency response and optical properties
  electromagnetic radiation, 189
  Fermi sphere, 188, 188f
  plasma frequency, 190
  refractive index, 190–191
  Maxwell-Boltzmann velocity distribution, 171–172
metallic behavior, 169–171
Pauli paramagnetism
  Bohr magneton, 185
  Curie-Weiss law, 187
  spin-up and spin-down electrons, 185, 186
photoelectric effect, 168
Sommerfeld model, 177–178
states, density of
  Fermi wave-vector, 180
  k-space, 179
  Sommerfeld expansion, 181, 182
Free electrons
  metals. see Free electrons in metals
  periodic potential
    Brillouin zone, 214, 215
    electronic states, 212
    reciprocal lattice vector, 213, 214
Frenkel defect, 98
Fundamental constants, 577–578

G

Giant magnetoresistance (GMR) effect, 376, 551
Gibbs free energy, 405
Ginzburg-Landau (GL) model
  Cooper pairs, 415
  order parameter, 413
Grain boundary, 121
Grüneisen constant, 157–158

H

Hall effect
  Hall coefficient, 256
  history, 254
  Lorentz force, 255
Hard-sphere depiction, 51
Heisenberg uncertainty principle, 11
Hexagonal close-packed structures, 51, 51f, 52
High resolution electron energy loss spectroscopy (HREELS), 502
High-temperature superconductivity (HTS)
  CuO_2 planes, 438, 439, 440
  YBCO compound, 437, 438, 440
Hooke's law, 133
Hydrogen bonding, 28–29

I

Internal friction, 121
Interstitial sites, 97
Intrinsic semiconductors
  conduction band, 280–281
  effective density of states, 282
  electrons and holes, 280
  parabolic bands, 281
  valence band, 281
Inverse effective mass tensor, 237
Ionic bonding
  cohesive energy, 22
  Madelung constant, 22
  NaCl, 21
Ionic conduction, 449–450
Ionic polarization, 462–467
Isotopes, 1

J

Josephson effects
   basic structure of, 430
   Cooper pairs, 431, 433
   Giaever tunneling, 434
   phase shifts, 435
   SQUID, 434, 435f, 436–437

K

Kramers-Kronig relations, 460

L

Lattice, 121
   vector, 37
Lattice vibrations
   anharmonic effects
      thermal conduction, 158–159
      thermal expansion, 157–158
      Umklapp processes, 160–162
   diatomic 1D lattice
      Brillouin zone, 145, 146, 148
      longitudinal and transverse, 148–149
      negative solution, 144
      phase and group velocities, 147
   monatomic lattice
      modes, number of, 139–143
      one-dimensional chain, 132–137
      three-dimensions, extension to, 138–139
   solids, thermal properties of
      Debye model, 152–155
      Dulong and Petit's law, 150
      Einstein's model, 150–152
Laue zone, 86
Law of mass action, 283–284
Linear combination of atomic orbitals (LCAO), 219
Lyddane-Sachs-Teller relation (LST), 465

M

Magnetic anisotropies
   magnetocrystalline anisotropy, 363–365
   shape anisotropy, 361-362
Magnetic force microscopy (MFM), 371
Magnetic materials
   AFM and MFM, 371
   anisotropy energy, 367
   antiferromagnetic order
      Curie-Weiss law, 355
      Néel temperature, 352, 353
      properties of, 355t
      spin-flip transition, 354
      sublattice, 352
   atomic magnetic moment
      Hund's rules and the ground state, 326–330
      moments and energies, 330–331
      orbital and spin angular momenta, 324–326
   Barkhausen effect, 370
   coercive field, 369
   diamagnetism, 331–334
   dipolar interaction, 339–341
   domain walls, 365
   exchange interactions, 341–344
      direct and indirect, 342
      DM interaction, 343–344
      Hamiltonian, 342
   ferrimagnetic order
      Brillouin function, 356
      classes of, 359
      inverse magnetic susceptibility, 358, 358f
      octahedral and tetrahedral sites, 360
      properties of, 360t
   ferromagnetic order

Index • 585

itinerant ferromagnetism, 349–351
mean field theory, 344–349
giant magnetoresistance and spintronics
  GMR, 379
  parallel and antiparallel states, 377
  RKKY interaction, 376
  spin-up and spin-down electrons, 376, 377
hysteresis loop, 369–370, 370f
Lorentz force, 371
magnetic anisotropies
  magnetocrystalline anisotropy, 363–365
  shape anisotropy, 361–362
magnetic domain closure, 368, 368f
magnetic domain wall parameters, 368, 368t
MOKE, 370
paramagnetism
  classical treatment, 334–336
  quantum mechanical treatment, 336–338
  Van Vleck paramagnetism, 338–339
spin dynamics
  experimental techniques, 385
  Kittel equation, 384
  LL and LLG equations, 382
  phenomenological description, 380
  relaxation process, 281
  SWR, 383
spin waves
  excitation energy, 375, 375f
  Hamiltonian, 371, 372
  phonons, 373, 374
  Reimann zeta function, 374
  Stoner excitation, 375
spontaneous magnetization, 323
transition regions, 366
virgin state, 369
Magnetic nanostructures
  antidots, 559
  ground state, 556
  magnetic dots structure, 555
  magnetic nanowires, 554
  vortex core, 556, 557
Magnetic quantum number, 14
Magnetic sub-bands/Landau levels, 264
Magnetic tunnel junction (MTJ), 522–523
Magnetocrystalline anisotropy, 361
Magneto-optic Kerr effect (MOKE), 370
Metallic bonding, 26–28
Molecular beam epitaxy (MBE), 310
Moseley's law, 81

N

Nanostructures
  Bohr model, 538
  conduction band, 535
  Coulomb interaction, 537
  Fermi's golden rule, 534
  matrix elements, 534
  photonics crystals, 539
  QW structure, 536, 537f, 538
Nanotechnologies and nanophysics
  electronic and optical properties
    insulating barrier and tunnel junctions, 519–523
    nanostructures, 534–539
    phase coherence length, 508
    quantum dots and coulomb blockade, 523–527
    quantum point contacts, 513–519

resonant tunneling, 527–529
SET, 529–533
size reduction and energy quantization, 509–513
IC, 487
LDS, 503–507
microscopic and macroscopic, 489
Moore's law, 488
nanomagnetism, aspects of
    magnetic length scales, 540–541
    magnetic nanostructures, 553–559
    magnetic thin films and multilayers, 550–553
    Stoner-Wohlfarth Model, 541–545
    superparamagnetism and ferromagnetic nanoparticles, 545–549
nanomedicine, 487
quantum mechanics, 488
STM, 488
surfaces, physics of
    Auger electron, 499, 500, 500f
    bulk crystals, 491
    EELS, 501, 502
    photoelectron spectroscopies, 502
    principal features, 499
    selvedge, 490
    UHV, 491
    UPS technique, 502
Neutron diffraction, 88–90
Non-crystalline materials, 124–125
Non-equilibrium distributions
    carrier injection/injection levels
        excess charge carriers, 293
        high-level carrier injection, 294
        low-level injection, 293–294
    continuity equations, 298–300
    generation and recombination processes, 294–298
        boundary condition, 297
        direct and indirect process, 294
        excess charge carriers, 295
        rates of, 296
        relaxation processes, 294
        trapping centers, 298
Nuclear magnetic resonance (NMR), 382

O

Object Oriented MicroMagnetic Framework (OOMMF), 369
One-dimensional chain, 132–137
    Brillouin zones, 137
    dispersion relation, 134, 136
    group velocity, 135
    Hooke's law, 133
    Newton's second law, 133
    non-dispersive regime, 135
    phase velocity, 135
    transversal modes, 137

P

Packing fraction, 55
Paramagnetism
    classical treatment, 334–336
    quantum mechanical treatment, 336–338
    Van Vleck paramagnetism, 338–339
Pauli exclusion principle, 14
Pauli paramagnetism, 185
    Bohr magneton, 185
    Curie-Weiss law, 187
    spin-up and spin-down electrons, 185, 186
Phonon emission, 104
Physical bonds, 21

Piezoelectrics, 472
Planar defects
  grain boundaries, 121–122
  tilt boundaries, 122–123
  twin boundaries, 123–124
Plasmon, 190
P-N junction
  breakdown voltage, 301
  current-voltage characteristics
    built-in potential, 307, 308
    ideal diode equation, 309
    leakage currents, 309
    n-region, 309
    p-region, 309
  depletion zone, 304-306
  junction capacitance, 306
  thermal equilibrium
    charge carriers, diffusion of, 301
    electron concentration, 302
    electrostatic potential, 303
    space-charge distribution, 302
Point defects
  color centers, 103–104
  crystals, diffusion in
    Fick's first law, 101
    Fick's second law, 102
    impurity diffusion, 101
    self-diffusion, 101
  defect density, thermodynamics of, 99–100
  types of
    antisite defects, 98
    Frenkel and Schottky defects, 98, 98f
Polaritons, 466
Polycrystalline systems, 58
Precipitation hardening, 121
Primitive unit cell, 38
  volume, 39
Principal quantum number, 13

Pseudopotential method, 224
Pyroelectrics, 469

Q

Quantum point contacts (QPC)
  DUT, 514
  Fermi-Dirac distribution function, 517
  Landauer formula, 518
  left-to-right transmission coefficient, 516
  QWR, 513
Quantum well (QW), 312
Quantum wires (QWR), 513
Quantum/quantized Hall effect (QHE), 256, 266, 268

R

Reciprocal lattice
  Bravais lattice, 64
  Fourier components, 63
  interplanar spacing, 65, 66f
  Kronecker delta, 64
  Miller indices, 65
  primitive lattice, 63
Resonant tunneling diode (RTD), 522, 527

S

Schottky defect, 98
Screw dislocations, 105
Semiconductors
  equilibrium statistics, electrons and holes
    compensated semiconductors, 290–292
    extrinsic, doping, 284–290
    intrinsic, 279–282
    law of mass action, 283–284
  heterostructures and quantum wells

band gap energy, 311
band structure engineering, 311
conduction and valence bands., 313, 314
GaAs and AlAs, 311
MBE, 310–311
QW structure, 312
materials
band gap, 277
direct band-gap, 277, 278
direct transitions, 278
III-V materials, 276
indirect band-gap, 277, 278
non-equilibrium distributions
carrier injection/injection levels, 292–294
continuity equations, 298–300
generation and recombination processes, 294–298
p-n junction
breakdown voltage, 301
current-voltage characteristics, 306–310
depletion zone, 304–306
junction capacitance, 306
thermal equilibrium, 301–304
Si, band structure for, 275
Silsbee effect, 397
Single electron transistor (SET)
Coulomb blockade, 531, 532
electron tunneling, 530, 531
QD, 529
Solid state physics
aspects, 2
atom, electronic structure of
angular momentum quantum number, 13–14
atomic orbital, 12–13
Bohr model, 6–11
eigenvalue equation, 12

electron filling, 15, 15f
electron orbits, 5–6
elemental particles, 4
Heisenberg uncertainty principle, 11
magnetic quantum number, 14
Pauli exclusion principle, 14
principal quantum number, 13
Spin quantum number, 14
electronic properties, 3
interatomic bonding
covalent bonding, 23–26
equilibrium separation, 19–20
hydrogen bonding, 28–29
ionic bonding, 21–23
metallic bonding, 26–28
Mie potential, 19
mixed covalent and Ionic bonding, 26
physical bonds, 21
Van der Waals bonding, 29
magnetism, 2
mechanical properties, 2
periodic table
electronic configuration, 16, 17f, 18, 18f
properties, 17
physical properties, 3
Space lattice/Bravais lattice, 38
Spin dynamics
experimental techniques, 385
Kittel equation, 384
LL and LLG equations, 382
phenomenological description, 380
relaxation process, 281
SWR, 383
Spin quantum number, 14
Spin wave resonance (SWR)
spectra, 383
Spin waves
excitation energy, 375, 375f

Hamiltonian, 371, 372
phonons, 373, 374
Reimann zeta function, 374
Stoner excitation, 375
Stacking fault, 120
Stoner-Wohlfarth Model, 541–545
Strain, 109
Structure factor, 62
   BCC structure, 78
   FCC structure, 79
   phase problem, 80
Superconducting quantum interference device (SQUID), 434, 435f
Superconductivity
   BCS theory
      Cooper pair, 417, 418, 421
      electrons, 429
      Fermi surface, 420
      ground state, 422–425
      isotope effect, 416
      outcomes of, 425–430
      phonon coupling, 419, 419f
      two-electron wave-function, 422
   BEC, 395
   critical fields and critical current, 401–404
   GL model
      Cooper pairs, 415
      order parameter, 413
   HTS
      CuO_2 planes, 438, 439, 440
      YBCO compound, 437, 438, 440
   Josephson effects
      basic structure of, 430
      Cooper pairs, 431, 433
      Giaever tunneling, 434
      phase shifts, 435
      SQUID, 434, 435f, 436–437
   London equations
      coherence length, 411
      first London equation, 409
      penetration depth, 410
      second London equation, 409
   Meissner-Ochsenfeld effect, 398–399
   perfect diamagnetism, 399–401
   thermodynamics
      heat capacity, 407–408
      phase stability of, 405–406
   zero-resistivity/infinite conductivity and persistent currents, 396–398
Superparamagnetic limit, 546, 548
Surface reconstruction, 87

T

Tight-binding model
   electronic wave functions, 218
   energy bands, 222, 222f
   FCC structure, 221, 221f
   Hamiltonian, 219
   Wannier function, 219, 220
Tilt boundary, 122
Transmission electron microscope (TEM), 86
Twin boundary, 123
Two-dimensional electron gas (2DEG), 265

U

Ultraviolet photoelectron spectroscopy (UPS), 502
Umklapp processes
   free path, 161, 162
   N and U–processes, 160, 161f, 161

V

Van der Waals bonding, 29
van Hove singularities, 143
Von Laue approach, 70–72

W

Wannier function, 219
Weiss molecular field theory, 344
Wigner-Seitz cell, 40
Work hardening, 109

X

X-ray diffraction
    Debye-Scherrer method, 83–85
    Laue method, 81–83
    rotating crystal method, 83